SEDIMENTOLOGY AND STRATIGRAPHY

Edited by

Antonella Galetto

Toronto Academic Press

SEDIMENTOLOGY AND STRATIGRAPHY

Editor: **Antonella Galetto**

Toronto Academic Press
4164 Lakeshore Road
Burlington ON L7L 1A4
Canada
www.tap-books.com
Email: orders@arclereducation.com

ISBN: 978-1-77956-336-1

ABOUT THE EDITOR

Antonella Galetto, is a PhD geologist with training in various earth sciences disciplines, including renewable energy exploration, geochronology, and geographic information systems. She graduated from the National University of Cordoba in Argentina. Her research projects have primarily focused on the exploration of geothermal energy and studying the evolution of the Andes through collaboration with multiple scientific groups worldwide. Currently, she's working on processing geospatial imagery for various exploration purposes and serve as a tutor of geographic information systems and geodata analysis at the university level. Additionally, she work as an academic editor for a company that offers revision, polishing, and translation services for scientific research papers in the geosciences field.

Table of Contents

List of Figures xi

List of Tables xvii

Preface xix

1 INTRODUCTION TO SEDIMENTOLOGY AND STRATIGRAPHY 1

Introduction 2

1.1. Sedimentology 3

1.1.1. Sedimentary Rock Types 3

1.1.2. Importance of Sedimentary Rocks 4

1.1.3. Basic Principles 4

1.1.4. Methodology 5

1.1.5. Recent Developments 5

1.2. Processes of Sedimentation 7

1.2.1. Physical Processes 8

1.2.2. Chemical Processes 11

1.2.3. Biological Processes 12

1.3. Sedimentary Environments and Facies 14

1.4. The Spectrum of Environments and Facies 16

1.5. Basic Principles of Stratigraphy 17

1.5.1. Order of Superposition 17

1.5.2. Original Horizontality 18

1.5.3. Lateral Continuity 18

1.5.4. Cross-Cutting Relationships 19

1.5.5. Inclusions 19

1.5.6. Unconformities 20

1.5.7. Fossil Succession 20

1.5.8. Uniformitarianism and Catastrophism 20

1.6. Stratigraphic Classification 21

1.6.1. Lithostratigraphy 22

1.6.2. Biostratigraphy 24

1.6.3. Chronostratigraphy 27

Summary 37

Review Questions 38

References 38

2 TERRIGENOUS CLASTIC SEDIMENTS: GRAVEL, SAND, AND MUD 41

Introduction 42

2.1. Classification of Sediments and Sedimentary Rocks 43

2.1.1. Clastic Sedimentary Rocks (Detrital) 43

2.1.2. Chemical, Biochemical, and Organic Sedimentary Rock 43

2.2. Gravel and Conglomerate 47

2.2.1. Gravel 47

2.2.2. Conglomerate 49

2.3. Sand and Sandstone 54

2.3.1. Origins 54

2.3.2. Components 55

2.3.3. Types of Sandstone 57

2.4. Clay, Silt, and Mudrock **61**
2.4.1. Definitions of Terms in Mudrocks 61
2.4.2. Silt and Siltstone 62
2.4.3. Clay Minerals 62
2.4.4. Petrographic Analysis of Clay Minerals 62
2.4.5. Clay Particle Properties 63
2.5. Textures and Analysis of Terrigenous Clastic Sedimentary Rocks **64**
2.5.1. Sorting 64
2.5.2. Rounding 64
2.5.3. Sediment Maturity 65
2.5.4. Cycles of Sedimentation 65
Summary **69**
Review Questions **70**
References **71**

3 BIOGENIC, CHEMICAL, AND VOLCANOGENIC SEDIMENTS 73

Introduction **74**
3.1. Limestone **75**
3.1.1. Description 75
3.1.2. Formation 78
3.1.3. Occurrence 80
3.1.4. Limestone Landscape 83
3.2. Evaporite Minerals **84**
3.2.1. Gypsum and Anhydrite 85
3.2.2. Halite 86
3.2.3. Other Evaporite Minerals 86
3.2.4. Depositional Environments 86
3.3. Cherts **90**
3.3.1. Description 91
3.3.2. Fossils 94
3.3.3. Prehistoric and Historic Uses 94
3.3.4. Varieties 95
3.4. Volcaniclastics **97**
3.4.1. Classification of Volcaniclastics 97
3.4.2. Types of Volcaniclastic Rocks 100
3.4.3. Nomenclature of Volcaniclastic Rocks 100
3.4.4. Recognition of Volcaniclastic Material 101
Summary **104**
Review Questions **105**
References **105**

4 PROCESSES OF TRANSPORT AND SEDIMENTARY STRUCTURES 107

Introduction **108**
4.1. Sediment Transport and Deposition **110**
4.1.1. Agents of Sediment Transport 111
4.1.2. Modes of Sediment Transport 112
4.1.3. Depositional Environments 114
4.1.4. Applications 117
4.2. Transport Media **118**
4.2.1. The Behavior of Fluids and Particles in Fluids 119
4.2.2. Transport of Particles in a Fluid 121
4.2.3. Entraining Particles in a Flow 121
4.2.4. Grain size and Flow Velocity 123
4.2.5. Clast-Size Variations: Graded Bedding 125
4.2.6. Fluid Density and Particle Size 126
4.2.7. Flows, Sediment, and Bedforms 126
4.3. Sedimentary Structures and Textures **128**
4.3.1. Importance of Sedimentary Structures and Textures in Sedimentology 129
4.3.2. Classification of Sedimentary Structures and Textures 129
4.3.3. Bedding Structures and Textures 130
4.3.4. Interpretation of Bedding Structures and Textures in Depositional Environments 132
4.3.5. Interpretation of Sedimentary Structures and Textures in Alluvial Systems 132
Summary **137**
Review Questions **137**
References **138**

5 FIELD SEDIMENTOLOGY, FACIES, AND ENVIRONMENTS 139

Introduction **140**

5.1. Field Sedimentology **141**

5.1.1. Field Equipment 141

5.1.2. Field Studies: Mapping and Logging 141

5.2. Graphic Sedimentary Logs **143**

5.2.1. Drawing a Graphic Sedimentary Log 144

5.2.2. Presentation of Graphic Sedimentary Logs 145

5.2.3. Other Graphical Presentations: Sketches and Photographs 146

5.3. Paleocurrent **148**

5.3.1. Paleocurrent Measurements 148

5.3.2. Structures for Paleocurrent Measurement 152

5.3.3. Presentation of Results and Calculation of Vector Means 154

5.3.4. Interpretation of the Paleocurrent Pattern 155

5.4. Collection of Rock Samples **158**

5.4.1. Provenance Studies 158

5.5. Description of Core **160**

5.6. Interpreting Past Depositional Environments **161**

5.6.1. The Concept of 'Facies' 161

5.6.2. Facies Analysis 162

5.6.3. Facies Associations 162

5.6.4. Facies Sequences/Successions 164

5.6.5. Facies Names and Facies Codes 165

Summary **171**

Review Questions **172**

References **172**

6 CONTINENTAL ENVIRONMENTS AND SOURCES OF SEDIMENT 175

Introduction **176**

6.1. Continental Environments **177**

6.1.1. Transitional Environments 178

6.1.2. Marine Environments 178

6.2. Continental Landform **180**

6.2.1. General Observations 181

6.2.2. Constraints on Modern Landform Theory 182

6.2.3. Landform Theories of the 18th and 19th Centuries 185

6.2.4. Davis's Erosion Cycle Theory 186

6.2.5. Climatic Morphogenesis 188

6.2.6. Tectonic Geomorphology 189

6.2.7. Orogenic and Epeirogenic Morphogenesis 191

6.2.8. A Unified Landform Theory 194

6.3. From Source of Sediment to Formation of Strata **198**

6.4. Mountain-Building Processes **199**

6.5. Global Climate **200**

6.6. Weathering Processes **202**

6.6.1. Physical Weathering 203

6.6.2. Chemical Weathering 204

6.6.3. The Products of Weathering 205

6.6.4. Soil Development 206

6.7. Erosion and Transport **207**

6.7.1. Erosion and Transport Under Gravity 207

6.7.2. Erosion and Transport by Water 209

6.7.3. Erosion and Transport by Wind 210

6.7.4. Erosion and Transport by Ice 210

6.8. Denudation and Landscape Evolution **211**

6.8.1. Topography and Relief 211

6.8.2. Climate Controls on Denudation Processes 211

6.8.3. Bedrock Lithology and Denudation 213
6.8.4. Soils and Denudation 213
6.8.5. Vegetation and Denudation 215
6.9. Tectonics and Denudation 216
6.10. Measuring Rates of Denudation 218
Summary 221
Review Questions 221
References 222

7 SOFT SEDIMENT DEFORMATION 223

Introduction 224
7.1. Soft-Sediment Deformation Structures 225
7.1.1. Classification of Soft-Sediment Deformation Structures 229
7.1.2. Types of Soft-Sediment Deformation Structures 230
7.2. Deformation Structures Related to Gravitationally Unstable Vertical Profiles of Bulk Density 232
7.2.1. Convolute Lamination 232
7.2.2. Load Casts 236
7.2.3. Pseudo-Nodules and Ball-and-Pillow 239
7.2.4. Heavy Mineral Sags and Drops 243
7.2.5. Deformation Involving an Ephemeral Layer: Passively Deformed Beds 245
7.2.6. Deformation Involving an Ephemeral Layer: Dish Structures 246
7.3. Deformation Structures Related to Non-Uniform Confining Loads 252
7.3.1. Load-Casted Ripple Marks 252
7.3.2. Folds Related to the Lateral Spread of Sand Mounds 255
7.4. Deformation Structures Related to a Downslope Body-Force Component 257
7.4.1. Sheet Slumps Formed from Multilayers 257
7.4.2. Deformed Cross-Bedding 260
Summary 263
Review Questions 264
References 264

8 BIOSTRATIGRAPHY 267

Introduction 268
8.1. Nature of Biostratigraphic Units 269
8.2. Fossils and Stratigraphy 270
8.2.1. Classification of Organisms 270
8.3. Evolutionary Trends 274
8.3.1. Population Fragmentation and Phyletic Transformation 274
8.3.2. Phyletic Gradualism and Punctuated Equilibrium 275
8.3.3. Speciation and Biostratigraphy 275
8.4. Biozones and Zone Fossils 276
8.4.1. Rate of Speciation 277
8.4.2. Depositional Environment Controls 278
8.4.3. Mobility of Organisms 278
8.4.4. Geographical Distribution of Organisms 279
8.4.5. Abundance and Size of Fossils 279
8.4.6. Preservation Potential 279
8.5. Taxa Used in Biostratigraphy 281
8.5.1. Marine Macrofossils 281
8.5.2. Marine Microfossils 282
8.5.3. Terrestrial Fossil Groups Used in Biostratigraphy 284
8.6. Biostratigraphic Correlation 285
8.6.1. Correlating Different Environments 285
8.6.2. Graphical Correlation Schemes 286
8.6.3. Biostratigraphy in Relation to Other Stratigraphic Techniques 287
8.7. Spores and Pollens 288
8.7.1. Distribution and Ecology 289
8.7.2. Dispersal and Sedimentation 290
8.7.3. Geological History 290
8.7.4. Applications of Fossil Spores and Pollen 292
8.8. Foraminifera 293
8.8.1. The Test 293
8.8.2. Foraminiferal Ecology 294
8.8.3. Planktic Foram Ecology 297
8.8.4. Temperature and Latitude 297

8.8.5. Calcite Compensation Depth (CCD) 299

8.8.6. Geological History of Foraminifera 300

8.8.7. Applications of Foraminifera 300

Summary **305**

Review Questions **306**

References **307**

9 DATING AND CORRELATION TECHNIQUES FOR STRATIGRAPHY 309

Introduction **310**

9.1. Radiometric Dating **311**

9.1.1. Fundamentals 311

9.1.2. Modern Dating Methods 315

9.1.3. Dating with Decay Products of Short-Lived Extinct Radionuclides 321

9.2. Other Isotopic and Chemical Techniques **323**

9.2.1. Strontium Isotopes 323

9.2.2. Thermochronological Techniques 324

9.2.3. Chemostratigraphy 325

9.3. Magnetostratigraphy **326**

9.3.1. The Magnetic Record in Rocks 326

9.3.2. Practical Magnetostratigraphy 327

9.3.3. Magnetostratigraphic Correlation 327

9.3.4. Technique 328

9.4. Dating in the Quaternary **330**

9.4.1. Carbon-14 Dating 330

9.4.2. Uranium-Series Dating 331

9.4.3. Oxygen Isotope Stratigraphy 331

9.4.4. Luminescence and Electron Spin Resonance Dating 331

9.4.5. Cosmogenic Isotopes 332

9.4.6. Amino-Acid Racemization 333

9.4.7. Annual Cycles in Nature 333

Summary **341**

Review Questions **342**

References **342**

10 SUBSURFACE STRATIGRAPHY AND SEDIMENTOLOGY 345

Introduction **346**

10.1. Basic Principles **347**

10.2. Seismic Reflection Data **348**

10.2.1. History 349

10.2.2. Theory 350

10.2.3. Applications 351

10.2.4. Environmental Impact 359

10.3. Borehole Stratigraphy and Sedimentology **363**

10.3.1. Borehole Cuttings 363

10.3.2. Core 364

10.3.3. Core Logging 364

10.4. Geophysical Logging **366**

10.4.1. Wireline Logging 366

10.4.2. Logging While Drilling 370

10.4.3. Coring 371

10.4.4. Mudlogging 371

10.4.5. Information Use 371

10.5. Subsurface Facies and Basin Analysis **372**

Summary **376**

Review Questions **377**

References **378**

INDEX 379

List of Figures

Figure 1.1. Sedimentology students examining Glacial Lake Oshkosh sediments

Figure 1.2. The rock cycle

Figure 1.3. Cambrian stramatolite

Figure 1.4. Very-fine-grained sediment from Glacial Lake Missoula, Montana, USA

Figure 1.5. Sedimentary layers

Figure 1.6. Monument Valley Navajo Tribal Park in Arizona shows The Mittens, i.e., East Mitten on the right and West Mitten on the left, with the matching of rock layers proving the law of lateral continuity

Figure 2.1. Sedimentary rock with sandstone in Malta

Figure 2.2. Carmelo FORMATION (conglomerate) at Point Lobos

Figure 2.3. A conglomerate at the base of the Cambrian in the Black Hills, South Dakota

Figure 2.4. Ternary plot showing the relative abundance of quartz, feldspar, and lithic grains in a sandstone

Figure 3.1. The proportions of the principal ions in seawater of normal salinity and 'average' river water

Figure 3.2. Ocean water contents

Figure 3.3. Marine evaporites the dead sea

Figure 3.4. (A) Ternary plot of the relative proportions of Lm:Lv:Ls. Lm: metamorphic lithics; Lv: volcanic lithics; Ls: sedimentary lithics (modified from [60]). (B) Relative proportions of Lv:M:S (e.g., [61]). Lv: All volcanic lithic textures (Lvl + ALvl + Algl + Lvf + Pm + Apm + Shard); M: mafic single crystal grains (pyroxene + amphibole + garnet + biotite + opaques); S: sialic single crystal grains (plagioclase + K-feldspar + monocrystalline and polycrystalline quartz without tectonic fabric)

Figure 4.1. Sediment transport in coastal lagoons

Figure 4.2. Six main domains/geomorphological units of sediment transport studies. (I) the watershed; (II) estuary/delta; (III) the estuary-littoral transition, including the river plume, the bottom nepheloid layer and the ROFI zone; (IV) the littoral zone; (V) the continental shelf; and (VI) canyon and deep sea

Figure 4.3. Depositional environments

Figure 4.4. Laminar and turbulent flow of fluids through a tube

Figure 4.5. Particles move in a flow by rolling and saltating (bedload) and in suspension (suspended load)

Figure 4.6. Flow of a fluid through a tapered tube results in an increase in velocity at the narrow end where a pressure drop results

Figure 4.7. The lift force resulting from the Bernoulli effect causes grains to be moved up from the base of the flow

Figure 4.8. The Hjülstrom diagram shows the relationship between the velocity of a water flow and the transport of loose grains. Once a grain has settled it requires more energy to start it moving than a grain that is already in motion. The cohesive properties of clay particles mean that fine-grained sediments require relatively high velocities to re-erode them once they are deposited, especially once they are compacted

Figure 4.9. Normal and reverse grading within individual beds and fining-up and coarsening-up patterns in a series of beds

Figure 4.10. Layers within a flow and flow surface roughness: The viscous sublayer, the boundary layer within the flow and the flow depth

Figure 4.11. The physical features and arrangements of sedimentary rocks

Figure 4.12. Sedimentary structures and textures

Figure 4.13. The surfaces that divide the layers of sedimentary rocks are known as bedding planes

Figure 4.14. Graded bedding

Figure 4.15. Cross bedding sedimentary structure

Figure 4.16. Ripple marks

Figure 4.17. Mud cracks

Figure 4.18. Biogenic structures

Figure 5.1. An example of a graphic sedimentary log: This form of presentation is widely used to summarize features in successions of sediments and sedimentary rocks

Figure 5.2. Examples of patterns and symbols used on graphic sedimentary logs

Figure 5.3. A proforma sheet for constructing graphic sedimentary logs

Figure 5.4. An example of an annotated sketch illustrating sedimentary features observed in the field

Figure 5.5. A field photograph of sedimentary rocks: An irregular lower surface of the thick sandstone unit in the upper part of the cliff marks the base of a river channel

Figure 5.6. (A) Local setting of the study area in the Western Mediterranean Sea and the Sahara–Sahel Dust Corridor. (B) Main trajectories of the water masses circulation in the Alboran Sea

Figure 5.7. Proforma for paleocurrent data. If dip of cross-beds is required, add another column. Corrections for tectonic tilt are required for linear structures such as groove marks if the strata are dipping more than 25° and for planar structures such as cross-bedding if the strata are dipping more than 10°. If this is the case, then add columns for the pitch and down-dip trend of linear structures or the dip and strike of cross-beds

Figure 5.8. An example of correcting the orientation of a linear structure for tectonic tilt. A bed dips at 50° to 225° and has a linear structure, such as a groove mark or parting lineation, with a pitch of 20° to the south-east. The bed is plotted as a great circle on a stereonet and the pitch of the structure is marked on that circle (X). The bed is restored to the horizontal and with it the structure along a small circle to give its original orientation at point Y. On rotating back, the azimuth of Y is given: 155°

Figure 5.9. Correction of cross-bedding for tectonic tilt using stereographic projection

Figure 5.10. Some of the heavy minerals that can be used as provenance indicators

Figure 5.11. When drilling through strata it is possible to recover cylinders of rock that are cut vertically to reveal the details of the beds

Figure 5.12. A graphic sedimentary log with facies information added. The names for facies are usually descriptive. Facies codes are most useful where they are an abbreviation of the facies description. The use of columns for each facies allows for trends and patterns in facies and associations to be readily recognized

Figure 5.13. A summary of the principal sedimentary environments

Figure 6.1. The pathway of processes involved in the formation of a succession of clastic sedimentary rocks, part of the rock cycle

Figure 6.2. The boundaries of the present-day principal tectonic plates

Figure 6.3. The present-day world climate belts

Figure 6.4. The principal weathering processes and their controls

Figure 6.5. Frost shattering of a boulder (50 cm across) in a polar climate setting

Figure 6.6. The relative stability of common silicate minerals under chemical weathering

Figure 6.7. An in situ soil profile with a division into different horizons according to presence of organic matter and degree of breakdown of the regolith

Figure 6.8. Mechanisms of gravity-driven transport on slopes. Rock falls and slides do not necessarily include water, whereas slumps, debris flows, and turbidity currents all include water to increasing degrees

Figure 6.9. A scree slope or talus cone in a mountain area with strong physical weathering

Figure 6.10. Erosion by solution in beds of limestone results in a karst landscape

Figure 6.11. Badlands scenery formed by the erosion of weak Mudrock beds

Figure 6.12. The development of land plants through time: Grasses, which are very effective at binding soil and stabilizing the land surface, did not become widespread until the mid-Cenozoic

Figure 6.13. Uplift due to thickening of the crust followed by erosion results in isostatic compensation as the load of the rock mass eroded is removed. If the erosion is uneven then locally the removal of mass from valleys can result in uplift of the mountain peaks between

Figure 6.14. The rain shadow effect in mountain belts: Moisture in air blown from the sea falls as rain as the air mass cools over a mountain range

Figure 7.1. Summary of the rheological history of unconsolidated sediments that become liquidized

Figure 7.2. Time required for the re-sedimentation of a bed of liquidized sand, as a theoretical function (Eqn. (1)) of initial thickness of the layer, particle falling velocity in the dispersion, and concentration difference between dispersion and deposit. Particle concentration in deposit assumed to be 0.60

Figure 7.3. Definition diagram for soft-sediment deformation at the interface between two horizontal sediment layers when fluidization is involved

Figure 7.4. Syndepositional convolute lamination (slightly schematic) in turbidite sandstone of Pico formation (Pliocene), Ventura basin, California, U.S.A

Figure 7.5. Syndepositional convolute lamination (note erosional truncations and evidence of earlier folding within fold limbs), River Ganges alluvium, Bhagalpur, Bihar, India. The deformed layer is normally graded, from sandy up to silty, and sandy sediment, in cross-laminated sets climbing at a low angle, reappears above the erosion surface truncating the folds toward the top of the section. Length of bed shown approximately 0.47 m

Figure 7.6. Types of convolute lamination

Figure 7.7. Vertical section through load casts beneath volcanogenic sand bed overlying mud, Langdale Slates (Ordovician), English Lake District

Figure 7.8. Photograph from below of load casts formed experimentally by collapse of a layer of soft sand mixed with Plaster of Paris into a layer of mud, when both became liquidized in response to a sequence of shocks. Specimen 0.3 m across

Figure 7.9. Bulk density of water-saturated sediment as a function of particle density (mineralogy) and fractional volume concentration

Figure 7.10. Schematic lateral passage of a continuous sand bed overlying mud into pseudonodules

Figure 7.11. A zone of ball-and-pillow structures (sandstone in mudstone) overlying shales and mudstones, and overlain (apparently erosively) by sheet sandstone and interbedded sandstone and mudstone. The ball-and-pillow zone is approximately 2 m thick. Coal measures (Carboniferous) east of Amroth Castle, Amroth, Dyfed, Wales

Figure 7.12. A single saucer-shaped sandstone pillow (note overturned edges) embedded in mudstone. Coin approximately 0.03 m across

Figure 7.13. Vertically flattened sigmoidally folded sandstone bed in mudstone. Coin approximately 0.03 m across

Figure 7.14. Drop and sag structures in heavy mineral bands (black) within sandstones, Torridon Group (Precambrian), northwest Scotland

Figure 7.15. Successive stages in the gravitational instability of a narrow cylinder (approximately 0.21 m long, 0.002 m diameter) of dyed glycerol mixed with very little water and introduced by means of a capillary tube into plain glycerol at room temperature

Figure 7.16. Vertical section through fine-grained sandstone with small dish structures. Specimen 0.035 m across. Note sharply defined bases and muddy lower zones to dishes, and also the relatively clean tops. Jackfork Group (Carboniferous), southeast Oklahoma

Figure 7.17. Dish structure developed after convolute lamination, Jackfork Group, Oklahoma, U.S.A. Numerous examples of laminae severed or distorted between successive dishes can be seen. Specimen approximately 0.15 m across

Figure 7.18. Dish structures. (a) Schematic vertical section through beds with dish structures, in which the dish-structured zone has a roof in sediment similar to that involved in the dishes. (b) Some schematic details of dish structures developed in a previously convolute-laminated bed

Figure 7.19. Vertical section in dish-structured sandstone (below) extending up to an irregular roof beneath parallel laminated rocks. The latter are penetrated by veins of destratified (? liquidized) sand. Tertiary deep-sea fan deposits, San Sebastian area, northern Spain. Width of bed 0.45 m

Figure 7.20. Schematic development of a row of load-casted ripple marks

Figure 7.21. Clusters of load-casted ripple marks. (a) A cluster of load-castcd ripple marks, Kliva Sandstone (Oligocene), Polish Carpathan Mountains. (b) Stages in development of cluster of load-casted ripple marks

Figure 7.22. Load-casted ripple marks, Kurkiselka Esker, Kiiminki, Finland

Figure 7.23. Schematic stages in the development of syndepositional load-casted ripple marks, according to Aario (1971)

Figure 7.24. Deformation by folding and faulting of layered plastic muds by a migrating sand body

Figure 7.25. Features of slump sheets, lacustrine muds and silts, Dead Sea, Israel. (a) Imbricately thrust folded layers in two zones. Hammer 0.33 m long. (b) Box folds with probable basal dicollernent. Hammer stuck into cliff below folded zone casts vertical shadow about 0.4 m long

Figure 7.26. Structures (slightly simplified) in slump sheets. Sandstone or marker sandstone beds shown stippled

Figure 8.1. The Linnaean hierarchical system for the taxonomy of organisms

Figure 8.2. An example of a macrofossil

Figure 8.3. Zonation schemes used in biostratigraphic correlation

Figure 8.4. Commonly used zone fossils (from the Cambrian onwards)

Figure 8.5. Meiosis and the production of bilaterally or radially symmetrical spores

Figure 9.1. Radioactive decay results in the formation of a new 'daughter' isotope from the 'parent' isotope

Figure 9.2. A graph showing the depleted-mantle model of DePaolo

Figure 9.3. A Concordia diagram as used in uranium–lead dating, with data from the Pfunze Belt, Zimbabwe [19]. All the samples show loss of lead isotopes, but the intercept of the errorchron (straight line through the sample points) and the Concordia (curve) shows the correct age of the rock

Figure 9.4. Isotope half-lives. The darker more stable isotope region departs from the line of protons (Z) = neutrons (N), as the element number Z becomes larger

Figure 9.5. Luminescence dating of the rock surface method

Figure 10.1. Sketch of seismic reflection energy propagation in the subsurface

Figure 10.2. Seismic testing in 1940

List of Tables

Table 2.1. Classification of sedimentary rocks

Table 3.1. Primary versus secondary volcaniclastic rocks

Table 3.2. Terms for Mixed pyroclastic-epiclastic rocks

Table 5.1. Paleocurrent patterns of principal depositional environments, together with best and other directional structures

Table 7.1. Classification of soft-sediment structures

PREFACE

Our understanding of the fascinating history of our planet primarily stems from deciphering the sedimentary record of events and trends across various systems, including biological, ecological, geochemical, climatic, and physical systems. This knowledge is derived through a combination of numerous tools and innovative techniques. One of the most critical approaches is integrated stratigraphy, which encompasses lithostratigraphy, biostratigraphy, chronostratigraphy/chronology, chemostratigraphy, magnetostratigraphy, and cyclostratigraphy. These methods collectively provide reliable timescales for reconstructing past events. Stratigraphy and sedimentology are essential fields of study that reveal significant information about the sources of energy present in different regions. Our country is abundant with resources, and it is crucial to understand the types of resources found in various kinds of soil. Sedimentary rocks are of particular importance as they are the primary locations where petroleum deposits are discovered.

Sedimentology, which involves the study of sedimentary rocks and the processes by which they are formed, is closely linked to stratigraphy. Stratigraphy examines the relationships between rock layers and how these layers can shift and move over time. This relationship is vital for determining where petroleum deposits are located and understanding how the extraction of petroleum can impact the surrounding sediment. One of the primary indicators of historical sea level changes can be found within coastal stratigraphy. Coastal sedimentary layers serve as natural archives, recording changes in the coast, sea levels, and water quality over time. According to Tooley (1992), these sedimentary layers provide crucial evidence of past environmental conditions. For instance, if shells from marine organisms are found on mountain tops, it suggests that these areas were once submerged under the ocean, indicating significant changes in sea levels.

Coal deposits, which are vital energy resources, are predominantly found in Europe. Coal is formed through the decay and decomposition of carboniferous vegetation. Interestingly, the present climate in these coal-rich areas does not support such luxuriant vegetation growth that could convert into vast quantities of coal. This discrepancy suggests that these regions once experienced a tropical climate, which supported thick vegetation growth, leading to the formation of extensive coal deposits.

Understanding the formation and distribution of these energy resources requires a comprehensive study of sedimentology and stratigraphy. By analyzing sedimentary records, scientists can uncover valuable information about past climatic conditions, biological activity, and geological processes. This knowledge not only helps in locating energy resources like petroleum and coal but also aids in understanding how their extraction might impact the environment.

Contents/Structure of the Book

This book is structured into 10 comprehensive chapters, designed to provide an in-depth understanding of sedimentological and stratigraphic principles. The content spans from basic concepts to advanced techniques, ensuring that readers gain a thorough knowledge of the subject.

Chapter 1 introduces the fundamental concepts of sedimentology and stratigraphy. It outlines the significance of these fields in geology, their historical development, and their applications in understanding Earth's history and resource exploration. This chapter serves as a foundation, defining key terms and providing an overview of the topics covered in the subsequent chapters.

Chapter 2 presents the coverage of clastic sediments, focusing on gravel, sand, and mud. It explores their origins, composition, and classification. The chapter discusses the processes that lead to the formation of clastic sediments, including weathering, erosion, and transportation, and examines the characteristics and depositional environments of these sediments.

Chapter 3 shifts the focus to biogenic, chemical, and volcanogenic sediments. It covers the formation and characteristics of sediments derived from biological activity, chemical precipitation, and volcanic processes. The chapter highlights the importance of these sediments in reconstructing past environmental conditions and their roles in the geologic record.

Chapter 4 addresses the processes responsible for sediment transport and the formation of sedimentary structures. It examines the mechanisms of fluid flow, sediment transport, and deposition, and describes various sedimentary structures such as crossbedding, ripple marks, and graded bedding. This chapter emphasizes the interpretation of these structures in understanding past depositional environments.

Chapter 5 focuses on field sedimentology, providing practical guidance on studying sediments in their natural settings. It introduces the concept of facies, which are distinctive bodies of sediment with specific characteristics reflecting particular depositional environments. The chapter discusses methods for identifying and interpreting facies and reconstructing ancient environments.

Chapter 6 explores sedimentary processes and environments on continents. It covers fluvial, lacustrine, eolian, and glacial environments, detailing the sources and characteristics of sediments in these settings. The chapter also discusses the impact of tectonics and climate on sedimentation in continental environments.

Chapter 7 examines the processes and significance of soft sediment deformation. It describes the types of deformation structures that can occur in unconsolidated sediments, such as slumps, convolute bedding, and load structures. The chapter highlights the importance of recognizing these features for interpreting sedimentary processes and environments.

Chapter 8 introduces biostratigraphy, the study of the distribution of fossils in sedimentary rocks to establish relative ages and correlations. It discusses the principles and methods of biostratigraphy, the use of index fossils, and the construction of biozones. The chapter emphasizes the role of biostratigraphy in constructing geological timescales and correlating strata.

Chapter 9 covers various techniques used for dating and correlating sedimentary rocks. It includes discussions on radiometric dating, magnetostratigraphy, chemostratigraphy, and cyclostratigraphy. The chapter explains how these methods are applied to establish absolute ages and correlate stratigraphic units across different regions.

Chapter 10 focuses on subsurface stratigraphy and sedimentology, which are essential for resource exploration and geological modeling. It covers methods for studying sedimentary rocks in the subsurface, including seismic stratigraphy, well logging, and core analysis. The chapter also discusses the interpretation of subsurface data for understanding sedimentary basin evolution and resource potential.

This book provides a detailed and comprehensive exploration of sedimentary processes, structures, and stratigraphic principles. Each chapter builds on the previous ones, guiding readers through the complexities of sedimentology and stratigraphy and equipping them with the knowledge needed to analyze and interpret sedimentary records.

Audience

This book is particularly suitable for students, offering invaluable insights for undergraduate and graduate students in geology, earth sciences, environmental science, and related disciplines. The book serves as a comprehensive textbook for courses in sedimentology, stratigraphy, and related subjects, providing both theoretical foundations and practical applications. Academics and researchers in sedimentology, stratigraphy, paleontology, and geosciences will also find this book beneficial, as it presents detailed explanations and the latest research findings. It offers a solid basis for teaching and conducting advanced research in these areas. Furthermore, professionals in environmental studies, engineering geology, and natural resource management can use this book to understand sedimentary environments and processes. It provides insights into soil and sediment analysis, environmental impact assessments, and land-use planning. Decision-makers and planners involved in natural resource management, environmental protection, and land-use planning can also benefit from this book. It provides a scientific understanding of sedimentary processes and their implications for sustainable development.

Learning Features:

- Start-of-chapter learning objectives;
- Key terms in each chapter;
- Summary points at the end of each chapter;
- Review Questions at the end of each chapter;
- Case Studies and Activities in each chapter for solving real problems;
- References at the end of each chapter.

CHAPTER

1 Introduction to Sedimentology and Stratigraphy

LEARNING OBJECTIVES

After studying this chapter, you will be able to:

- Understand the fundaments of sedimentology;
- Explain the processes of sedimentation;
- Describe sedimentary environments and facies;
- Define the spectrum of environments and facies;
- Discuss the basic principles of stratigraphy;
- Deal with stratigraphic classification.

KEY TERMS FROM THIS CHAPTER

Aeolian sandstones
Chronostratigraphy
Fossils
Lithification
Sediment transportation
Sedimentary sequence
Stratigraphy
Biostratigraphy
Exogenic processes
Hydrocarbon exploration
Lithostratigraphy
Sedimentary rock
Sedimentary structures
Stromatolites

INTRODUCTION

The physical and chemical characteristics of sedimentary rocks as well as the processes that go into their formation, such as sediment transport, deposition, and lithification (conversion to rock), are the focus of the scientific field of sedimentology. A major goal of most sedimentological research is the interpretation of ancient environmental conditions in sediment source areas and depositional sites. Sedimentologists study the composition, structure, textures, and fossil content of deposits that were made in different geographical settings. In the geologic record, they are able to discriminate between continental, littoral, and marine deposits.

The arrangement and succession of strata, or layers, as well as the origin, composition, and distribution of these geological strata, are the subjects of stratigraphy, a branch of geology and the Earth Sciences. Thus, time and space are evaluated in the study of natural and archaeological stratification. In particular, the horizontal (lateral) distribution of layers represents the space dimension (i.e., spatial variations in the stratigraphy), whereas the vertical arrangement of layers represents the time dimension (oldest at the bottom, youngest at the top).

1.1. SEDIMENTOLOGY

The study of contemporary sediments, such as sand, silt, and clay, as well as the processes that lead to their formation (weathering and erosion), transport, deposition, and diagenesis, is known as sedimentology. Sedimentologists use their knowledge of contemporary processes to analyze sedimentary rocks and sedimentary structures to interpret the history of the Earth.

Up to 75% of the Earth's surface is made up of sedimentary rocks, which also preserve the fossil record and record a large portion of the planet's history. Stratigraphy, the study of the spatial and temporal relationships between rock layers or strata, and sedimentology are closely related fields. (Juergen et al., 2007).

Determining how sedimentary features in the rock record were formed is predicated on the idea that the processes affecting the earth today are the same as those affecting it in the past. Geologists reconstruct past environments by comparing similar features found today with features found in the rock record. For example, they compare modern sand dunes to dunes preserved in ancient aeolian sandstones (Figure 1.1).

Figure 1.1. Sedimentology students examining Glacial Lake Oshkosh sediments.

Source: https://www.uwosh.edu/faculty_staff/hiatt/Teaching/314/314_Lab.html.

1.1.1. Sedimentary Rock Types

Sedimentary rocks fall into four main categories: clastics, carbonates, evaporites, and chemical.

- Particles from the weathering and erosion of precursor rocks make up clastic rocks, which are mainly made up of fragments. Clastic rocks are categorized based on the composition and predominant grain size. Although there have been instances of clastic carbonate rocks, the term "clastic sedimentary rocks" was originally used to refer to silica-rich clastic sedimentary rocks. The phrase "siliciclastic sedimentary rocks" is more appropriate.

- o Basins of clastic sedimentary rocks are usually home to significant deposits of coal and oil shale, which are produced by the buildup of biological detritus. These deposits are known as organic sedimentary rocks.
- Carbonates consist of different types of carbonate minerals, primarily calcium carbonate ($CaCO_3$), which are formed through a range of inorganic and organic processes. Reef material usually makes up the majority of carbonate rocks.
- Evaporites, which are mostly composed of gypsum or halite, are created when water evaporates at the Earth's surface.
- Chemical sedimentary rocks, such as certain carbonates, form when minerals precipitate out of water. Examples of these rocks include jaspilite and chert.

1.1.2. Importance of Sedimentary Rocks

Ancient and modern societies have come to use the many products that sedimentary rocks provide.

- Art: Marble is an example of the use of sedimentary rocks for aesthetic and creative purposes, even though it is a metamorphosed limestone.
- Uses in Architecture: Dimension stone and architectural uses of stone derived from sedimentary rocks include the use of slate (shale that has undergone metamorphosis) for roofing and sandstone for load-bearing buttresses.
- Ceramics and Industrial Materials: Clay for pottery and ceramics including bricks; cement and lime derived from limestone.
- Economic Geology: Sedimentary rocks contain significant SEDEX ore deposits (lead, zinc, and silver), copper ore deposits, gold, tungsten, and uranium deposits, as well as numerous other industrial minerals, gemstones, and valuable minerals, such as heavy mineral sands ore deposits.
- Energy: The ability of sedimentary rocks to produce petroleum oil deposits is the foundation of petroleum geology. Rocks that are sedimentary contain coal and oil shale. The majority of the uranium energy resources on Earth are found in sedimentary successions.
- Groundwater: A substantial portion of Earth's groundwater aquifers are found in sedimentary rocks. Our knowledge of the rocks that contain these aquifers—the reservoir—is essential to our comprehension of their extent and the amount of water that can be extracted from them.

1.1.3. Basic Principles

The study of sediments, or sedimentology, aims to provide a coherent understanding of the evolution of sedimentary sequences and basins, and consequently, the Earth's geological history overall, by obtaining information on the depositional conditions that acted to deposit the rock unit and the relationships between the individual rock units in a basin.

The uniformitarianism principle, which holds that sediments deposited at the Earth's surface today are deposited in the same manner as those found in ancient sedimentary rocks, provides the scientific foundation for this.

Sedimentological circumstances are recorded in the sediments as they are deposited; the current form of the sediments is a reflection of past events, and all events influencing the sediments, from the origin of the sedimentary material to the stresses applied to them following diagenesis, can be studied.

The interpretation of sedimentary sequences depends on the principle of superposition. In older metamorphic terrains, or fold and thrust belts, where sediments are frequently intensely folded or deformed, the interpretation of the sedimentary section, and frequently the deformation and metamorphic structure of the area, depends on the recognition of younging indicators or graded bedding.

The concept of original horizontality, which holds that sediments are deposited at their angle of repose—which is essentially horizontal for the majority of sediment types—is used to analyze folding in sediments. As a result, the rocks can be unfolded and interpreted using the sedimentary information they contain once the younging direction is known.

According to the cross-cutting relationships principle, anything that penetrates or cuts through strata is younger than the strata itself.

1.1.4. Methodology

The methods employed by sedimentologists to gather data and evidence on the nature and depositional conditions of sedimentary rocks include;

- Measuring and describing the outcrop and distribution of the rock unit.
- Describing the rock formation, a formal process of documenting thickness, lithology, outcrop, distribution, and contact relationships to other formations.
- Mapping the distribution of the rock unit, or units.
- Descriptions of rock core (drilled and extracted from wells during hydrocarbon exploration).
- Sequence stratigraphy.
- Describes the progression of rock units within a basin.
- Describing the lithology of the rock.
- Petrology and petrography; particularly measurement of texture, grain size, grain shape (sphericity, rounding, etc.), sorting and composition of the sediment.
- Analyzing the geochemistry of the rock.
- Isotope geochemistry, including the use of radiometric dating, to determine the age of the rock, and its affinity to source regions.

1.1.5. Recent Developments

Geologists at Massachusetts Institute of Technology and Indiana University (Bloomington) have questioned the conventional wisdom regarding the formation of some mudstones. The findings, which are published in the December 14, 2007, edition of science, refute the widely held belief among geologists that mud only settles in situations where water is motionless or slow-moving and instead demonstrate that mud can accumulate even in the presence of swift currents. According to the research, mudstones can be deposited under more energetic conditions than previously thought, necessitating a reevaluation of many geologic records. Some mudstones may have formed in swiftly moving waters.

Reviewing Schieber et al.'s research were Macquaker & Bohacs declare that all mudstones that were previously thought to have been continuously deposited under still waters need to be critically reevaluated in light of these findings. These rocks are frequently used to deduce historical ocean conditions, temperatures, and orbital fluctuations.

The recent push to economically extract hydrocarbons from mudstones as unconventional reservoirs for both shale gas and tight oil (or light tight oil) plays has sparked a great deal of recent research into these formations.

Australian sedimentologist Dutkiewicz has recently published research that explains the connection between geocirculation and changes in global temperatures and climate. The study discussed the circulation of carbon and water, as well as the effects of heat on the ocean's present and potential ability to absorb carbon.

1.2. PROCESSES OF SEDIMENTATION

Different types of sedimentary rocks are formed through physical, chemical, and biological processes known as sedimentation. The primary goal is to comprehend and relate the sedimentary system's processes and products.

Sediments are created by mechanical and chemical processes, and then they undergo cementation, recrystallization, and overburden pressure to become sedimentary rocks. Diagenesis, also known as diagenetic changes, is the term for the physical and chemical changes that occur during the consolidation process from sediments to sedimentary rocks. In contrast to igneous or metamorphic rocks, which require higher temperatures and pressures, the entire process of rock decay, sediment transportation, deposition, precipitation, and diagenesis occurs at or near the earth's surface under normal temperature and pressure conditions (Figure 1.2).

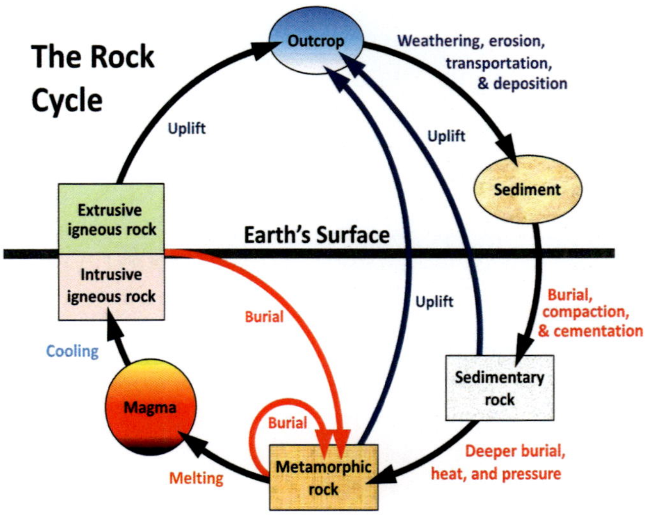

Figure 1.2. The rock cycle.

Source: https://opentextbc.ca/geology/chapter/3-1-the-rock-cycle/.

Despite making up only 5% of the crust, sedimentary rocks are found throughout much of the earth's surface. On the continental crust, the average thickness of sediment cover is around 2 km, while on the oceans, it is about 1 km.

Evidence of the physical, chemical, and biological processes that led to the formation of sedimentary rocks is preserved in these materials (Victor et al., 2006).

1.2.1. Physical Processes

Water and wind are two natural forces that dominate sedimentary processes. As a result, the cycles of water and wind are essential to the sedimentation process. Weathering, erosion, and material transportation and deposition by wind, water, and glaciers are examples of physical processes. Understanding the physical circumstances under which sediments are deposited is the goal of a discussion of various physical processes. (Blatt, Harvey; Middleton, Gerard et al., 980)

1.2.1.1. Weathering

The breakdown of minerals, rocks, and soil by atmospheric agents is known as weathering. There is no transportation involved; weathering happens in place. Both chemical and physical weathering processes can occasionally include a biological component. The disintegration of rocks and soils caused by direct contact with atmospheric elements like heat, water, ice, and pressure is known as mechanical or physical weathering. When rocks, soils, and minerals break down, chemical weathering—also referred to as biological weathering—involves the direct impact of chemicals in the atmosphere or chemicals produced by biological processes. Chemical reactions are most intense in wet and hot climates, while physical weathering is more prominent in extremely cold or dry environments. But the two kinds of weathering happen simultaneously, and one tends to hasten the other. For instance, rubbing together causes physical abrasion, which reduces particle size and increases surface area, increasing the particle's susceptibility to quick chemical reactions. Together, the different agents transform feldspars and micas, which are primary minerals, into secondary minerals like carbonates and clays, which liberate plant nutrient elements that are soluble. Soil is created by the remnants of broken-down rock mixed with organic matter (Murton et al., 2006).

1.2.1.1.1. Physical Weathering

- Freeze-Thaw: Water seeps through cracks and in between rock grains in this kind of weathering. The force of ice crystallization caused by freezing water fractures the rock. It works best during the spring thaw and is most active in polar regions.
- Insolation: In regions with significant daily temperature fluctuations, this kind of weathering takes place. This is common in hot, dry climates. In the desert, the wintertime diurnal temperature range is about 25°C. Rocks react to temperature by expanding and contracting. Temperature variations create differential stresses that eventually lead to the rock breaking apart because some minerals expand more than others. Some rocks weather by exfoliation, or the peeling away of outer layers, because the outer surface of a rock is frequently warmer or colder than the more protected inner portions. If ice forms in the surface cracks, this process could accelerate dramatically.
- Hydration and Dehydration: Clays and lightly indurated shales in areas with alternating wet and dry seasons expand when exposed to water and shrinkage cracks when

they become dehydrated. As a result, the rock's physical strength decreases. Waterlogged clays can cause landslides, but shrinkage cracks increase permeability, facilitating chemical weathering from rainfall.

- Stress Release: Rocks are compressed by the overburden at depth and have elastic properties, albeit to varying degrees. Rocks expand and occasionally fracture as a result of weathering and erosion gradually removing the overburden pressure. Lateral downslope creep frequently facilitates this kind of fracturing. Stress-release fractures are prone to enlargement by precipitation-related solutions and other processes once they are opened.

1.2.1.1.2. Chemical Weathering

The primary process of rock disintegration is chemical weathering. In tropical, humid climates where the rainy season comes after the summer, this process works better. Rainwater is ionized with dissolved carbon dioxide to produce carbonic acid. Because of this, rainwater typically has an acidic pH of 4 to 7. Natural carbonate is dissolved in the solution by acidic water. In addition to making limestone unstable, humus in the soil releases carbon dioxide when organic matter oxidizes, resulting in an acidic environment with a pH as low as 2.

$$H_2O + CO_2 = H_2CO_3 = HCO_3^- + H^+$$

Chemical weathering of silicate minerals: In carbonated water, feldspar breaks down into clay minerals, silicic acid and carbonates.

$$6H_2O+CO_2+2KAlSi_3O_8= Al_2Si_2O_5(OH)_4+4SiO(OH)_2+K_2CO_3$$

Orthoclase Kaolinite Silicic acid

The first mineral to form as a result of chemical weathering is typically kaolinite. The weathering process produces smectite, illite, and chlorite clay minerals, depending on the makeup of the source rock. The degree to which a mineral is stable depends on how well it can withstand the significant changes that weathering brings about in the source sediments.

1.2.1.1.3. Biological Weathering

While biological weathering generally doesn't contribute much on its own, it frequently speeds up and helps chemical weathering. Chemical weathering can be produced by a variety of plants and animals by releasing acidic compounds. Soil microorganisms can initiate and/or accelerate the weathering of minerals. It is believed that lichens on rocks speed up chemical weathering. The most prevalent type of biological weathering occurs when plants release organic acids into the soil below them, which breaks down compounds that contain iron and aluminum. The organic acids produced by the decomposing remnants of dead plants in the soil can lead to chemical weathering when dissolved in water.

1.2.1.2. Erosion

Erosion is the process by which rock and soil are moved and deposited in different locations after being removed from one location by exogenic processes like wind, water, and glaciers. Weathered material is left in its original location when it weathers. The crucial element of sedimentary processes is erosion. The sediments needed for final deposition and transportation are provided by erosion.

Although erosion is a natural process carried out by natural forces, human activity has increased erosion rates globally by 10–40 times. Problems brought on by excessive erosion include the expansion of desert areas, deterioration of land, obstruction of waterways, and ecological imbalance from the loss of nutrient-rich soil.

The main physical processes responsible for erosion are:

- Rivers and streams;
- Wind;
- Glaciers;
- Rainfall;
- Floods;
- Coastal erosion;
- Freezing and thawing;
- Gravitational;
- Exfoliation.

Controlling factors of erosion are:

- Rainfall and wind speed;
- Strength of water current;
- Rock structure and composition;
- Topography;
- Vegetation cover.

Human activities responsible for erosion-

- Deforestation;
- Land use pattern;
- Climate change.

Environmental implications:

- Ecological imbalance;
- Land degradation;
- Tectonic effects.

1.2.1.3. Transportation and Deposition

The movement of solid particles, or sediments, across the surface of the earth caused by water, wind, ice, or gravity is known as sediment transportation. It is caused by both the movement of the fluid in which the sediment is entrained and the force of gravity acting on the sediment. Because of currents and tides, sediment is transported by fluid motion in rivers, lakes, and other bodies of water. Sediments flow in glaciers, and wind carries them on land surfaces. On sloping surfaces such as hill slopes, scarps, and cliffs, as well as on the boundary between the continental shelf and the continental slope, sediment transport can only happen as a result of gravity.

Brief mechanisms of sediment transportation and resultant products are narrated here:

- Aeolian/Eolian: Sand dunes and ripples are the end products of this process, which involves the wind carrying sediments. Because air is a fluid with low density and viscosity and cannot exert much shear on its bed, transported sediment is typically fine sand (<1 mm) and smaller. In the terrestrial near-surface environment, aeolian sediment transport creates bed forms. Because vegetation does not impede the presence and motion of sand fields in these environments, aeolian sediment transport is frequent on beaches and in arid regions of the world. Loess is the term for deposits of fine-grained glacial silt carried by the wind.
- Fluvial: Rivers move the sediments during this process. This is the sediment transportation process that is most significant. Because water has a higher density and viscosity than air, sediment moved by water can be larger than sediment moved by air. Depending

on the gradient and strength of the river's current, sediments ranging in size from large boulders and cobbles to as fine as clay are carried downstream.

Rivers carry sediment, which causes the creation of flood plains, deltas, and other sedimentary structures. Grain sediments are carried downstream by the flow once they start to move (above the critical erosional velocity). The energy of flow and grain size determine the mode of grain transportation.

- Coastal: The movement of waves and currents causes coastal sediment transport in near-shore environments. Deltas are formed at river mouths by the interaction of fluvial and coastal sediment transport processes. Additionally, it creates distinctive landforms like barrier islands and beaches. In marine depositional realms, sediments are also dispersed by waves and currents.
- Glacial: All sizes of material are entrained and moved by glaciers as they pass over their beds. Areas with glacial deposition frequently have a high concentration of glacial erratic, many of which are several meters in diameter. Glaciers are also capable of carrying the largest amounts of sediment. Additionally, glaciers crush rock into tiny fragments that are frequently carried thousands of kilometers away by winds to form "loess" deposits. The sediment that is entrained in glaciers frequently flows roughly along the lines of glacial flow, appearing at the surface in the ablation zone.
- Debris Flow: Massive amounts of material, consisting of mud mixtures, boulder-sized clasts, and water, are carried by debris flows. Granular debris flows down steep mountain valleys and washes in the form of debris flows. Their transport mechanisms and capacities scale differently from those of fluvial systems because they move sediment as a granular mixture. A similar mechanism applies to depositional systems in deep water as well.
- Hill Slopes: Regolith is transported downslope by a number of processes in the sediment transport of hill slopes.
- Soil Creep–Tree Throw: Movement of soil by burrowing animals – Slumping and land sliding of the hill slop.

1.2.2. Chemical Processes

The physical processes of weathering, erosion, material transportation, and deposition are all involved in sedimentation. Chemical changes occur in solutions as a result of sedimentation processes, which precipitate sediments out of the solution.

The calcium carbonate dissolves more readily in seawater due to its increased dissolved carbon dioxide content. Precipitation would result from the decrease in CO_2. Lime precipitation is encouraged, and bicarbonate ions are converted to carbonate ions by the processes that remove CO_2 from normal water (pH = 8.4). When the bicarbonate ions in seawater react with the Ca^{++} ions, $CaCO_3$ is produced, and CO_2 is released.

$$Ca^{++} + 2HCO_3^{-} <-> CaCO_3 + CO_2 + H_2O$$

Put another way, when carbon dioxide is dissolved in water, carbonic acid is created, and calcium carbonate (either aragonite or calcite) is dissolved. On the other hand, calcium carbonate can precipitate in the presence of enough bicarbonate and calcium.

Common and widely distributed sedimentary rocks called limestone are created chemically in shallow marine depositional environments. The majority of the calcium carbonate found in limestone is derived from biological sources, which include the hard, shelly portions of invertebrates like mollusks and the minuscule calcite and aragonite particles produced by algae. Sediment accumulation in carbonate-forming environments is primarily regulated by variables that affect the kinds and numbers of living things. Carbonate deposition and the accumulation of limestone successions are influenced by several factors, including water depth, temperature, salinity, nutrient availability, and the availability of terrigenous clastic material. Certain depositional environments are formed by living things; coral reefs, for instance, are the result of sedentary colonial organisms like these. Through the stratigraphic record, changes in the biota over geological time have also been crucial in defining the properties of shallow-marine sediments. Evaporite successions, which are created when gypsum, anhydrite, and halite chemically precipitate from seawater evaporation, are linked to carbonate sedimentation in arid environments. Among the other rocks created through chemical processes are phosphorites, chert, and dolomite.

1.2.3. Biological Processes

Numerous biochemical reactions—reactions that are driven by organisms—occur in the sediments. Important activities in the biological processes are (Figure 1.3):

1. Secretion of Calcium Carbonate Skeletons by Organisms: As a result of their metabolic processes, many living things cover their soft parts with calcium carbonate ($CaCO_3$), acting as external skeletons that support and shield them. These kinds of organisms are more prevalent in tropical and shallow waters. The skeleton of an organism can be composed of either low- or high-magnesian aragonite and/or calcite. A given chemical variety's secretion is influenced by temperature, growth rate, and the presence of specific inorganic or organic compounds.
2. Degradation of Calcium Carbonate Skeletons into Skeletal Debris: Lime sand, lime mud, and skeletal debris are the products of altered calcium carbonate skeletons. Skeletal materials are primarily used by predators to create sedimentary debris. Similar to corals, over 90% of the skeletal debris produced by reef degradation is the size of sand.
3. Trapping and Baffling by Organisms: The processes of baffling and trapping that organisms of lime mud perform result in deposits such as beds, laminae, and mounds. Filamentous blue-green algae are the most common indicator of trapping in shallow waters. Lime mud-like fine-grained foreign particles are captured by these algae. Stromatolites are laminated lithified deposits made up of alternating layers of material produced by microorganisms, primarily algae. Precambrian and other geologically older rocks are common places to find strombolites.

Figure 1.3. Cambrian stramatolite.

Source: https://commons.wikimedia.org/wiki/File:Stromatolites_%28Hoyt_Limestone,_Upper_Cambrian;_Lester_Park_Road_outcrop,_W_of_Saratoga_Springs,_New_York_State,_USA%29_11_%2815200834042%29.jpg.

4. Pelletization: Pelletizing soft lime mud, deposit feeding organisms excrete particles the size of sand. These pellets typically contain higher levels of organic matter.
5. Effects of Microorganisms: Microbial activities that propel different chemical reactions that result in the precipitation of pyrite, sulfur, and calcite.

1.3. SEDIMENTARY ENVIRONMENTS AND FACIES

The physical and chemical processes that are occurring there as well as the organisms that are currently residing there can be used to describe the environment at any given location on land or under the sea. A channel that restricts the flow of fresh water that carries and deposits sand or gravel on the channel's bars is an example of a fluvial (river) environment. Thin layers of relatively fine sediment are deposited on the floodplain by the river's floodwaters. In the area of the floodplain, vegetation grows and soils form. A lens of sandstone or conglomerate that displays internal structures created by deposition on the channel bars may serve as a representation of the channel in a succession of sedimentary rocks. Sandstone with roots and other signs of soil formation, as well as thinly bedded mudrock, will serve as representations of the floodplain environment.

The word "facies" is frequently used when describing sedimentary rocks in terms of their depositional environments. A body of rock with particular qualities that represent the circumstances of its formation is called a rock facies. A body of sediment's facies is described by recording all the features of its lithology, texture, sedimentary structures, and fossil content that can be used to infer formation processes. Facies associations and environments of deposition are related because the features of a depositional environment are shaped by the processes that exist within it. This can be achieved by identifying the combinations of processes that dominated the facies. If the deposits from the floodplain are discovered to be connected to the sandstone lens in Figure 1.4, it could be demonstrated to be a river channel. However, because channels filled with sand exist in a variety of settings, such as deltas, tidal environments, and the deep-sea floor, recognizing a channel form alone is not a sufficient basis to determine the depositional environment. Rather, the full picture of a depositional environment is provided by the association of various processes.

Figure 1.4. Very-fine-grained sediment from Glacial Lake Missoula, Montana, USA

Source: By Matt Affolter (talk) (Transferred by Citypeek/Original uploaded by Matt Affolter) – I (Matt Affolter (talk)) created this work entirely by myself. (Original uploaded on en.wikipedia), CC BY-SA 3.0, https://commons.wikimedia.org/w/index.php?curid=14711882.

1.4. THE SPECTRUM OF ENVIRONMENTS AND FACIES

Every depositional environment has a different set of processes, and the sedimentary rocks that result from these processes are an equally distinct assemblage. Depositional environments are categorized as, for example, deltas, estuaries, or shorelines, and subcategories of each are established, such as wave-dominated, tide-dominated, and river-dominated deltas, for ease of description and interpretation. This book uses the same methodology that sedimentary geologists generally employ. Nonetheless, it's critical to understand that these deposition environments are merely handy classifications or "pigeonholes," and that descriptions of them typically focus on "typical" instances. Although we categorize deltas into a number of types, the reality is that every delta, for instance, differs from its neighbor in space or time. Every deltaic deposit will also be unique. Our deposit, however, is likely to fall somewhere in between these "pigeonholes." When dealing with past events, incomplete data sets are inevitable, so it may not always be possible to determine with certainty whether deposits are from an estuary or a delta. Nonetheless, it is always feasible to offer some indication of the location and formation process of a sedimentary rock by evaluating each bed objectively in terms of physical, chemical, and biological processes.

1.5. BASIC PRINCIPLES OF STRATIGRAPHY

The foundation of stratigraphy is a set of rules that control the sedimentation processes. Rock material that has been deposited in a basin by glaciers, wind, and water has accumulated and is the cause of sedimentation. The fundamentals of stratigraphic studies are determined by the processes of sedimentation and accumulation of other stratified rocks, such as lava flows.

There are eight basic principles or laws of stratigraphy, which are listed below:

- Order of superposition;
- Original horizontality;
- Lateral continuity;
- Cross-cutting relationships;
- Inclusions;
- Unconformities;
- Fossil succession;
- Uniformitarianism and catastrophism.

1.5.1. Order of Superposition

It is evident that sediments settle layer by layer to the bottom of a basin where they are deposited. Thus, the first layer to deposit is the lowest layer. As more sediments are deposited in the ensuing layers, this process keeps going throughout the deposition process. Because they were deposited first, the beds or layers at the bottom of a sedimentary sequence are the oldest. The beds above them are more recent. This means that it is possible to identify the older and younger beds in a sedimentary sequence. Because the beds are layered on top of one another, a sedimentary sequence indicates the relative time of deposition. The idea is that, as one moves up a sedimentary sequence from bottom to top, the beds get progressively younger. One of the fundamental ideas of stratigraphy is the order of superposition, which is known as this principle (Figure 1.5).

Figure 1.5. Sedimentary layers.

Source: https://www.geologyin.com/2023/02/the-principle-of-superposition.html.

1.5.2. Original Horizontality

Sedimentary rocks can be found in all orientations, but they are rarely horizontal; instead, they are frequently angled, dipping, or even folded. They could not have been deposited in that state, that much is clear. No matter how the basin is shaped, sediments are always deposited in a horizontal fashion. After the rocks are deposited and consolidated, structural alterations like tilting and folding occur. Therefore, to study the rocks, you must imagine their nature rather than how they are currently arranged, presuming that they were originally deposited horizontally. This is the original horizontality principle.

1.5.3. Lateral Continuity

According to the lateral continuity principle, sediment layers are initially laterally continuous, extending laterally in all directions. Therefore, it is reasonable to assume that otherwise similar rocks that are currently divided by a valley or other erosional feature were once continuous.

Sediment layers do not go on forever; rather, their boundaries are defined by the kind and quantity of sediment that is available as well as the dimensions of the sedimentary basin. Sediment will eventually be deposited anywhere it is transported. Nevertheless, the layer of that material will get thinner as the amount of material decreases away from the source (Figure 1.6).

Figure 1.6. Monument Valley Navajo Tribal Park in Arizona shows The Mittens, i.e., East Mitten on the right and West Mitten on the left, with the matching of rock layers proving the law of lateral continuity.

Source: https://geologybase.com/lateral-continuity/.

Coarser-grained materials are frequently incapable of being moved to a certain area because the transporting medium cannot move them there with enough energy. There will be a lateral transition from coarser to finer-grained material, and finer-grained particles from the transporting medium will settle in its place. Sedimentary facies refers to the lateral variation in sediment within a stratum.

The sedimentary material will be deposited as far as the boundaries of the sedimentary basin if there is enough of it available. The sedimentary basin is frequently found inside rocks that differ greatly from the sediments being deposited. In those circumstances, a sudden shift in the type of rock will identify the lateral limits of the sedimentary layer.

1.5.4. Cross-Cutting Relationships

According to the geological principle of cross-cutting relationships, the younger of two geologic features is the one that cuts through the other. In geology, it's a relative dating method. The concept was initially introduced by Nicholas Steno, a Danish geological pioneer, in his Dissertationis prodromus. James Hutton later developed it in Theory of the Earth, and Charles Lyell refined it in Principles of Geology.

1.5.4.1. Types

There are several basic types of cross-cutting relationships:

- Structural relationships may be faults or fractures cutting through an older rock.
- Intrusional relationships occur when an igneous pluton or dike is intruded into pre-existing rocks.
- Stratigraphic relationships may be an erosional surface (or unconformity) cuts across older rock layers, geological structures, or other geological features.
- Sedimentological relationships occur where currents have eroded or scoured older sediment in a local area to produce, for example, a channel filled with sand.
- Paleontological relationships occur where animal activity or plant growth produces truncation. This happens, for example, when animal burrows penetrate into pre-existing sedimentary deposits.
- Geomorphological relationships may occur where a surficial feature, such as a river, flows through a gap in a ridge of rock. In a similar example, an impact crater excavates into a subsurface layer of rock.

Cross-cutting connections could have a complex structure. For instance, if a dike were to cut off an unconformity that caused a fault to be truncated. It is evident from these compound cross-cutting relationships that the fault predates the unconformity, which predates the dike. Such a justification helps to clarify the order of geological events.

1.5.4.2. Scale

Megascopically, microscopically, and cartographically, cross-cutting relationships can be observed. Put differently, there are different scales for these relationships. A large fault dividing the landscape on a large map could be an example of a cartographic crosscutting relationship. Megascopic cross-cutting relationships are features that would be found on an outcrop or in a restricted geographic area, similar to the previously mentioned igneous dikes. Microscopic cross-cutting relationships are ones that need to be examined closely or under magnification. An illustration of such a relationship is the drilling action of a boring organism penetrating a fossil shell.

1.5.5. Inclusions

Clasts or pieces of older rocks that are carried by river water or any other medium and deposited in a basin are frequently what make up sedimentary rocks. Following deposition, the clasts unite to form a rock. These clasts, which are inclusions in sedimentary rocks, vary in size from very fine silts to gravel fragments. There may also be inclusions in lava flows; these are referred to as xenoliths. The inclusions all have an age greater than that of the rock in which they are deposited. Since they are formed from the detritus of pre-existing rocks, this is evident.

1.5.6. Unconformities

A sedimentary sequence is made up of sediments that are continuously deposited. Frequently, a break in the sedimentation process occurs, known as an unconformity. The lack of sediment or the basin filling up could be the cause of the break in sedimentation. Alternatively, the basin might be elevated, in which case sedimentation would be impossible. The break could be extended, or it could just be temporary. Finding an unconformity in the former scenario might be challenging, but in long breaks, signs of erosion or a shift in the beds' angle of inclination can help locate the unconformity.

1.5.7. Fossil Succession

Fossils, which are the skeletal remains of extinct organisms, are found in most sedimentary rocks. However, the kind and species of organisms are constantly shifting. The fossils in various strata are therefore distinct from one another. Thus, it is possible to distinguish between beds at different levels based on the kind of fossils present in each bed. Fossils are useful in classifying the order of rocks and creating a chronology of deposition.

1.5.8. Uniformitarianism and Catastrophism

The uniformitarianism principle holds that current Earth processes, such as erosion and sedimentation, are natural, were also the same in the past. The changes that have occurred on Earth can be attributed to these processes over an extended duration. According to James Hutton, the adage "study of the present is the key to the past" serves as a compass for comprehending Earth's historical evolution. Uniformitarianism is a gradual process that results in changes over an extended length of time. However, it has been noted that a few catastrophic events have caused abrupt major changes on Earth. Thus, the history of Earth is made up of extended stretches of uniform, slow change (uniformitarianism) interspersed with sporadic, brief bursts of catastrophic activity (catastrophism). Hence, even though catastrophism and uniformitarianism have both been active historical processes on Earth, catastrophism is ultimately responsible for all of the major changes (Hakim et al., 2021; Gong et al., 2022).

1.6. STRATIGRAPHIC CLASSIFICATION

Geology's field of stratigraphy studies rock layers, or strata, and layering, or stratification. It is mostly applied to the study of layered volcanic and sedimentary rocks. Three related subfields of stratigraphy are called chronostratigraphy (stratigraphy by age), biostratigraphy (biologic stratigraphy), and lithostratigraphy (lithologic stratigraphy).

Although the term "stratigraphy" was first used by d'Orbingy in 1852, the idea that strata—or layers of rocks—represent a chronological order of past events is much older. The superposition principle was created by Steno in 1667 and states that "any layer in a sequence of layered rocks is older than the layer next above it." The relationship between rocks and time is known as stratigraphy, and a stratigrapher's job is to observe, describe, and interpret direct, tangible evidence found in rocks to learn about Earth's history. We all understand that the Earth is a dynamic environment, with mountains and oceans formed by plate tectonics and climate fluctuations possibly occurring on a human time scale due to changes in the atmosphere. The study of stratigraphy provides us with the historical behavior records we need to analyze the functioning of these global systems.

The geological sciences' temporal framework is provided by stratigraphy. Simple stratigraphic relationships (younger rocks typically lie on top of older, as Steno recognized), fossils preserved in strata, and measurements of processes like the radioactive decay of elements that allow us to date some rock units can all be used to determine the relative ages of rocks and, consequently, the events that are recorded in those rocks. Stratigraphy serves two purposes: first, it provides a global nomenclature for rock units of all ages and allows for their correlation; second, it looks for evidence of past climate change or plate tectonic movements. The record of local and global changes over periods of thousands to millions of years found in the rock strata is one of the most potent tools we have for forecasting future climate change. Furthermore, the study of fossils discovered in rocks of various ages has contributed to our understanding of evolutionary processes by providing information about the evolution of life forms over time. Additional features of stratigraphy offer the means to discover new resources. For instance, "sequence stratigraphy" is a forecasting method that is frequently employed in the oil and gas sector and can assist in locating new oil and gas reserves.

We are able to construct and connect images of the Earth's surface at various times and locations by combining sedimentology and stratigraphy. For example, the type of sedimentary rocks that have been deposited in a particular area may suggest that it was once an arid landscape with desert dunes and gravel washes from a nearby mountain range. In the same location, but at a different time, circumstances led to the development of coral reefs in a shallow sea far from any landmass. By analyzing the rocks according to their deposition processes and environments, we can discover the record of this change. Additionally, we could prove that while shallow tropical seas existed in one location, a few tens of kilometers away lay a deep ocean that was being deposited with fine sediment by ocean currents. As a result, we are able to construct images of the paleogeography, the past appearance of a region, and the evolution of paleogeography over Earth's history. To finish the picture, plate tectonics can be used to explain the distribution of various environments and how they have changed over time. This is because the majority of sediment is sourced from mountain building, and plate movements also produce the sedimentary basins where sediment is deposited.

1.6.1. Lithostratigraphy

Stratigraphy, the branch of geology that studies rock layers or strata, includes lithostratigraphy as a subdiscipline. Three main areas of study are petrology, comparative geology, and geochronology.

Generally speaking, strata are mainly igneous or sedimentary, depending on the process of rock formation. Sedimentary layers are formed by chemical precipitation, biogenic decomposition of organic matter, or sediment deposition linked to weathering processes. These layers are significant for the study of biostratigraphy and are frequently identifiable as having a large number of fossils. Igneous layers can be found in layered intrusions formed deep beneath the Earth's surface, stacks of lava flows, and layers of lava fragments (called tephra) erupted onto the surface by volcanoes. Igneous layers, which show magmatic or volcanic activity that took place during an area's geologic history, are typically fossil-free.

Relationships between strata are explained by a variety of principles. It is possible to determine the age of an igneous intrusion by observing its cutting through a sedimentary rock formation. A sedimentary rock layer in a tectonically undisturbed stratum is younger than the layer below it and older than the layer above it, according to the principle of superposition. According to the original horizontality principle, sediments deposit as essentially horizontal beds.

1.6.1.1. Types of Lithostratigraphic Units

The Danish naturalist Nicolas Steno first articulated the fundamentals of lithostratigraphy in his 1669 Dissertationis prodromus. A lithostratigraphic unit complies with the modern law of superposition, which states that younger rocks are positioned above older rocks in any succession of strata that hasn't been disturbed or turned over since deposition. According to the lateral continuity principle, a set of beds can be traced over a sizable area.

Physical characteristics of the rock that can be observed are the basis for defining and identifying lithostratigraphic units. A unit's lithology comprises elements like its primary depositional structures, texture, color, and chemical and mineralogical composition. It may also contain other organic materials like coal or kerogen, as well as fossils that are thought to be rock-forming particles. A lithostratigraphic unit cannot be defined lithologically using the taxonomy of fossils. Facies are defined by the descriptions of strata based on physical characteristics.

A lithostratigraphic unit's formal description comprises a stratotype, which is typically a type of section. An ideal exposure of the unit that displays its full thickness is a type of section. Further reference sections might be defined if the unit exhibits significant lateral variation or is never fully exposed. As their stratotype, long-standing lithostratigraphic units that predate the current stratigraphic codification or that lack tabular form (like volcanic domes) may use a type locality rather than a type of section. It is expected of the geologist defining the unit to provide enough information about the stratotype so that other geologists can identify the unit with certainty.

Lithosome: Masses of rock of essentially uniform character and having interchanging relationships with adjacent masses of different lithology, e.g., shale lithosome, limestone lithosome.

The formation is the basic Lithostratigraphic unit. A formation is a large enough mappable and traceable stratigraphic unit with unique lithology. Formations can be combined with other formations to form groups and supergroups, or they can be further divided into members and beds.

1.6.1.2. Stratigraphic Relationship

Two types of contact: conformable and unconformable:

- Conformable: Unbroken deposition, no break or hiatus (break or interruption in the continuity of the geological record). The surface strata resulting is called a conformity.
- Unconformable: Period of erosion/ non-deposition. The surface stratum resulting is called an unconformity.

Four types of unconformities:

- Angular Unconformity: Younger sediment lies upon an eroded surface of tilted or folded older rocks. The older rock dips at a different angle from the younger.
- Disconformity: The contact between younger and older beds is marked by visible, irregular erosional surfaces. Paleosol might develop right above the disconformity surface because of the non-deposition setting.
- Paraconformity: The bedding planes below and above the unconformity are parallel. A time gap is present, as shown by a faunal break, but there is no erosion, just a period of non-deposition.
- Nonconformity: Relatively young sediments are deposited right above older igneous or metamorphic rocks.

1.6.1.3. Lithostratigraphic Correlation

Geologists define facies and search for key beds or key sequences that can be used as a datum to correlate lithostratigraphic units.

- Direct correlation: based on lithology, color, structure, thickness.
- Indirect correlation: electric log correlation (gamma-ray, density, resistivity…)

The primary method for reassembling the geometry of layering in sedimentary basins is geological correlation. The process of lithological correlation determines which strata (layers) in geological cross-sections at various locations currently (or historically) belong to the same geological body. The identification process relies on general presumptions known as the Steno's principles as well as a comparison of the rocks' physical and mineralogical characteristics:

- The sedimentary strata occurred sequentially in time: the youngest at the top.
- The strata are originally horizontal.
- The stratum extends in all directions until it thins out or encounters a barrier.

Numerous challenges exist in practical correlation, such as hazy layer borders, differences in the makeup and structure of the rocks within the layer, irregularities in the layer sequence, etc. For this reason, mistakes in correlation schemes happen frequently. The quality of correlation improves as the distances between available cross-sections get closer together (for instance, by drilling new wells). However, in the interim, poor geological decisions may be made, driving up the cost of geological projects.

1.6.1.4. Lithodemic Stratigraphy

Rock bodies that are intrusive, highly deformed, or metamorphic and do not exhibit observable stratification are not subject to the law of superposition. Based on the properties of the rock, these groups of rock are identified and categorized as lithodemic. The formal terms "lithodeme," which is equivalent to a formation, "suite," which is equivalent to a group, and "supersuite," which is equivalent to a supergroup, were adopted by the North American Stratigraphic Code in 1983. The basic unit, known as a lithodeme, should have unique and consistent lithological characteristics, such as a single rock type or a combination of two or more types that set the unit apart from the surrounding units. Similar to formations, a lithodemic unit is named after a location along with a term that describes its form or a name for the rock. The phrase "suite" is out of use. The term "complex," which refers to a group of rocks that belong to two or more genetic classes (igneous, metamorphic, or sedimentary), has also been formalized (Hakim et al., 2021).

1.6.2. Biostratigraphy

The area of stratigraphy known as biostratigraphy focuses on using the fossil assemblages that are present in rock strata to correlate and determine the relative ages of those strata. Correlation, or proving that a specific horizon in one geological section represents the same period of time as another horizon at a different section, is the main goal of biostratigraphy. Because local variations in the sedimentary environment can result in sediments of the same age looking completely different, fossils found within these strata are valuable. For instance, there may have been more chalky limestones in one area and clays and marls in another. On the other hand, the two sediments are probably from about the same time if the fossil species found in them are similar. As the fundamental biostratigraphy units, these fossils are ideally used to define geological time periods based on the fossil species found within each section and to aid in the identification of biozones.

The foundational ideas of biostratigraphic principles date back several centuries to the early 1800s. One of the first geologists to notice that rock layers correlate to the Law of Superposition was Nicolas Steno, a Danish scientist and bishop. Science and technology progressed, and by the 18th century, the idea that fossils were the remains of extinct species preserved in the rock record started to gain traction. The process was well-established prior to Charles Darwin's explanation of the underlying mechanism, evolution. The scientists William Smith, George Cuvier, and Alexandre Brongniart concluded that fossils then revealed a sequence of chronological events, defining rock strata layers as a kind of unit that was subsequently named a biozone. From this point on, researchers started to establish boundaries and time periods within significant faunal changes by connecting the alterations in strata and biozones to distinct geological eras. These discoveries led to the international recognition of the Cambrian and Carboniferous periods by the late 18th century. Technological developments in the early 1900s allowed scientists to investigate radioactive decay. By using the isotopes found in fossils through radioactive decay, scientists were able to determine geological time, the boundaries of the various eras (Paleozoic, Mesozoic, Cenozoic), as well as Periods (Cambrian, Ordovician, Silurian). The oil and gas industries mainly use biostratigraphy to determine the age of rock layers for drilling workflows and resource allocations in the 21st century.

1.6.2.1. Fossils as a Basis for Stratigraphic Subdivision

Traditionally, periods were determined by the composition of their fossils. The majority of the periods that are now recognized end with a major extinction event or faunal turnover because early stratigraphers needed a significant change in the fauna to create a new period.

1.6.2.1.1. Concept of Stage

A stage is a significant division of strata that bears a distinct collection of fossils and follows each other in a systematic manner. Thus, a set of strata with similar major fossil assemblages can be referred to as a stage. This idea is credited to the French paleontologist Alcide d'Orbigny. He gave the stages their names based on geographical locations that had especially good portions of rock strata bearing the distinctive fossils that serve as the basis for the stages.

1.6.2.1.2. Concept of Zone

The concept of zones, also known as biozones or Oppel zones, was introduced by German paleontologist Albert Oppel in 1856. A zone consists of layers of rock containing fossils that have overlapping ranges. These zones represent the time period between the first appearance of a specific species at the base of the zone and the subsequent appearance of another chosen species at the base of the next zone. Oppel's zones are named after a unique fossil species, known as an index fossil, which is one of the species that define the zone.

Biostratigraphy utilizes zones as the primary unit of measurement, with their thickness and range varying from a few meters to hundreds of meters. The horizontal extent of these zones, which can be local or worldwide, is influenced by tectonic plates and activity. Metamorphic folding and subduction are two tectonic processes that can potentially alter the range of these zones. Additionally, biostratigraphic units are classified into six main types of biozones: Taxon range biozone, Concurrent range biozone, Interval biozone, Lineage biozone, Assemblage biozone, and Abundance biozone.

The known stratigraphic and geographic range of occurrence for a single taxon is represented by the Taxon Range Biozone. The concurrent, coincident, or overlapping portion of the ranges of two designated taxa is included in the concurrent range biozone. Based on either the lowest or highest occurrences, interval biozones comprise the strata that lie between two distinct biostratigraphic surfaces. Species belonging to a particular evolutionary lineage segment are found in strata known as lineage biozones. Strata that have a distinct association of three or more taxa within them are known as assemblage biozones. Strata known as abundance biozones are those where a given taxon or group of taxa is substantially more abundant than it is in the section's neighboring section.

1.6.2.1.3. Index Fossils

The fossilized remains or traces of specific plants or animals that are typical of a given period of geologic time or environment are known as index fossils (also called guide fossils, indicator fossils, or dating fossils), and they can be used to identify and date the rocks that contain them. Index fossils need to have a broad geographic distribution, a restricted vertical time range, and quick evolutionary trends to be useful. It is therefore known that two rock formations that are geographically far apart but contain the same index fossil species formed during the species' brief existence.

Index fossils served as the foundation for the initial definition and identification of geologic units, as well as the definition of geologic periods and faunal stages and zones.

Numerous species from the animal groups known as ammonites, graptolites, archeocyathids, inoceramids, and trilobites have been identified as index fossils, which are extensively utilized in biostratigraphy. Acritarchs, chitinozoans, conodonts, dinoflagellate cysts, ostracods, pollen, spores, and foraminiferans are a few other species of microfossils that are commonly employed. Different fossils are more beneficial for different ages of sediments; trilobites, for instance, are especially beneficial for Cambrian-aged sediments. A vast array of ammonite and inoceramid species are especially helpful in linking global environmental events that occurred during the Late Cretaceous super-greenhouse.

The fossils that are used must be widely distributed geographically to be found in a variety of locations. They also need to be a species with a limited life span, meaning that there is a limited amount of time for them to become incorporated into the sediment. Ammonites are favored over forms that evolve much more slowly, like nautiloids, because the longer a species lived, the less accurate the stratigraphy was.

A faunal assemblage, as opposed to a single species, is frequently the basis for biostratigraphic correlations because it allows for more precision because the time span during which every species in the assemblage existed collectively was shorter than the time spans of any individual members. Additionally, the presence of a single species in a sample may indicate that: (1) the strata formed within the organism's known fossil range; or (2) the organism's known fossil range was only partially understood and the strata extended it. For example, the Cambrian period's base was determined by the presence of the trace fossil Treptichnus pedum, which has since been discovered in older strata. It is possible to estimate the stratigraphic layers' times more precisely if the fossil is simple to identify and preserve.

1.6.2.1.4. Faunal Succession

William Smith first proposed the idea of faunal succession at the start of the 19th century. William first noticed that rock outcrops held a distinct collection of fossils while researching rock strata. Smith was able to arrange rock formations all over

England because of the theory that these far-off rock outcrops held comparable fossils. Smith noticed there were small variations within or between these fossil groups, although some rock beds may contain primarily similar species as a result of his work on these outcrops and his mapping of England. This variation in initially seemingly identical assemblages gave rise to the principle of faunal succession, according to which fossil organisms follow one another in a specific and predictable order, allowing any period of time to be classified according to the extent of its fossil record.

1.6.3. Chronostratigraphy

One aspect of stratigraphy that deals with the age and relative time relations of rock bodies is called chronostratigraphy. It should be mentioned that chronostratigraphy is an abstract concept that primarily relies on relative time as deduced from biostratigraphy. As the biostratigraphic classification indicates, the time must be inferred from the fossil content rather than being visible within the rocks. The primary goal of chronostratigraphy is to globally arrange the rock sequence into chronostratigraphic units, allowing for the correlation of all local, regional, and global occurrences with a single, accepted geological scale. Thus, the age of the rock sequence and their temporal relationships are the primary concerns of chronostratigraphy. Thus, the arrangement of rocks into units according to their age or time of origin is known as chronostratigraphic classification.

1.6.3.1. Methodology

To determine the hard dating of known and well-defined rock units that contain the particular fossil assemblages defined by the stratigraphic system, chronostratigraphy mainly relies on isotope geology and geochronology. In actuality, assumptions must be made to arrive at an age date that represents the start of the interval because it is exceedingly difficult to directly date most fossils and sedimentary rocks using isotope dating.

The cross-cutting relationship principles and the law of superposition serve as the foundation for the employed methodology.

The formation of a chronostratigraphic column mainly depends on intrusive and extrusive igneous rocks because they are nearly instantaneous on a geologic time scale, occur at specific intervals in time, and contain mineral assemblages that can be more precisely and accurately dated using isotopic techniques.

Metamorphism can be utilized to demarcate depositional intervals in a chronostratigraphic column. It is frequently connected to faulting. The age range in which a bed could have been formed may be limited by the occasional dating of metamorphic rocks. For instance, dating the crystalline basement will determine the maximum age of the fossil assemblage if a bed containing graptolites has ever covered the crystalline basement.

This procedure calls for a significant amount of work, as well as the verification of age dates and field relationships. The estimate of age must necessarily be made between the oldest cross-cutting intrusive rock in the fossil assemblage and the youngest rock upon which the fossil assemblage rests. For example, there may be many millions of years between a bed being laid down and an intrusive rock cutting it.

1.6.3.2. Chronostratigraphic Units

A body of rocks that includes all rocks formed during a particular interval of geologic time, as well as those formed during that time span, is known as a chronostratigraphic unit. In chronostratigraphy, a biozone's duration is referred to as a "biochron," and the different chronostratigraphic units can be distinguished using biochrons. The basic operational units of the geological time scale are called chronostratigraphic units.

Chronostratigraphic units, with examples:

- Eonothem – Phanerozoic;
- Erathem – Paleozoic;
- System – Ordovician;
- Series – Upper Ordovician;
- Stage – Ashgill.

We identify the stage that makes up the lowest unit of chronostratigraphy based on biochron. The remaining units, erathem, eonothem, series, and system, are arranged in ascending order. These are all relative in nature and are all identified based on the amount of fossils present. The sequence of the sediments and distinctive fossils of various beds that establish their relative ages are determined by the order of superposition.

Note that the fundamental goal of both geochronologic and chronostratigraphic classifications is to systematically group Earth's rock strata into units called units (geochronologic units) and units that correspond to intervals of geologic time (chronostratigraphic units). In addition to making, it easier to correlate time and determine the age of rock layers, this arrangement acts as a reference system for documenting geologic history events.

1.6.3.3. Differences from Geochronology

It's critical to distinguish between chronostratigraphic and geochronologic units. Since chronostratigraphic units are composed of geological material, it is accurate to state that Upper Cretaceous Series fossils of the Tyrannosaurus rex species have been discovered. Periods of time known as geochronological units share the same name as standard stratigraphic units; however, late/early is used in place of upper/lower. Therefore, it is also accurate to state that the Late Cretaceous Epoch is when Tyrannosaurus rex lived.

Because the age correlations obtained from chronostratigraphy are essential for creating precise cross-sections of the spatial organization of rocks and for creating precise paleogeographic reconstructions, chronostratigraphy is a significant subfield of stratigraphy (Gong et al., 2022).

CASE STUDY

- **A Case Study on Rock–Soil–Sediment System in Wanquan Area of Zhangjiakou, North China** (An, Y., Yin, X., Gong, Q., Li, X., & Liu, N., (2023). Classification and provenance on geochemical lithogenes: A case study on rock–soil–sediment system in Wanquan Area of Zhangjiakou, North China. Applied Sciences, 13(2), 1008. https://doi.org/10.3390/app13021008).

Geochemical gene is a new technique that can be used for composition classification and source traceability of geological samples, and the proposal of lithogene has initiated the study of geochemical lithogenes in recent years. The advantage of geochemical lithogenes is that they express trends among chemical elements in geological materials, such as the elemental spider diagram, and are applicable to rocks, soils, sediments, and plants, etc. Thus, it is a key link between geology and ecology using the relationship of elements rather than their absolute concentrations and overcomes the limitations of the applicability of graphical tracing methods and isotopic tracing methods in the traditional study of geochemistry (Figure 1; Tables 1 and 2).

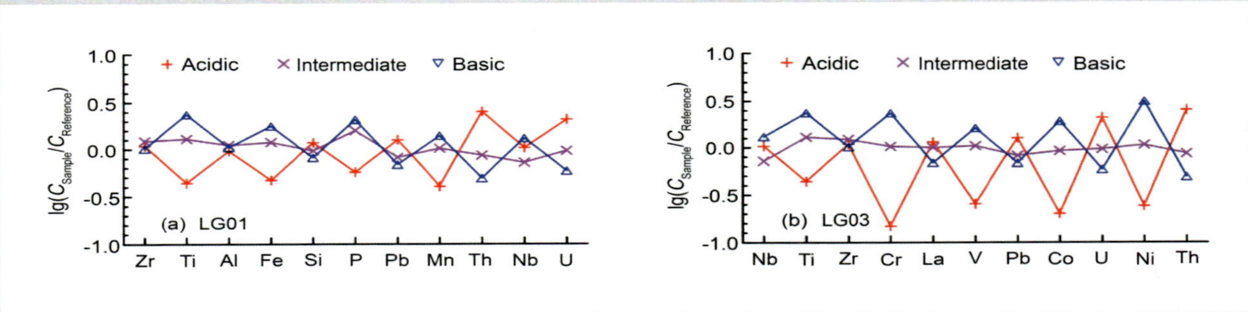

Figure 1. *Spectral lines of LG01 (a); and LG03 (b) lithogenes for the ideal rocks in China. Al, Fe, and Si are the abbreviations of Al_2O_3, TFe_2O_3, and SiO_2 on the horizontal axis.*

Source: An, Y., Yin, X., Gong, Q., Li, X., & Liu, N., (2023). Classification and provenance on geochemical lithogenes: A case study on rock–soil–sediment system in Wanquan Area of Zhangjiakou, North China. Applied Sciences, *13(2), 1008. https://doi.org/10.3390/app13021008.*

Table 1. *Elemental Sequences and Their Reference Values of LG01 and LG03 Geochemical Lithogenes*

10	Sequence No.	1	2	3	4	5	6	7	8	9	10	11
LG01	Gene elements	Zr	Ti	Al_2O_3	TFe_2O_3	SiO_2	P	Pb	Mn	Th	Nb	U
–	Reference values	147	4,016	14.8	6.4	60	756	19	940	5.72	14.5	1.2
LG03	Gene elements	Nb	Ti	Zr	Cr	La	V	Pb	Co	U	Ni	Th

–	Reference values	14.5	4,016	147	81	35	130	19	24	1.2	32	5.72

Source: An, Y., Yin, X., Gong, Q., Li, X., & Liu, N., (2023). Classification and provenance on geochemical lithogenes: A case study on rock–soil–sediment system in Wanquan Area of Zhangjiakou, North China. Applied Sciences, 13(2), 1008. https://doi.org/10.3390/app13021008.

Table 2. *The Elemental Abundances of Rocks in China*

Rocks	Zr	Ti	Al_2O_3	TFe_2O_3	SiO_2	P	Pb	Mn	Th	Nb	U	Cr	La	V	Co	Ni
Acidic rock	160	1,770	14.20	3.00	70.85	430	24	380	14.5	15	2.5	12	40	33	4.8	7.7
Intermediate rock	180	5,200	16.42	7.62	57.79	1,200	15.5	960	4.9	10.4	1.15	83	35	135	22	34
Basic rock	150	9,470	15.54	11.33	48.68	1,570	13	1,310	2.8	19	0.7	190	24	210	46	100

Source: An, Y., Yin, X., Gong, Q., Li, X., & Liu, N., (2023). Classification and provenance on geochemical lithogenes: A case study on rock–soil–sediment system in Wanquan Area of Zhangjiakou, North China. Applied Sciences, 13(2), 1008. https://doi.org/10.3390/app13021008.

The advantage of geochemical lithogenes is that they cannot only classify geological materials (rocks, weathered debris, soils, stream sediments, etc.) compositionally, breaking through the technical bottleneck that rocks, debris, soils, and sediments cannot be classified uniformly due to different criteria but also can establish a uniform classification method. The ideal acidic rock in China is a virtual rock sample represented by the elemental abundances of acidic rock in China, whose LG01 and LG03 lithogenes are both coded as 10202020202. The gene similarity between a sample and the ideal acidic rock is called the acidic similarity of the sample, which is labeled as R_{Acidic}. The R_{Acidic} of samples can be used as a tool to classify geological materials into three types compositionally, namely acidic-like ($R_{Acidic} \geq 80\%$), intermediate-like ($25\% \leq R_{Acidic} \leq 75\%$), and basic-like ($R_{Acidic} \leq 20\%$) components.

To further integrate the classification results of LG01 and LG03 lithogenes and make the sample classification results more accurate, Wu et al. proposed the concept of LG_CR (classification results on lithogenes). Firstly, samples were classified into three types: acidic-like component with $R_{Acidic} \geq 80\%$ labelled as 1-type, intermediate-like component with $75\% \geq R_{acidic} \geq 25\%$ labeled as 2-type, and basic-like component with $R_{acidic} \leq 20\%$ labeled as 3-type on LG01 and LG03, respectively. Subsequently, the classification results were expressed as double-digit numbers on the sequence of placing the classification results of LG01 in the first digit and LG03 in the second digit. Therefore, a total of 9 types (or LG_CR types) could be classified theoretically as 11, 12, 13, 21, 22, 23, 31, 32, and 33 types, although the 13 type and the 31 types may rarely occur in nature. The advantage of this classification method is that it is applicable to geological materials, such as fresh and weathered rocks, debris, soils, and sediments on a uniform criterion.

- **Geographical and Geological Settings:** The study area is located in the northern part of Wanquan district, Zhangjiakou city, Hebei province, China, with an area of about 1,080 km^2 ranging from E 114°16′33″ to 114°38′36″ and N 40°44′14″ to 41°04′07″ and is located in the transition zone between the North China Plain and the Inner Mongolia Plateau, with high topography in the northwest and low topography in the southeast, undulating hills, and a mainly shallow-cut landform type. The landform type is mainly shallowly cut stripped-erosion low hills, with altitudes between 1,000 and 1,300 m. The northern part is adjacent to the Zhangbei Dam Plateau. The area has a temperate semi-humid continental monsoon climate, with an average precipitation of about 400–500 mm. The soil type is mainly chestnut brown soil and brown soil, and the mountainous land is covered by larch, white birch, and mountain poplar, etc., while bulk crops and vegetables grow on the plains (Figure 2).

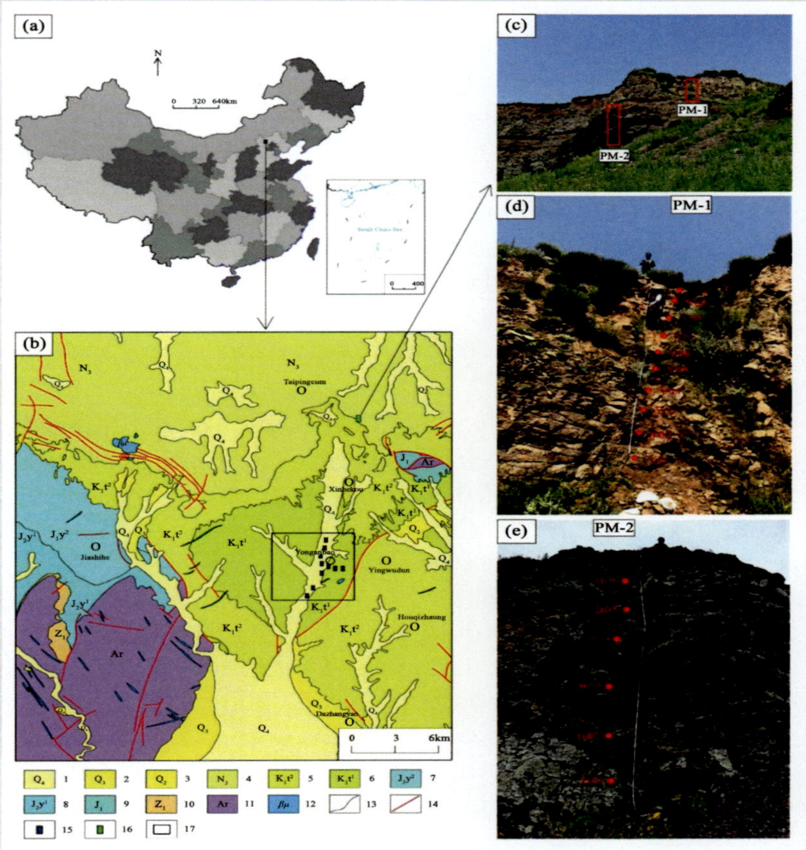

Figure 2. *Location of the study area in the Chinese mainland.*

Source: An, Y., Yin, X., Gong, Q., Li, X., & Liu, N., (2023). Classification and provenance on geochemical lithogenes: A case study on rock–soil–sediment system in Wanquan Area of Zhangjiakou, North China. Applied Sciences, *13(2), 1008. https://doi.org/10.3390/app13021008.*

The strata in the study area belong to the Archean, Sinian, Jurassic, Cretaceous, Neogene, and Quaternary periods, respectively, of which petrological descriptions are illustrated briefly in Figure 2. The spatial distribution has the following characteristics: the northern region is dominated by basaltic rocks of Hannuoba Formation (N_3), the central and southeast region is dominated by Jurassic, Cretaceous, and Quaternary rocks or sediments, and the southwest region is dominated by Archean and Sinian rocks. Faults mainly trend northwest and northeast. Intrusions in the study area are less developed, except diabase stock in the northwest.

- **Materials and Methods:**
  - **Materials:** A weathering profile was collected, with 16 samples (including 3 rock samples, 5 weathered debris samples, and 8 soil samples), which included 2 sub-profiles labeled as PM-1 and PM-2. The sampling length or regolith depth of PM-1 is ca. 7.5 m and that of PM-2 is ca. 7.6 m. There is a boundary or platform between the two parts and the total depth of the weathering profile is ca. 15.1 m. The profile samples are divided into three parts from the top to the bottom sequentially as soil, debris, and rock. When collecting soil or debris samples from the weathering profile, the lateral exposed surface soil should be removed at a depth of about 10–20 cm, while rock samples should be collected by removing the external weathering surface and taking the fresh part. The location of this weathering profile is just near the boundary between the first section of the Lower Cretaceous Tujingzi Formation (K_1t^1) and the upper Neogene Hannuoba Formation (N_3).

To trace soil migration scientifically according to the topography of the terrain, which can be initially recognized as two pathways, it is necessary to determine the main direction of migration first. In total, 10 vertical sediment profiles were laid out along these two migration directions. The first pathway profile CJPM-1 along the primary channel from N-S, with a vertical elevation difference of about 260 m and a transverse migration distance of about 5 km, includes 7 profiles of TP1, TP2, TP3, TP4, TP5, TP7, and TP8, which were spaced almost equally except the distance between TP5 and TP7. The second pathway profile CJPM-2 includes 4 profiles, TP14, TP15, TP16, and TP4 (used repeatedly), along the E-W tertiary channel, which were spaced almost equally with an interval of ca. 2 km and with a height difference of ca. 150 m. With respect to these 10 profiles, 5 samples in each profile were collected sequentially from the surface soil to the bottom soil. Each sample is collected continuously within a depth of 0.2 m and, therefore, the depth of each soil profile is 1 m. Thus, 50 soil samples in these 10 profiles were collected.

  - **Methods of Analyses:** Soil samples were put into clean white cotton bags after removing gravel, grass roots, animal dung, insect shells, etc., in the field. After collecting on the same day, they were placed in a ventilation room for air-drying to avoid breakage or mold caused by prolonged wet accumulation and then sent to the laboratory in time to complete preprocessing, such as grinding and sieving (200 mesh). The weight of bedrock and weathered debris samples was ca. 500 g. The field collection and processing of the samples were in accordance with the requirements of the Specification for Geochemical Evaluation of Soil Quality.

Rocks, weathered debris, and soils were analyzed for the major oxides of SiO_2, Al_2O_3, TFe_2O_3, K_2O, Na_2O, CaO, MgO, TiO_2, P_2O_5, and MnO and trace elements of Zr, Nb, Th, U, La, Pb, V, Cr, Co, and Ni. The major oxides were determined using a wavelength dispersive X-ray fluorescence spectrometer (ARL Advant XP + 2413) with detection limits of 0.05%, except Al_2O_3 of 0.03% and MgO, CaO of 0.02%. The trace elements were determined by high-resolution plasma mass spectrometry (X Serise2/SN01831C) where the detection limits were 5 for Zr, 2 for Nb, Th, V, Cr, 1 for La, Ni, 0.2 for Co, and 0.1 for Pb, U in μg/g. The accuracy and precision of the analytical method were controlled by the national standard substances (GSB-1, GSB-5, GSS-20, GSS24, GSS34) by adding 10% blank samples and parallel samples. The accuracy of the analysis of the first-grade standard substances was more than 98%, the repeatability of the sample test was more than 100%, and the relative standard deviation was less than 5%. The analytical methods, precision, accuracy, and detection limits were all in accordance with the requirements of the specification for a multi-purpose regional geochemical survey.

- **Methods of Weathering Indices:** The weathering degree of a sample is commonly depicted on weathering indices. The commonly used indices are CIA (chemical index of alteration) and WIG (weathering index of granite). The CIA was developed by Nesbit and Young in reconstructing paleoclimate from Early Proterozoic sediments and the WIG (weathering index of granite) was proposed by Gong et al. to describe the weathering degrees of granitic weathered products in the absence of CO_2 contents. Their calculation methods used here were detailed by Wu et al. and briefly illustrated as

$$CIA = [Al_2O_3/(Al_2O_3 + CaO^* + Na_2O + K_2O)] \times 100 \quad (1)$$

$$WIG = [Na_2O + K_2O + (CaO\text{-}10/3P_2O_5)]/(Al_2O_3 + TFe_2O_3 + TiO_2) \times 100 \quad (2)$$

where the oxide content is expressed in moles, CaO * represents CaO in silicates (i.e., removing CaO in carbonates and apatite), and ($CaO\text{-}10/3P_2O_5$) is taken as a non-negative value (i.e., 0 when its value is less than 0). It has been shown that CIA values are divided into <60, 60–80, and >80, and WIG values are divided into >60, 60–20, and <20, which represent the incipient, intermediate, and extreme weathering degree, respectively. Values of CIA increase with the weathering degree, while WIG values decrease with the weathering degree according to their definitions (Equations (1) and (2)).

- **Results and Discussion:**
  - **Weathering Profile:** The results on weathering indices of CIA and WIG of the weathering profile are shown in Figure 3. The values of CIA in the profile vary in a range of 44.6 to 74.0, and WIG values vary from 19.5 to 78.3, which indicates an incipient to intermediate weathering degree

mostly. Although the weathering degree is increasing gradually from fresh bedrock at the bottom to the weathered debris in the middle parts and then to topsoil, an abrupt variation at the boundary of PM-1 and PM-2 is recognized at the platform.

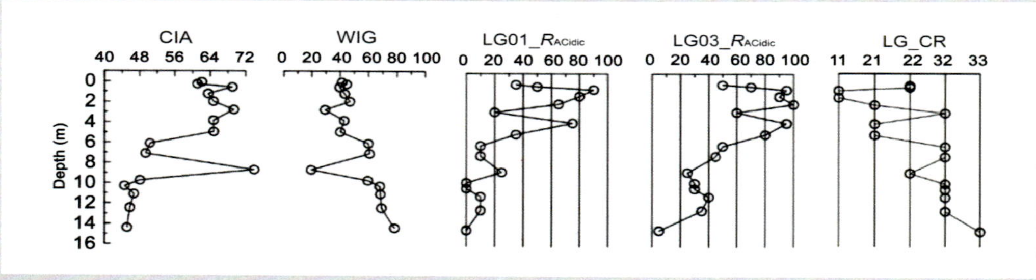

Figure 3. *Weathering indices, acidic similarities of lithogenes, and the LG_CR results in the weathering profile.*

Note: The geochemical lithogenes of LG01 and LG03 of samples from the weathering profile were coded on GGC (Geochemical Gene Coding) software, and then the acidic similarities (R_{Acidic}) were calculated and are illustrated in Figure 3.

Source: An, Y., Yin, X., Gong, Q., Li, X., & Liu, N., (2023). Classification and provenance on geochemical lithogenes: A case study on rock–soil–sediment system in Wanquan Area of Zhangjiakou, North China. Applied Sciences, *13(2), 1008. https://doi.org/10.3390/app13021008.*

With respect to the bottom rock, the R_{Acidic} values of LG01 and LG03 are all ≤20%, which indicates that the bottom sample is the 33 types of LG_CR compositionally. With respect to the other rock and debris (below 5 m in profile depth), they are basic-like materials on LG01 classification and intermediate-like on the LG03 classification; therefore, they belong to the 32 type of LG_CR compositionally, except the top material in the sub-profile of PM-2, which is the 22 type of LG_CR and is located near the boundary of the platform or sub-profiles, while the soils (above 5 m in profile depth) are 11, 21, 22, or 32 types of LG_CR, which indicates the lithology of the weathering profile is heterogeneous compositionally or soils were mixed with different sources compositionally. The lithological variation in the upper soils in the profile may be explained by its location near the petrological boundary of the conglomerate (K_1t^1) and the basalt with clay and shale in the middle and upper part (N_3).

- **Gully Sediment Profiles:** The LG01 and LG03 lithogenes of the 50 soils in CJPM-1 and CJPM-2 pathway profiles were coded by GGC software, and then the acidity similarity (R_{Acidic}) of each sample was calculated. The LG_CR results are derived based on the acidic similarities and illustrated in Figure 4.

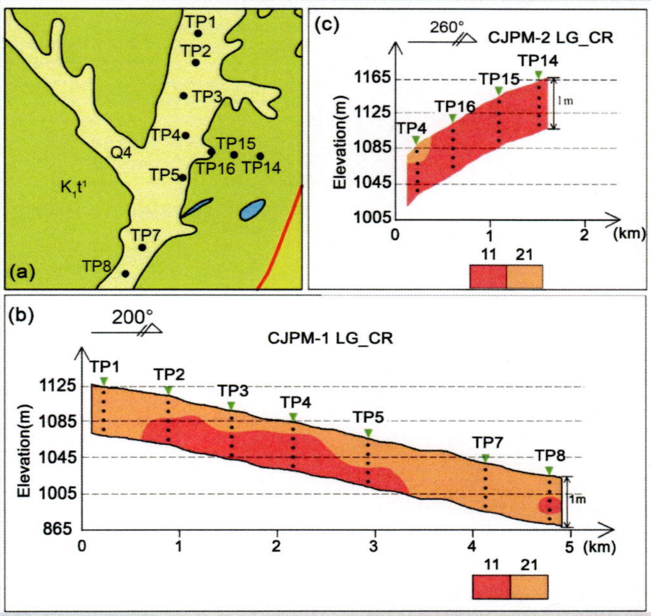

Figure 4. *Locations of the pathway profiles (a) and LG_CR results of soils from the CJPM-1; (b) and CJPM-2; and (c) profiles.*

Source: An, Y., Yin, X., Gong, Q., Li, X., & Liu, N., (2023). Classification and provenance on geochemical lithogenes: A case study on rock–soil–sediment system in Wanquan Area of Zhangjiakou, North China. Applied Sciences*, 13(2), 1008. https://doi.org/10.3390/app13021008.*

- o **Stream Sediments from RGNR Project:** Based on the 278 stream sediments' data retrieved from the RGNR project in the study area of Wanquan district of Zhangjiakou city in Hebei province, the codes and acidic similarities of LG01 and LG03 lithogenes are calculated and the LG_CR results are derived. The geochemical maps of RAcidic of LG01, RAcidic of LG03, and LG_CR values are contoured and classified compositionally like the method of the geochemical map.

- **Conclusions:** (1) The geochemical lithogenes are applied in the Wanquan area of Zhangjiakou city in Hebei province, China, on samples from a weathering profile, 10 sediment profiles, and 278 stream sediments, covering an area of ca. 1,080 km^2 to classify geological materials.

(2) The soils in the gully have undergone mixing with the upstream soils and the lithological results on LG_CR of stream sediments are basically consistent with the petrological results derived from the regional geological survey.

(3) The LG_CR method on geochemical lithogenes can be viewed as a useful tool to recognize and trace geological materials. Furthermore, it will have great potential in lithological mapping in petrological overburdened areas.

CLASS ACTIVITY

EXPLORING SEDIMENTARY ROCKS:

- **Objective:** Understand the formation, classification, and characteristics of sedimentary rocks through hands-on investigation and analysis.
- **Materials Needed:**
  - Samples of different sedimentary rocks (sandstone, shale, limestone, conglomerate);
  - Magnifying glasses;
  - Small plastic containers;
  - Sand, clay, small pebbles, and shells (for creating mini sedimentary layers);
  - Water spray bottles;
  - Handouts with a sedimentary rock classification chart;
  - Pencils and notebooks;
  - Poster board and markers.

SUMMARY

- Sedimentology is a scientific discipline concerned with the physical and chemical properties of sedimentary rocks as well as the processes involved in their formation, such as sediment transportation, deposition, and lithification (conversion to rock).
- Sedimentology encompasses the study of modern sediments such as sand, silt, and clay, and the processes that result in their formation (erosion and weathering), transport, deposition, and diagenesis.
- Sedimentary rocks cover up to 75% of the Earth's surface, record much of the Earth's history, and harbor the fossil record.
- Folding in sediments is analyzed with the principle of original horizontality, which states that sediments are deposited at their angle of repose which, for most types of sediment, is essentially horizontal.
- Processes of sedimentation include physical, chemical, and biological processes leading to the formation of different types of sedimentary rocks.
- Weathering is the breaking down of rocks, soil and minerals through the atmospheric agencies. Weathering occurs in situ and no transportation is involved.
- Sediment transportation is the movement of solid particles, i.e., sediments across the earth's surface by water, wind, ice or gravity.
- In the physical processes of sedimentation weathering, erosion, transportation, and deposition of material is involved. In the chemical processes of sedimentation, sediments are precipitated from the solution, i.e., chemical changes take place in solutions.
- Every depositional environment has a unique combination of processes, and the products of these processes, the sedimentary rocks, will be a similarly unique assemblage.
- Stratigraphy is based on a set of principles that govern the processes of sedimentation. The sedimentation is a result of accumulation of rock material carried by agencies like water, wind, and glaciers and deposited in a basin.
- The principle of lateral continuity states that layers of sediment initially extend laterally in all directions; in other words, they are laterally continuous.
- Lithostratigraphy is a sub-discipline of stratigraphy, the geological science associated with the study of strata or rock layers.
- Biostratigraphy is the branch of stratigraphy that focuses on correlating and assigning relative ages of rock strata by using the fossil assemblages contained within them.
- The chronostratigraphy is defined as an element of stratigraphy that deals with the relative time relations and age of rock bodies.

REVIEW QUESTIONS

1. What are the major classes of sedimentary rocks?
2. What do you mean by weathering?
3. Discuss the significance of the sedimentary environment.
4. What is stratigraphy? Discuss the concept and historical development of stratigraphy.
5. Describe the various principles of stratigraphy.

REFERENCES

1. Blatt, H., Middleton, G., & Murray, R. (1980). Origin of Sedimentary Rocks (2nd ed.). Prentice-Hall.
2. Bufe, A., Cook, K. L., Galy, A., Wittmann, H., & Hovius, N. (2022). The effect of lithology on the relationship between denudation rate and chemical weathering pathways—Evidence from the eastern Tibetan Plateau. Earth Surface Dynamics, 10, 513–530.
3. Carcaillet, J., Bourles, D. L., Thouveny, N., & Arnold, M. (2004). A high-resolution authigenic 10Be/9Be record of geomagnetic moment variations over the last 300 ka from sedimentary cores of the Portuguese margin. Earth and Planetary Science Letters, 219(3–4), 397–412. https://doi.org/10.1016/s0012-821x(03)00702-7.
4. Gaillardet, J., Dupré, B., Louvat, P., & Allègre, C. J. (1999). Global silicate weathering and CO_2 consumption rates deduced from the chemistry of large rivers. Chemical Geology, 159, 3–30.
5. Gong, Q., Yan, T., Wu, X., Li, R., Wang, X., Liu, N., Li, X., Wu, Y., & Li, J. (2022). Geochemical gene: A promising concept in discrimination and traceability. Applied Geochemistry, 136, 105133. https://doi.org/10.1016/j.apgeochem.2022.105133.
6. Hakim, K., Bower, D. J., Tian, M., Deitrick, R., Auclair-Desrotour, P., Kitzmann, D., Dorn, C., Mezger, K., & Heng, K. (2021). Lithologic controls on silicate weathering regimes of temperate planets. The Planetary Science Journal, 2(5), 1–29.
7. Hodson, A., Porter, P., Lowe, A., & Mumford, P. (2002). Chemical denudation and silicate weathering in Himalayan glacier basins: Batura Glacier, Pakistan. Journal of Hydrology, 262, 193–208. https://doi.org/10.1016/S0022-1694(02)00036-7.
8. Jiang, H., Liu, W., Xu, Z., Zhou, X., Zheng, Z., Zhao, T., Zhou, L., Zhang, X., Xu, Y., & Liu, T. (2018). Chemical weathering of small catchments on the Southeastern Tibetan Plateau I: Water sources, solute sources, and weathering rates. Chemical Geology, 500, 159–174. https://doi.org/10.1016/j.chemgeo.2018.09.030.
9. Ková⊠, M., Hók, J., Minár, J., Vojtko, R., Bielik, M., Pipík, R., Rakús, M., Kráå, J., Šujan, M., & Králiková, S. (2011). Neogene and Quaternary development of the Turiec Basin and landscape in its catchment: A tentative mass balance model. Geological Carpathica, 62(4), 361–379. https://doi.org/10.2478/v10096-011-0027-6.

10. Li, X., Ding, Y., Liu, Q., Zhang, Y., Han, T., Jing, Z., Yu, Z., Li, Q., & Liu, S. (2019). Intense chemical weathering at glacial meltwater-dominated Hailuogou Basin in the Southeastern Tibetan Plateau. Water, 11(6), 1–22. https://doi.org/10.3390/w11061209.

11. Liu, B., Li, Y., Eckstein, D., Zhu, L., Dawadi, B., & Liang, E. (2013). Has an extending growing season any effect on the radial growth of Smith fir at the timberline on the southeastern Tibetan Plateau? Trees, 27, 441–446. https://doi.org/10.1007/s00468-012-0819-z.

12. McLennan, S. M. (2001). Relationships between the trace element composition of sedimentary rocks and upper continental crust. Geochemistry, Geophysics, Geosystems, 2(4). https://doi.org/10.1029/2000GC000109.

13. Merchel, S., Beutner, S., Opel, T., Rugel, G., Scharf, A., Tiessen, C., Weiß, S., & Wetterich, S. (2019). Attempts to understand potential deficiencies in chemical procedures for AMS. Nuclear Instruments and Methods in Physics Research Section B: Beam Interactions with Materials and Atoms, 456, 186–192. https://doi.org/10.1016/j.nimb.2019.05.005.

14. Moore, D. M., & Reynolds, R. C. (1997). X-Ray Diffraction and the Identification and Analysis of Clay Minerals. Oxford University Press.

15. Murton, J. B., Peterson, R., & Ozouf, J. C. (2006). Bedrock fracture by ice segregation in cold regions. Science, 314(5802), 1127–1129.

16. Pipík, R., Bodergat, A. M., Briot, D., Kováč, M., Král, J., & Zielinski, G. (2012). Physical and biological properties of the late Miocene, long-lived Turiec Basin, Western Carpathians (Slovakia) and its paleobiotopes. Journal of Paleolimnology, 47(2), 233–249. https://doi.org/10.1007/s10933-011-9573-2.

17. Schieber, J., Southard, J., & Thaisen, K. (2007). Accretion of mudstone beds from migrating floccule ripples. Science, 318(5854), 1760–1763.

18. Sun, S. S., & McDonough, W. F. (1989). Chemical and isotopic systematics of oceanic basalts: Implications for mantle composition and processes. In Magmatism in the Ocean Basins (Vol. 42, pp. 313–345). Geological Society.

19. Torres, M. A., West, A. J., Clark, K. E., Paris, G., Bouchez, J., Ponton, C., Feakins, S. J., Galy, V., & Adkins, J. F. (2016). The acid and alkalinity budgets of weathering in the Andes–Amazon system: Insights into the erosional control of global biogeochemical cycles. Earth and Planetary Science Letters, 450, 381–391.

20. Victor, S., Neth, L., Golbuu, Y., Wolanski, E., & Richmond, R. H. (2006). Sedimentation in mangroves and coral reefs in a wet tropical island, Pohnpei, Micronesia. Estuarine, Coastal and Shelf Science, 66(3–4), 409–416.

21. Wu, Y., Li, X., Gong, Q., Wu, X., Yao, N., Peng, C., Chao, Y., Wang, X., & Pu, X. (2021). Test and application of the geochemical lithogene on weathering profiles developed over granitic and basaltic rocks in China. *Applied Geochemistry, 128*, 104958. https://doi.org/10.1016/j.apgeochem.2021.104958.

22. Zhang, Y., Han, R., Wang, L., & Wei, P. (2022). Zn–S isotopic fractionation effect during the evolution process of ore-forming fluids: A case study of the ultra-large Huize rich Ge-bearing Pb–Zn deposit. *Applied Geochemistry, 140*, 105240. https://doi.org/10.1016/j.apgeochem.2022.105240.

CHAPTER

2

Terrigenous Clastic Sediments: Gravel, Sand, and Mud

LEARNING OBJECTIVES

After studying this chapter, you will be able to:

- Understand the classification of sediments and sedimentary rocks;
- Explain gravel and conglomerate;
- Discuss sand and sandstone;
- Identify clay, silt, and mudrock;
- Know textures and analysis of terrigenous clastic sedimentary rocks.

KEY TERMS FROM THIS CHAPTER

Bedrock	Breccia
Calcite	Chemical weathering
Clastic sedimentation	Claystone
Coal	Conglomerates
Eogenesis	Glaciers
Gravel	Ion concentration
Lithification	Matrix

INTRODUCTION

Clastic terrigenous sedimentary rocks are made up of rock and mineral grains, or clasts, that range in size from materials the size of pebbles, cobbles, and boulders to clay, silt, and sand. These clasts are eventually deposited in diverse locations (e.g., in desert dunes, on alluvial fans, across continental shelves, and in river deltas) after being carried by wind, gravity, mudflows, flowing water, glaciers, and running water.

Terrigenous clastic sedimentary rocks are further divided based on average clast diameter because the agents of transportation typically sort discrete particles by clast size. Cobbles, boulder-sized gravels, and coarse pebbles lithify to form conglomerates and breccia; sand transforms into sandstone; clay and silt combine to form mudrock, shale, siltstone, and claystone. Siliciclastic sedimentary rocks are another name for them because the majority of the clasts are silica-rich. Based on the diameter of the clasts, silica rocks are further separated into breccia and conglomerate, sandstone, siltstone, and mudrock which is finer than silt (shale, claystone, and mudstone).

2.1. CLASSIFICATION OF SEDIMENTS AND SEDIMENTARY ROCKS

Particles washed from eroding rocks can be carried by rivers, oceans, winds, and rain runoff. These materials, referred to as detritus, are made up of broken rocks and minerals. The particles fall out during the sedimentation process when the energy of the transporting current is insufficient to carry them. Clastic sedimentation is the term used to describe this kind of sedimentary deposition. A different kind of sedimentary deposition happens when something dissolves in water and separates chemically from the liquid. Chemical sedimentation is the term used to describe this kind of sedimentation. A third process is possible, in which living things draw out ions that have been dissolved in water to form structures like shells and bones. We refer to this kind of sedimentation as biochemical sedimentation. Organic sedimentation is the term used to describe the buildup of plant matter, such as at a swamp's bottom. Clastic sedimentary rocks, chemical sedimentary rocks, biochemical sedimentary rocks, and organic sedimentary rocks are the four main categories of sedimentary rocks.

2.1.1. Clastic Sedimentary Rocks (Detrital)

Clastic sedimentary rock, sometimes referred to as "detrital" sedimentary rock, is composed of sediment fragments, or "clasts" or rock detritus, which have been weathered bedrock and cemented together to form a new rock. Due to its origin in the rocks of the crust, which are primarily made up of silicate minerals from igneous rock, the clastic sediment is frequently referred to as "siliciclastic." The size and makeup of the clasts determine how clastic (also known as detrital) rock is categorized and given names. The term "sediment grain size" describes their range of sizes, from boulders to clay. The top portion of the chart below lists the different clastic rocks along with their classification and description. Depending on the source rock and how long the sediment has been moving through the system, the clasts' size, shape, and composition can vary greatly.

2.1.2. Chemical, Biochemical, and Organic Sedimentary Rock

Chemical weathering and erosion are not the only processes that create chemical sedimentary rock. The ions that dissolve in water as a result of chemical weathering are what eventually form the different types of rock. The minerals found in inorganic chemical sedimentary rock are formed by ions that have dissolved in solution. When ion concentration, dissolved gases, temperatures, or pressures change and cause minerals to crystallize—for example, through evaporation—inorganic chemical sedimentary rock is created (Figure 2.1).

Figure 2.1. Sedimentary rock with sandstone in Malta.

Source: By Marco Almbauer – Own work, CC0, https://commons.wikimedia.org/w/index.php?curid=41805278.

Biochemical sedimentary rocks are composed of the bodies and shells of submerged organisms, like coral. The chemical elements that the living things take from the water are used to construct their shells and other body parts. Silica, mostly from the single-celled microorganisms known as radiolarians, or calcite, from a variety of marine life, make up biochemical sedimentary rocks.

The remnants of organic material that has gone through the lithification process are found in organic sedimentary rock. The raw materials are the remains of plants and animals, which are converted into coal, oil, and methane (natural gas) by burial and heat (Table 2.1).

Table 2.1. Classification of Sedimentary Rocks

Step 1: Determine Makeup	Step 2: Determine Grain Size	Step 3: Rock Description	Step 4: Rock Name
CLASTIC: Composed of pieces of rocks and minerals.	Granule (> 2 mm)	Rounded rock or mineral fragments; usually poorly sorted.	Conglomerate
–	–	Angular rock or minerals fragments; usually poorly sorted.	Breccia
–	Sand (0.06 mm to 2 mm).	Mostly quartz grains; sorting and rounding variable.	Quartz sandstone

–	–	Mostly quartz with at least 25% feldspar; rock fragments common; usually poorly sorted with angular grains.	Arkose
–	–	Clay, quartz, feldspars, and rock fragments; usually poorly sorted with angular grains; often dark color.	Graywacke
–	Silt (< 0.06 mm)	Silt-sized particles are too small to identify; no layering making it not fissile.	Siltstone
–	–	Silt and clay-sized particles are too small to identify; layering makes rock break in planes making if fissile.	Shale
–	Clay (< 0.004 mm)	Clay-sized particles are too small to see.	Claystone
Step 1: Determine makeup	Step 2: Composition	Step 3: Rock description	Step 4: Rock name
Chemical, biochemical, and organic: Made of minerals that have crystallized together, or biological fragments of shells of plants.	Calcite – All fizz with acid.	Abundant fossils. Possibly in micrite.	Fossilliferous Limestone[1]
–	–	Abundant ooids – coarse, sand sized spheres with concentric internal layers.	Oolitic limestone
–	–	Microcrystalline[2]; breaks with conchoidal fracture.	Micrite
–	–	Shell fragments loosely cemented with a high porosity.	Coqina
–	–	Microscopic fossil fragments; chalky; white color; soft.	Chalk

–	–	Microcrystalline[2]; color banding of browns, grays, whites, and blacks.	Travertine
–	–	Coarse crystals easily visible.	Crystalline limestone
–	Dolomite	Fizzes with acid only when powdered; fine to coarsely crystalline; may contain fossils.	Dolostone
–	Microcrystalline quartz	Microcrystalline[2]; conchoidal fracture; hardness of 7 so steel nail leaves a metal streak on the chert surface; sharp edges used for pre-historic spearheads and knives; synonymous with flint.	Chert
–	Gypsum	Crystalline; very soft with a hardness <2, white, gray, or pink color.	Rock gypsum
–	Halite	Crystalline; salty taste, white or gray color; fairly soft hardness of 2.5.	Rock salt
–	Plant material	Plant fragments; low density; brown to black color; often crumbly.	Bituminous coal

[1]Fossiliferous limestone can be further subdivided into Grainstone, Packstone, Wackestone, and Mudstone depending on the amount of mud with the lithified lime mud. Grainstone = less than 10% mud →→ Mudstone = less than 10% fossil organisms.

[2]Microcrystalline – crystals that are visible only through a high-powered microscope.

Source: https://geo.libretexts.org/Bookshelves/Geology/Historical_Geology_(Bentley_et_al.)/05%3A_Earth_Materials_-_Rocks/5.05%3A_Classification_of_Sedimentary_Rocks.

2.2. GRAVEL AND CONGLOMERATE

An unconsolidated collection of rounds, larger-than-sand fragments—more than 2 mm—is referred to as gravel. Content between 2- and 4-mm range has been referred to as very fine gravel or granule gravel. Conglomerates are silica-selective sedimentary rocks that are primarily made up of consolidated accumulation gravel-size (>2 mm) clasts. These rocks are sometimes referred to by their Latin-derived term, rudite. Conglomerates are common rocks found in all stratigraphic sequences, but they account for less than 1% of the sedimentary rock mass overall. Conglomerates, as opposed to sandstones, have a sizable portion of gravel-size particles. The percentage of gravel-size particles needed to differentiate a conglomerate from mudstone (shale) or sandstone.

2.2.1. Gravel

A loose collection of broken rock is called gravel. Because of erosive and sedimentary geological processes, gravel is a naturally occurring material on Earth. It is also widely produced commercially as crushed stone.

Gravel is categorized according to the size range of its particles, ranging from tiny granules to large boulder-sized pieces. Granular gravel (2–4 mm or 0.079–0.157 in) and pebble gravel (4–64 mm or 0.2–2.5 in) are the two categories for gravel according to the Udden-Wentworth scale. With ranges of 2–6.3 mm (0.079–0.248 in) for fine and 20–63 mm (0.79–2.48 in) for coarse, ISO 14688 classifies gravels as fine, medium, and coarse. Gravel usually weighs approximately 1,800 kg (4,000 lb) per cubic meter or approximately 3,000 lb (1,400 kg) per cubic yard. Gravel plays a crucial role in various industries due to its versatile applications. Approximately half of all gravel produced is utilized as an aggregate in concrete production. The remaining gravel is commonly used in road building, either as a base or surface material (with or without additional binders like asphalt). Gravel deposits found in nature are known for their high hydraulic conductivity, making them valuable sources of groundwater.

2.2.1.1. Definition and Properties

Informally, "gravel" refers to a blend of various-sized stone fragments combined with sand and potentially clay. The American construction industry distinguishes between gravel (a natural material) and crushed stone (produced artificially by mechanical crushing of rock).

Gravel has different technical definitions depending on the area and application. Many geologists do not specify an upper size limit when defining gravel; they simply define it as loose, rounded rock particles larger than 2 mm (0.079 in). Rubble, which is loose rock fragments in the same size range but with an angular shape, can occasionally be distinguished from gravel. Granular gravel is defined as particles with a size between 2 and 4 mm (0.079 to 0.157 in) and pebble gravel as particles with a size between 4 and 64 mm (0.16 to 2.52 in) by the Udden-Wentworth scale, which is commonly used by geologists in the US. This is true for all particles ranging in size from cobbles to coarse sand.

The U.S. Department of Agriculture and the Soil Science Society of America defines gravel as particles from 2 to 80 mm (0.079 to 3.150 in) in size, while the German scale (Atterburg) defines gravel as particles from 2 to 200 mm (0.079 to 7.874 in) in size. The U.S. Army Corps of Engineers defines gravel as particles under 3 in (76 mm) in size that are retained by a number 4 mesh, which has a mesh spacing of 4.76 mm (0.187 in). ISO 14688 categorizes gravel into fine, medium, and coarse grades based on particle size ranges.

Gravel ranges in bulk density from 1,460 to 1,920 kg/m^3 (2,460 to 3,240 lb/cu yd). The hydraulic conductivity of natural gravel can occasionally be more than 1 cm/s.

2.2.1.2. Origin

The majority of gravel is produced by bedrock weathering and disintegrating. Due to its hardness, lack of chemical inertness, and lack of cleavage planes that would facilitate easy rock splitting, quartz is the most frequently occurring mineral in gravel. Since few rocks contain mineral grains coarser than about 8 millimeters (0.31 in.), the majority of gravel particles are composed of multiple mineral grains. Pegmatites, high-grade metamorphic rock, deep intrusions, and quartz veins are examples of exceptions. Rivers carry the rock fragments, which are quickly rounded, frequently to within a few tens of kilometers of their original outcrops.

In stream channels, alluvial fans, near-shore marine environments where streams or coastal erosion supply the gravel, and in the deltas of swift-moving streams, gravel is deposited as gravel blankets or bars. There are large chert gravels in the upper Mississippi embayment that are estimated to have originated fewer than 100 miles (160 km) from the embayment's edge. There have been suggestions that the aeolian gravel "megaripples" that form in Argentina are similar to those on Mars.

2.2.1.3. Types

The composition, place of origin, and applications of various types of gravel set them apart. There are various kinds of gravel:

- Bank gravel is the naturally occurring gravel that is found in and near rivers and streams, mixed with sand or clay. Also referred to as "river run" or "bank run."
- Bench gravel is a gravel bed that marks the former location of the stream bed when it was at a higher level. It is found on the side of a valley above the current stream bottom. The Yukon Territory and Alaska are where the term is most frequently used.
- Stone that has been crushed, graded by screens, and combined with fines to create a mixture of fines and stones. It is frequently

used as a driveway and road surface, occasionally with tar applied on top. Rocks such as granite, limestone, dolomite, and others can be crushed to create crushed stone. Also referred to as "crusher run," shoulder stone, DGA (dense grade aggregate), and QP (quarry process). Crushed stone is distinguished from gravel by the U.S. Geological Survey.

- Fine gravel is made up of particles that range in diameter from 2 to 6.83 millimeters (0 to 248 in).
- Lag gravel is an accumulation of coarse gravel on the surface that is created when finer particles are removed.
- "Pay dirt," or "pay gravel," is another term for gravel that has a high concentration of gold and other precious metals. Gold panning is the method used to recover the metals.
- Pea gravel, sometimes called "pea shingle," is fine, tidy gravel that resembles garden peas in size, utilized as a substrate in home aquariums as well as on concrete surfaces, driveways, and walkways.
- Piedmont gravel is a coarse gravel that is transported by mountain streams from high altitudes to comparatively level areas where water moves more slowly.
- Plateau gravel is a layer of gravel that is found on a plateau or in another area that is higher than the normal location of stream-terrace gravel.
- Shingle is a loose, coarse, well-rounded, waterworn sediment that is mostly made up of flattened or smoothed pebbles, cobbles, and occasionally small boulders. It is typically between 20 and 200 millimeters (0.79 and 7.87 in) in diameter. Shingles are found in alluvial and beach environments.

2.2.1.4. Production and Uses

One important basic raw material used in construction is gravel. Official statistics typically do not distinguish between sand and gravel, but they do treat crushed stone as a different category. In 2020, sand and gravel together made up 23% of all industrial mineral production in the U.S., with a total value of about $12.6 billion. There were produced about 960 million tons of gravel and sand for construction. This is far more than the 68 million tons of industrial sand and gravel produced, the majority of which is sand rather than gravel.

Roughly 50% of gravel and sand used in construction are thought to be utilized as aggregate for concrete. Additional significant applications include filler for construction, road base or blacktop; and numerous smaller applications. Because gravel is so widely and abundantly distributed—mostly as river deposits, river flood plains, and glacial deposits—quality and environmental factors determine whether using crushed stone or other alternatives is more cost-effective. In the eastern United States, recycled gravel is also becoming more significant, and crushed stone is already replacing natural gravel.

2.2.2. Conglomerate

Conglomerate is a type of clastic sedimentary rock that consists primarily of gravel-sized clasts that are rounded to subangular in shape. The spaces between the clasts in a conglomerate are usually filled with a matrix of finer-grained sediments, such as clay, silt, or sand. Usually, hardened clay, silica, iron oxide, or calcium carbonate are used to cement the clasts and matrix.

Gravel solidifies and becomes lithified to form conglomerates. Although they are present in sedimentary rock sequences of all ages, their weight is most likely less than 1% of that of all sedimentary rocks. They share many characteristics with sandstones, including similar types of sedimentary structures, and are closely related to them in terms of origin and depositional mechanisms, e.g., graded bedding, tabular and trough crossbedding (Friedman, 2003).

The largest gravel accumulations in the geologic record are probably found in fanglomerates, which are poorly sorted, matrix-rich conglomerates that formed as debris flows on alluvial fans (Leeder & Mike, 2011).

2.2.2.1. Classification of Conglomerates

Conglomerates may be named and classified by the:

- Amount and type of matrix present;
- Composition of gravel-size clasts they contain;
- Size range of gravel-size clasts present.

The classification method depends on the type and detail of research being conducted. The naming of a sedimentary rock that is primarily made up of gravel is based on how round the gravel is. It is a conglomerate if the majority of the gravel clasts that make it up are well-rounded to subrounded. It is a breccia if the majority of the gravel clasts that make it up are angular. Such breccias can be called sedimentary breccias to differentiate them from other types of breccia, e.g., volcanic and fault breccias. Breccio-conglomerate is a term occasionally used to describe sedimentary rocks that have a mixture of angular and round gravel clasts (Figure 2.2).

Figure 2.2. Carmelo FORMATION (conglomerate) at Point Lobos.

Source: By Brocken Inaglory – Own work, CC BY-SA 3.0, https://commons.wikimedia.org/w/index.php?curid=10004410.

2.2.2.1.1. Texture

At least 30% of rounded to subangular clasts with a diameter greater than 2 mm (0.079 in) are found in conglomerates, e.g., boulders, cobbles, pebbles, and grains. Conglomerates, however, are rarely made up totally of clasts the size of gravel. The material that fills the spaces between the gravel-sized clasts is called matrix and is typically made up of different proportions of silt, sand, and clay. A conglomerate is referred to as a paraconglomerate if the individual gravel clasts are kept apart from one another by an abundance of matrix, allowing them to float within the matrix and avoid contact with one another. In addition to frequently being unstratified, paraconglomerates can have higher matrix contents than gravel clasts. An orthoconglomerate is created when the gravel clasts in a conglomerate come into contact with one another. Orthoconglomerates, in contrast to paraconglomerates, are frequently cross-bedded and well-cemented, with lithification often achieved by calcite, hematite, quartz, or clay (Nichols, 2009).

The variations in how paraconglomerates and orthoconglomerates are deposited are reflected in their differences. Paraconglomerates are typically deposits left over from debris flows or glacial tills. Aqueous currents are commonly linked to orthoconglomerates (Figure 2.3).

Figure 2.3. A conglomerate at the base of the Cambrian in the Black Hills, South Dakota.

Source: By Wilson44691 – Own work, Public Domain, https://commons.wikimedia.org/w/index.php?curid=4319174.

2.2.2.1.2. Clast Composition

Conglomerates are further categorized based on the make-up of their clasts. A conglomerate is referred to as monomict, monomictic, oligomict, or oligomictic conglomerate if it is any clastic sedimentary rock composed of a single rock or mineral. A conglomerate is referred to as polymict or polymictic if it contains two or more distinct types of rocks, minerals, or a combination of both. A polymictic conglomerate is referred to as either a petromict or petromictic conglomerate if it includes a variety of the clasts of metastable and unstable rocks and minerals.

Furthermore, conglomerates are categorized according to their source, which is indicated by the lithology of the gravel-sized clasts. An extraformational conglomerate is one that is made up of rocks and minerals that differ significantly from the surrounding matrix in terms of lithology, indicating that it is older and originated from outside the deposition basin. An intraformational conglomerate is created when the rocks and minerals found in these clasts are identical to or consistent with the lithology of the surrounding matrix, meaning they are penecontemporaneous and originated from within the deposition basin.

Shale-pebble and flat-pebble conglomerates are the two recognized forms of intraformational conglomerates. Shale-pebble conglomerates are formed by erosion in places like river channels and lake edges, and they are primarily made up of clasts of rounded mud chips and pebbles held together by clay minerals. Conglomerates known as flat-pebble conglomerates (also known as edgewise conglomerates) are made up of comparatively flat clasts of lime mud that are eroded by tidal currents along a shoreline or by storms or tsunamis that erode a shallow sea bottom.

2.2.2.1.3. Clast Size

Last but not least, conglomerates are frequently identified and distinguished based on the predominant clast size that makes up the group. According to this classification, a conglomerate made up primarily of clasts the size of grains would be referred to as a granule conglomerate, a conglomerate made up primarily of clasts the size of pebbles would be referred to as a pebble conglomerate, and a conglomerate made up primarily of clasts the size of cobbles would be referred to as a cobble conglomerate.

2.2.2.2. Sedimentary Environments

Conglomerates are deposited in a variety of sedimentary environments.

2.2.2.2.1. Deepwater Marine

The base of a bed in turbidites is usually coarse-grained, occasionally conglomeratic. Conglomerates in this context are typically highly sorted, round, and frequently exhibit a strong A-axis type imbrication of the clasts.

2.2.2.2.2. Shallow Marine

Basal conglomerates are conglomerates that are typically found at the base of sequences that were laid down during marine transgressions above an unconformity. They are diachronous and depict the location of the shoreline at a specific moment in time.

2.2.2.2.3. Fluvial

It is common for conglomerates deposited in fluvial environments to be poorly sorted and well-rounded. This size of clast is only transported as bedload during periods of high flow rate. Conglomerates are more typical of young river systems because the maximum clast size reduces as the clasts are transported farther owing to attrition. Conglomerates are typically limited to the lowest portion of a channel fill in the sediments left behind by established rivers, where they are referred to as pebble lags. Conglomerates that are deposited in a riverine environment frequently exhibit imbrication of the AB-plane type.

2.2.2.2.4. Alluvial

High relief regions are where alluvial deposits form, and they are usually coarse-grained. The thickest conglomerate deposits are found in the braidplains formed by the merging of individual alluvial fans at mountain fronts. The majority of conglomerates that are deposited in this environment have a strong AB-plane imbrication and are clast-supported. Due to debris-flow deposition, matrix-supported conglomerates are frequently linked to a large number of alluvial fans. When these clusters build up in an alluvial fan, in quickly eroding (e.g., desert) settings, the resulting unit of rock is commonly referred to as a fanglomerate.

2.2.2.2.5. Glacial

A large amount of material with coarse grains is carried by glaciers, and many glacial deposits are conglomeratic. The sediments that a glacier directly deposits are called tillites, and they are usually poorly sorted, matrix-supported conglomerates. Rock fragments that have been finely ground make up the fine-grained matrix in most cases. Glacier-related waterlaid deposits frequently form conglomeratic structures called eskers.

2.2.2.3. Conglomerate Uses and Application

Conglomerate is not very useful because of its unreliable fine particle breakdown and lack of cleanliness. Only in situations where low-performance material is desired can it be used as a crush. Conglomerate is rarely used for business purposes. Its inconsistent composition renders it a rock with unpredictable physical strength and durability, and its inability to break cleanly makes it a poor choice for dimension stone. When a low-performance material is appropriate, conglomerate can be crushed to create a fine aggregate. Even though conglomerates are frequently beautiful and colorful rocks, they are not frequently used as interior ornamental stones.

Conglomerate analysis is occasionally employed as a prospecting technique. For instance, kimberlite hosts the majority of diamond deposits. If kimberlite clasts are present in a conglomerate, the kimberlite's source must be upstream (Nichols, 2009; Leeder, 2011).

2.3. SAND AND SANDSTONE

When a geologist refers to "sand" in sandstone, they are referring to the size of the grains within the rock, not the substance that makes up the rock. Particles of sand vary in diameter from 1/16 millimeters to 2 millimeters. Sand-sized grains make up the majority of the composition of sandstones.

Sand-sized particles of mineral, rock, or organic material make up sandstone, a sedimentary rock. A matrix of silt- or clay-sized particles may be present to fill in the voids between the sand grains, in addition to a cementing substance that holds the sand grains together. Sandstone is a prevalent form of sedimentary rock that can be found worldwide in sedimentary basins.

2.3.1. Origins

Unlike either organic materials like chalk and coal or chemical materials like gypsum and jasper, sandstones are clastic in origin. Bedrock has weathered both chemically and physically, producing the silicate sand grains that give rise to them. In regions of high relief, such as volcanic arcs, areas where continents are rifting, and orogenic belts, weathering, and erosion occur most quickly.

Sand that has been eroded is carried by rivers or wind from its original locations to depositional settings where tectonics has made room for sediments to build up. Sand that is high in plagioclase and lithic grains is typically accumulated in forearc basins. Sand deposition also frequently occurs in intracontinental basins and grabens along continental margins.

The process of diagenesis occurs when older sand is buried by younger sediments as sediments continue to build up in the depositional environment. The sand is lithified and compacted primarily in this process. Early diagenesis, also known as eogenesis, is characterized by bioturbation and minor mineralogical changes in the sands, with only minor compaction, and occurs at shallow depths (a few tens of meters). Red bed sandstones get their color from red hematite, which is most likely formed during eogenesis. Mesogenesis, which occurs with deeper burial, is when most of the lithification and compaction happens.

As the pressure from the sediments above the sand increases, compaction occurs. Reduced pore space, deformation of ductile grains (like mica grains), and movement of sediment grains into more compact arrangements are all observed. Pressure solutions can also be used to achieve chemical compaction in addition to physical compaction. The most strained mineral is more soluble than the remainder of the grain at the points of contact between grains. The result is the dissolution of the contact points, which permits the grains to come into closer contact.

Compaction is closely followed by lithification because higher temperatures at depth accelerate the cement's deposition, which holds the grains together. Because the mineral dissolved from strained contact points is redeposited in the unstrained pore spaces, pressure solution aids in cementing.

The majority of mechanical compaction occurs at depths of less than 1,000 meters (3,300 feet). The majority of cementation occurs between 2,000 and 5,000 meters (6,600–16,400 feet), where chemical compaction continues down to 2,000 meters (6,600 feet). The third and last stage of diagenesis, telogenesis, occurs concurrently with the uncovering of buried sandstone. Re-exposure to meteoric water causes additional alterations to the sandstone, such as the dissolution of some of the cement to create secondary porosity, as erosion lowers the depth of burial.

2.3.2. Components

2.3.2.1. Framework Grains

The majority of sandstone is composed of framework grains, which are sand-sized (0.0625-to-2-millimeter (0.00246 to 0.07874 in) diameter) detrital fragments. As demonstrated by the Goldich dissolution series, the common minerals most resistant to weathering processes at the Earth's surface are quartz and feldspar, which make up the majority of framework grains. Based on their mineral makeup, framework grains can be divided into many categories:

- The majority of clastic sedimentary rocks are dominated by quartz framework grains due to their remarkable physical characteristics, including chemical stability and hardness. These physical characteristics enable the quartz grains to exhibit some degree of rounding and to withstand several recycling cycles. Quartz grains are formed by the recycling of older sandstones and plutonic rock, which is made of felsic material.
- In sandstones, feldspathic framework grains are typically the second most common mineral. A petrographic microscope can be used to distinguish between alkali feldspars and plagioclase feldspars, two subtypes of feldspar.
- Alkali feldspar range in chemical composition from $KAlSi_3O_8$ to $NaAlSi_3O_8$.
- Plagioclase feldspar range in composition from $NaAlSi_3O_8$ to $CaAl_2Si_2O_8$.
- Lithic framework grains are fragments of old source rock that haven't weathered into individual mineral grains. They're also referred to as lithic fragments or lithic clasts. Any fine- or coarse-grained igneous, metamorphic, or sedimentary rock can contain lithic fragments; however, volcanic rock clasts are the most frequently found lithic fragments in sedimentary rocks.
- All other mineral grains found in sandstone are considered accessory minerals. In most cases, these

minerals only comprise a small portion of the sandstone's grain content. Corundum, olivine, pyroxene, and micas (muscovite and biotite) are examples of common accessory minerals. Compared to the silicates that make up the majority of the rock, many of these accessory grains are denser. Zircon, tourmaline, rutile (hence ZTR), garnet, magnetite, and other dense, resistant minerals that are extracted from the source rock are examples of common heavy minerals. By using the ZTR index, these heavy minerals—which are frequently resistant to weathering—can be used as a gauge of the maturity of sandstone.

2.3.2.2. Matrix

The very fine material known as matrix is found in the spaces between the framework grains' interstitial pores. Two classifications arise from the nature of the matrix in the interstitial pore space:

- Arenites are texturally clean sandstones that are free of or have very little matrix.
- Wackes are texturally dirty sandstones that have a significant amount of matrix.

2.3.2.3. Cement

The siliciclastic framework grains are joined by cement. A secondary mineral called cement is created when the sandstone is buried, and deposition occurs. These cementing materials could be non-silicate minerals like calcite or silicate minerals like limestone.

- Minerals used to make silica cement may include either opal or quartz. The most prevalent silicate mineral used as cement is quartz. When silica cement is present in sandstone, the quartz grains are bonded to the cement, forming an overgrowth rim around the quartz grains. The overgrowth preserves the quartz framework grain's crystallographic continuity, which is being cemented. Opal cement is extremely rarely found in other sandstones; it is typically found in sandstones that are rich in materials formed by volcanoes.
- The most popular carbonate cement is calcite cement. A mixture of smaller calcite crystals is known as calcite cement. The framework grains are cemented together by the cement's adherence to them.
- Hematite, limonite, feldspars, anhydrite, gypsum, barite, clay minerals, and zeolite minerals are additional minerals that function as cements.

When weathering strips sandstone of its cement binder, it progressively becomes brittle and unstable. Applying tetraethyl orthosilicate ($Si(OC_2H_5)_4$), which will deposit amorphous silicon dioxide between the sand grains, can partially reverse this process. The response is given below:

$$Si(OC_2H_5)_4\ (l) + 2\ H_2O\ (l) \rightarrow SiO_2\ (s) + 4\ C_2H_5OH\ (g)$$

2.3.2.4. Pore Space

The pore spaces found in rocks and soil are referred to as pores. A rock's porosity and permeability are directly correlated with its pore space. The way the sand grains are arranged has a direct impact on the porosity and permeability.

- The percentage of a rock's bulk volume that is made up of interstitial spaces is known as porosity. The packing of even-sized spherical grains, which are rearranged in sandstones from loosely packed to tightest packed, directly affects porosity.
- The rate at which liquids, such as water, pass through rock is known as its permeability. Work permeability for groundwater can be expressed in gallons per day through a cross-section of one square foot at a hydraulic gradient of one unit.

2.3.3. Types of Sandstone

The standard method for classifying sandstones is to use a technique similar to the Gazzi-Dickinson Method to point-count a thin section. This produces the amount of clay matrix as well as the relative percentages of lithic, quartz, and feldspar grains. In conjunction with a triangular Quartz, Feldspar, Lithic fragment (QFL diagram), the composition of a sandstone can yield significant information regarding the genesis of the sediments. However, geologists haven't been able to come to a consensus on a set of borders that would divide the QFL triangle into different regions (Figure 2.4).

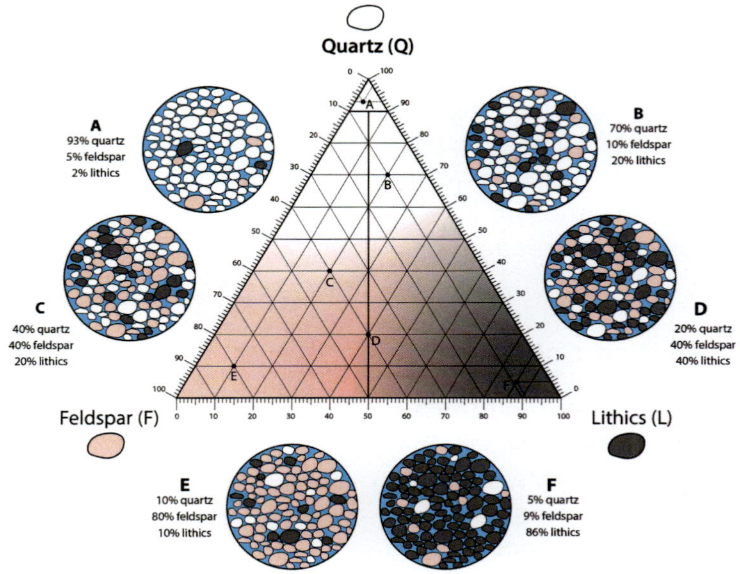

Figure 2.4. Ternary plot showing the relative abundance of quartz, feldspar, and lithic grains in a sandstone.

Source: By Michael C. Rygel – Own work, CC BY-SA 4.0, https://commons.wikimedia.org/w/index.php?curid=141773863.

Diagrams are visual aids that help geologists understand various sandstone properties. A provenance model, for instance, can be used to indicate on a QFL chart the probable tectonic origin of sandstones with different framework grain compositions. The various stages that a sandstone goes through as the degree of kinetic processing of the sediments increases are also shown by the stage of textural maturity chart.

- The framework grains and matrix found in sandstone are depicted in a QFL chart. This chart is akin to those employed in the field of igneous petrology. This analytical model produces a meaningful quantitative classification of sandstones when it is plotted correctly.
- Although it is usually based on a QFL chart, a sandstone provenance chart enables geologists to visually interpret the various locations that sandstones can originate from.
- The distinctions between immature, submature, mature, and supermature sandstones are displayed on a textural maturity stage chart. The sandstone's grains round out and the amount of clay in the rock matrix decreases as it ages.

2.3.3.1. Dott's Classification Scheme

Geologists utilize a variety of schemes, including Dott's (1964) classification scheme, to categorize sandstones. Dott's scheme is a modification of Gilbert's classification of silicate sandstones, and it incorporates R.L. Folk's dual textural and compositional maturity concepts into one classification system (Robert, 1964). The philosophy behind combining Gilbert's and R. L. Folk's schemes is that it is better able to "portray the continuous nature of textural variation from mudstone to arenite and from stable to unstable grain composition." Dott's classification system is predicated on the mineralogy of the framework grains as well as the kind of matrix that exists in between them.

Dott has established a 15% matrix boundary as the separation between arenite and wackes in this particular classification scheme. Furthermore, Dott divides the various kinds of framework grains that can be found in sandstone into three main groups: lithic, feldspar, and quartz grains.

- Sandstones classified as arenites have less than 15% clay matrix between the framework grains.
    - o Sandstones with more than 90% siliceous grains are known as quartz arenites. Rock fragments made of quartz or chert are examples of grains. The textural maturity of quartz arenites reaches supermature sandstone status. The extensive weathering that took place both before and during transport is what produced these pure quartz sands. Everything was gone except for the most stable mineral, quartz grains, due to weathering. They are frequently associated with rocks that were deposited in shelf or aeolian environments, which are stable cratonic environments. The multiple recycling of quartz grains results in quartz arenites, which are typically found as sedimentary source rocks or, less frequently, as first-cycle deposits that are derived

from primary igneous or metamorphic rocks.

- o Sandstones classified as feldspathic arenites have less than 90% quartz, more feldspar than unstable lithic fragments, and trace amounts of accessory minerals. Sandstones that are feldspathic are typically immature or sub-mature. These sandstones are found in stable or cratonic shelf environments. The primary crystalline rocks of the granitic type are the source of feldspathic sandstones. The sandstone is of igneous origin if it is primarily plagioclase.
- o The predominant feature of lithic arenites is the elevated concentration of unstable lithic fragments. Volcanic and metamorphic clasts are two examples; however, stable clasts, like chert, are frequently found in lithic arenites. This kind of rock has more unstable rock fragments than feldspars and less than 90% quartz grains. Texturally, they are typically immature to submature. They are found in marine conglomerates in deeper water or in conjunction with other fluvial deposits and conglomerates. They originate from fine-grained rocks, primarily shales, volcanic rocks, and metamorphic rock, and are formed in environments that generate significant amounts of unstable material.

- Sandstones known as wackes have more than 15% clay matrix in between the framework grains.
  - o Because quartz arenites are supermature and texturally mature, quartz wackes are rare.
  - o Felspathic wackes are composed of feldspathic sandstone with a matrix content higher than 15%.
  - o A sandstone with a matrix percentage of more than 15% is called lithic wacke.
- More than 25% of feldspar is present in arkose sandstones. Compared to pure quartz sandstones, the grains are typically less well sorted and less well-rounded. These quickly eroding granitic and metamorphic terrains, where physical weathering supersedes chemical weathering, are the source of these feldspar-rich sandstones.
- A fine-grained clay matrix encircles angular quartz and feldspar grains, forming a heterogeneous mixture known as greywacke sandstones. After the sandstone formation is deeply buried, comparatively soft fragments, like shale and some volcanic rocks, are physically and chemically compressed to form a large portion of this matrix.

2.3.3.2. Quartzite

The metamorphic rock known as quartzite is created when the intense heat and pressure of regional metamorphism cause the individual quartz grains in sandstone to recrystallize together with the previous cementing substance. The metamorphism erases most or all of the sandstone's original sedimentary structures and texture. Because the grains are so closely spaced, an irregular or conchoidal fracture results from the rock breaking through the grains.

By 1941, geologists had established that certain rocks exhibit the macroscopic features of quartzite even in the absence of high temperature and pressure metamorphism. Only the much lower temperatures and pressures associated with sedimentary rock diagenesis have been applied to these rocks; however, the rock has been cemented so thoroughly by diagenesis that only a microscopic examination can differentiate it from metamorphic quartzite. Such sedimentary rock is referred to as orthoquartzite to differentiate it from metaquartzite, which is the result of metamorphism. As a result, any quartz-cemented quartz arenite has occasionally been referred to more broadly by the term orthoquartzite. In the narrow sense, orthoquartzite is typically 99% SiO_2, with trace amounts of iron oxide and minerals like magnetite, rutile, and zircon. Normal fossil concentrations are low, but the original sedimentary structures and texture are retained.

A true orthoquartzite is usually distinguished from an ordinary quartz sandstone by its extremely high degree of cementation, which causes it to fracture across grains rather than around them. In the field, this distinction is identifiable. The start of the existing grains' recrystallization is what separates an orthoquartzite from a metaquartzite. A polarizing microscope can be used to identify the mortar texture in thin sections, which can be determined by marking the point at which strained quartz grains start to be replaced by new, unstrained, small quartz grains. Further recrystallization results in foam texture, which is characterized by polygonal grains coming together at triple junctions, and porphyroblastic texture, which is characterized by coarse, irregular grains, including some larger grains (porphyroblasts), as the metamorphism grade increases (Nichols, 2009; Leeder, 2011).

2.4. CLAY, SILT, AND MUDROCK

Despite being the most common type of sedimentary rocks in terms of volume, fine-grained terrigenous clastic sedimentary rocks tend to get less attention than any other group of deposits. Until scanning electron microscopes and X-ray diffraction analysis techniques were developed, little was known about the constituents of these sediments because the grain size is typically too small for optical methods of mineral determination. Mudrocks in the field rarely exhibit the distinct sedimentary and biogenic structures found in limestone and coarser clastic rocks. Because they typically do not form steep cliffs and because the soils support vegetation that covers the outcrop, exposure is typically poor. Because of this, this group of sediments is often disregarded, but as we will see in later sections pertaining to stratigraphy and depositional environments, they can offer just as much information as any other type of sedimentary rock.

2.4.1. Definitions of Terms in Mudrocks

The term "silt" refers to material with grains that range in diameter from 4 to 62 microns. There are four categories within this size range: coarse, medium, fine, and very fine. Only with a hand lens or the unaided eye can one see the larger silt grains. When a small amount of finer silt is ground between teeth, it will feel "gritty," whereas clay feels smooth. This is how finer silt is most easily distinguished from clay by touch. The finest grade of clastic sedimentary particles, those with a diameter of less than 4 microns, are referred to as clay in textural terminology. Particles are too small to be seen with the unaided eye and can only be slightly resolved using a high-power optical microscope. The primary components of clay-sized particles are phyllosilicate minerals, or clay minerals.

Mud is the term used to describe the mixture of unidentified proportions of silt and clay-sized particles that make up unconsolidated sediment. Any indurated silty or clay-based sediment can be referred to as mudrock in general. Mixtures of more than one-third of each component are referred to as mudstone. A rock may be referred to as a claystone or a siltstone depending on whether it can be shown that the majority of the particles (more than two-thirds) are clay-sized (Folk, 1974; Blatt et al., 1980). The term shale is sometimes applied to any mudrock (e.g., by drilling engineers) but it is best to use this term only for mudrocks that show a fissility, which is a strong tendency to break in one direction, parallel to the bedding. Keep in mind that slate, as opposed to shale, refers to fine-grained metamorphic rocks that fracture along one or more cleavage planes.

2.4.2. Silt and Siltstone

Because of the small particle size, silt is more difficult to determine in terms of textural parameters and mineralogy than sandstone. Optical microscope techniques are only easily applicable to coarser silt grains for analysis. Because other minerals have frequently undergone chemical breakdown prior to physical breakdown to this size, resistant minerals are more prevalent at this size. The most prevalent mineral in silt deposits is quartz. In addition to many other minute elements, this sediment grade also contains feldspars, muscovite, calcite, and iron oxides. The only place where silt-sized lithic fragments are common is in the "rock flour" created by glacial erosion.

Silt remains suspended in aqueous currents until the flow becomes extremely slow; as a result, deposition is typically observed in low-velocity flows or in standing water with minimal wave action. Silt-sized particles have the ability to float in the air as dust for extended periods of time and even travel very high into the atmosphere. Strong, continuous winds have the ability to transport silt-sized dust thousands of kilometers and deposit it in widely spaced lateral sheets; wind-blown silt appears to have played a significant role in the formation of loess deposits during glacial periods.

2.4.3. Clay Minerals

Commonly, feldspars and other silicate minerals break down to produce clay minerals. They are composed of aluminosilicates and are phyllosilicates with a layered crystal structure resembling micas. The silica, magnesium, and aluminum ions that make up the crystal layers are joined by oxygen atoms. There are two different patterns of layering: the smectite group, which has three layers, and the kandite group, which has two layers. The four most prevalent clay minerals—of which there are many—that are found in sedimentary rocks are examined here.

The most prevalent member of the kandite group, kaolinite, typically forms in warm, humid soil profiles where acidic waters strongly leach bedrock lithologies like granite. The swelling or expandable clay minerals of the smectite group, like montmorillonite, are capable of absorbing water into their structure. More temperate temperature conditions in soils with neutral to alkaline pH produce montmorillonite. In arid climates, it also forms in alkaline conditions. Illite, a member of the mica group and the most prevalent clay mineral in sediments, is another example of a three-layer clay mineral that forms in temperate soils with little leaching. A three-layer clay mineral called chlorite is most frequently found in arid soils and soils with moderate leaching under somewhat acidic groundwater conditions. Volcanic rocks, especially volcanic glass, weather to produce montmorillonite, illite, and chlorite.

2.4.4. Petrographic Analysis of Clay Minerals

A more advanced technological approach is necessary for the identification and interpretation of clay minerals compared to coarser sediment. X-ray diffraction pattern analysis and scanning electron microscopy are the two main methods. A fine electron beam that scans the sample's surface produces secondary electrons, which are used to create an image of the material under a scanning electron microscope (SEM). With this technique, features as small as microns can be imaged at a much higher resolution than can be achieved with an optical microscope. As a result, it's employed to look into the composition of clay minerals and how they relate to other rock grains. An SEM is the most practical tool for differentiating between clay minerals that are formed diagenetically within the sediment and those that are deposited as detrital grains.

The process of using an X-ray diffractometer (XRD) involves aiming an X-ray beam at a mineral powder or broken-up clay powder and measuring the angles at which the radiation is diffracted by the crystal lattice. Certain minerals have distinct patterns of diffracted X-ray intensity at different angles, which can be used to identify the mineral or minerals that are present. A reasonably quick and simple technique for semi-quantitatively figuring out the mineral composition of fine-grained sediment is X-ray diffractometer analysis. Additionally, some carbonate minerals with extremely similar optical properties can be distinguished using it.

2.4.5. Clay Particle Properties

Because clay minerals are small and have a platy shape, they stay suspended in relatively weak fluid flows and only settle out in very slow or stationary flows. As a result, clay particles are suspended in most water and air currents and only deposit when the flow stops.

When clay particles come into contact with each other, they stick together due to their cohesive nature. This cohesion can be attributed to both a thin film of water between the particles and an electrostatic effect caused by incomplete bonds in the mineral structure. The result is the formation of small aggregates of clay particles, known as flocculated groups, which settle out of suspension more quickly than individual particles. Flocculation is promoted by saline water conditions, leading to clay deposition in areas like estuaries and deltas. Once deposited, the cohesive nature of clay particles makes them resistant to being remobilized by flowing water, allowing for the preservation of fine sediment in environments with intermittent flows, such as tidal areas (Nichols, 2009; Leeder, 2011).

2.5. TEXTURES AND ANALYSIS OF TERRIGENOUS CLASTIC SEDIMENTARY ROCKS

Sediment transport and deposition leave behind markers indicating the mode of transport and deposition. For instance, the deposits that are left behind when material is transported by sliding down a slope are typically chaotic in nature and exhibit a large range of particle sizes. The texture of the resulting sediment is determined by the size and interactions of the grains. Consequently, we can deduce information about the mode of transport and deposition from the texture of the resulting deposits.

2.5.1. Sorting

The extent of grain size homogeneity. The energy of the transporting medium causes particles to be sorted according to density. Larger fragments can be carried by high-energy currents. Lighter fragments continue to be transported while heavier particles are deposited as the energy levels drop. Density causes sorting as a result.

The heavier particles will also be larger if the particles have the same density, so size will be the determining factor in the sorting process. This size sorting can be divided into two categories: well-sorted and poorly sorted. Sorting provides information about the energy state of the transport medium the sediment was deposited in.

Examples:

- Beach and wind-blown deposits typically exhibit good sorting because the transporting medium's energy is typically constant.
- Because the energy (velocity) in a stream fluctuates with time and position within the stream deposits are typically not well sorted.

2.5.2. Rounding

Grain abrasion during transit may cause grains to shrink in size. The sharp edges and corners of grains eventually round off due to random abrasion. As a result, grain rounding provides information about how long a sediment has been involved in the transportation cycle. Additionally, rounding is categorized using relative terms.

2.5.3. Sediment Maturity

The amount of time that sediment has been a part of the sedimentary cycle is referred to as sediment maturity. Texturally mature sediment is well-sorted and well-rounded, with better sorting occurring when smaller clasts are carried away and larger clasts are left behind. Rounding increases with transport distance and time. Mineral grains that are unstable near the surface become less common as the distance of transport or time in the cycle increases because the weathering processes continue during sediment transport. Thus, only the most stable minerals make up compositionally mature sediment.

An example of a texturally and compositionally immature sediment would be one with glassy angular volcanic fragments, olivine crystals, and plagioclase. The angular shape of the fragments suggests they have not been transported far, while the poor sorting indicates minimal time for the separation of larger and smaller fragments. Furthermore, the presence of unstable glass and minerals not typically found near the surface like olivine and plagioclase add to its immature composition.

Conversely, well-sorted beach sand with a majority of well-rounded quartz grains is texturally mature because the grains' roundness suggests a lengthy transportation cycle, and the sediment's well-sorting indicates a lengthy process needed to separate the sand's finer and coarser-grained components. Because the only material present in beach sand is quartz, which is extremely stable at the earth's surface, it has reached a mature composition.

2.5.4. Cycles of Sedimentation

Mineral grains and lithic clasts eroded from an igneous rock, like granite, are carried by various processes to a point where they are deposited to create a collection of clastic sediment. This material, known as a first cycle deposit, has undergone one cycle of erosion, transportation, and deposition. Once this sediment has solidified into sedimentary rock, it may be uplifted by tectonic activities and endure further erosion, transportation, and re-deposition. The re-deposited material is classified as a second cycle deposit, as the individual grains have experienced two cycles of sedimentation. Clastic sediment can undergo multiple cycles of sedimentation, with the mineralogical and textural maturity of the detritus increasing each time. Only resistant minerals such as quartz and lithic fragments of chert can endure repeated weathering, erosion, transportation, and re-deposition. Some heavy minerals, like zircon, are also highly resistant, with the roundness of zircon grains indicating the number of sedimentation cycles the material has undergone (Nichols, 2009; Leeder, 2011).

CASE STUDY

SWANSEA SAND DUNES – NATURE-BASED SOLUTION

Natural Resources Wales is a devolved body responsible for maintaining environment and resources sustainably in Wales.

- **The Idea:** Rising sand levels in Swansea Bay, coupled with high winds off the sea, meant a seafront road had to be regularly cleared. Swansea Council spent tens of thousands of pounds a year sweeping the road clear. There were also disposal costs, as contamination with road debris meant sand had to go to landfill, instead of back on the beach. A seafront promenade and cycleway were made dangerous by the sand.

Natural Resources Wales suggested encouraging a sand dune system to form. This would stabilise the sand and contribute to the local environment.

- **What Happened?** As part of new requirements to consult on plans for Welsh river catchments, NRW undertook a very broad consultation with residents and businesses around the Tawe. The wind-blown sand problem came out of this consultation. NRW recognized that they could use this as an opportunity to showcase the potential of nature-based solutions, rather than expensive hard infrastructure such as groynes or rock-filled gabions. They had to get the Council on board though.

Luckily, Swansea Council had already been advised that a dune system might be the answer, but they didn't have the up-front cash to pay for it. Joint funding, use of NRW's internal works team and a reduction in the length of the dune, made the project affordable.

Volunteers from local school and organizations planted the grasses required to help the dunes form. NRW experimented with different types of grass to see what effect they had on dune formation.

The sand dune project has produced multiple benefits, beyond reducing maintenance costs and reducing risks to cyclists. The aesthetics of the beach are improved, and habitats are more diverse. As dunes have started to form naturally, the protection they provide from flooding and erosion of the foreshore is only going to increase.

- **Insights:**
  - Up-front costs can put off public bodies from projects with public benefit and savings. Invest to Save and Innovate to Save programs can help kick start projects like these.
  - Bringing diverse sets of expertise together around shared goals can enrich ideas and bring about broader ranges of benefits – as well as allowing pooled funding.

- **What's Next?**

Natural Resources Wales has begun using the dune project as a case study of how nature-based solutions can outperform traditional hard infrastructure when it communicates with Councils and other bodies in Wales.

The whole promenade wasn't protected by the scheme, so windblown sand still affects the promenade. Swansea Council hasn't said it plans to extend the sand dune project across the whole area, and since this was a demonstration project, Natural Resources Wales doesn't intend on joint funding the rest of the dune system. Swansea Council has begun using the contaminated sand to supply its cold weather road gritting program.

CLASS ACTIVITY

- **Activity:** Sedimentary rocks are composed of different layers. These layers can be pebbles, sand, silt, or fossil fragments. Your students will understand the characteristics of sedimentary layers.
- **Objective:** Students will understand that most sedimentary rocks such as coal are formed under the water and in swamps, lakes, seas, and oceans.
- **Materials:**
  - Glass jars with lids;
  - Water;
  - Rocks;
  - Pebbles;
  - Sand;
  - Soil.

Source: https://wiki.ubc.ca/index.php?curid=98656.

This activity can be done individually, in small groups or as a demonstration by the teacher for the class.

1. Have students collect rocks, pebbles, sand, and soil at school or home.
2. Fill a jar about one-third full of the rocks, pebbles, sand, and soil. Add water and shake the jar carefully.
3. Have students write a paragraph/statement on what they predict will happen and why.
4. Let the jar stand undisturbed. Materials will settle into layers with the coarser, heavier material at the bottom and the finer, lighter layers at the top.
5. Tell the students that these sedimentary layers usually form in lake beds and near running water. After a long period of time, they will turn to rock.
6. Have students measure and graph the various sizes of the layers.
7. Students will enjoy re-shaking the jar and watching the materials resettle.

SUMMARY

- Rivers, oceans, winds, and rain runoff all have the ability to carry the particles washed off of eroding rocks. Such material, called detritus, consists of fragments of rocks and minerals.
- Clastic sedimentary rock (also called 'detrital' sedimentary rock) consists of sediment pieces (clasts/rock detritus) that come from weathered bedrock, cemented together to make a new rock. The clastic sediment is often referred to as "siliciclastic" because it is derived from the rocks of the crust which, as we have learned, is largely composed of silicate minerals of igneous rock.
- Chemical sedimentary rock is formed by processes that do not directly involve mechanical weathering and erosion. Chemical weathering contributes to the dissolved ions in water that ultimately form the various rock types.
- Biochemical sedimentary rock is formed from shells and bodies of underwater organisms (such as coral). The living organisms extract chemical components from the water and use them to build shells and other body parts. Biochemical sedimentary rocks are commonly composed of calcite, from a wide variety of sea life, or silica, largely from the single-celled microorganisms called radiolarian.
- Organic sedimentary rock contains the remains of organic material that has undergone the lithification process. The source materials are plant, and animal remains that are transformed through burial and heat, and end up as coal, oil, and methane (natural gas).
- Gravel is a loose aggregation of rock fragments. Gravel occurs naturally on Earth as a result of sedimentary and erosive geological processes; it is also produced in large quantities commercially as crushed stone.
- Conglomerate is a clastic sedimentary rock that is composed of a substantial fraction of rounded to subangular gravel-size clasts. A conglomerate typically contains a matrix of finer-grained sediments, such as sand, silt, or clay, which fills the interstices between the clasts. The clasts and matrix are typically cemented by calcium carbonate, iron oxide, silica, or hardened clay.
- Sandstone is a sedimentary rock composed of sand-size grains of mineral, rock, or organic material. It also contains a cementing material that binds the sand grains together and may contain a matrix of silt- or clay-size particles that occupy the spaces between the sand grains.
- Sandstones are typically classified by point-counting a thin section using a method like the Gazzi-Dickinson Method. This yields the relative percentages of quartz, feldspar, and lithic grains and the amount of clay matrix.
- Fine-grained terrigenous clastic sedimentary rocks tend to receive less attention than any other group of deposits although they are volumetrically the most common of all sedimentary rock types.
- Silt is defined as the grain size of material between 4 and 62 microns in diameter. This size range is subdivided into coarse, medium, fine and very fine. The coarser grains of silt are just visible to the naked eye or with a hand lens.

- Clay minerals commonly form as breakdown products of feldspars and other silicate minerals. They are phyllosilicates with a layered crystal structure similar to that of micas and compositionally they are aluminosilicates. The crystal layers are made up of silica with aluminum and magnesium ions, with oxygen atoms linking the sheets.
- Identification and interpretation of clay minerals requires a higher technology approach than is needed for coarser sediment. There are two principal techniques: scanning electron microscopy and X-ray diffraction pattern analysis.
- The small size and platy shape of clay minerals means that they remain in suspension in quite weak fluid flows and only settle out when the flow is very sluggish or stationary. Clay particles are therefore present as suspended load in most currents of water and air but are only deposited when the flow ceases.
- When sediment is transported and deposited, it leaves clues to the mode of transport and deposition. For example, if the mode of transport is by sliding down a slope, the deposits that result are generally chaotic in nature and show a wide variety of particle sizes. Grain size and the interrelationship between grains give the resulting sediment texture. Thus, we can use the texture of the resulting deposits to give us clues to the mode of transport and deposition.

REVIEW QUESTIONS

1. Describe the three main types of sedimentary rocks.
2. How can different types of limestone be bioclastic, chemical, and organic?
3. List sedimentary rock types by grain size, from small to large.
4. List the three types of sedimentary rocks.
5. Where and how are clastic rocks found?
6. How do clastic rocks form?
7. Contrast conglomerates and breccia rocks.
8. What can be found in clastic rocks?
9. Explain the difference between layers and bands.
10. What can we learn from sedimentary rocks?
11. How do chemical rocks form?
12. What are bioclastic rocks?
13. List the two types of bioclastic rocks.
14. What is the significance of red colored sediment? What is the significance of black colored sediment with high amounts of organic material?
15. How does sediment turn into hard sedimentary rock?

REFERENCES

1. Blatt, H. (1982). Sedimentary Petrology. W. H. Freeman and Co.
2. Blatt, H., Middleton, G. V., & Murray, R. C. (1980). Origin of Sedimentary Rocks (2nd ed.). Prentice-Hall.
3. Boggs, S. (2006). Principles of Sedimentology and Stratigraphy (2nd ed.). Prentice Hall.
4. Dott, R. H. (1964). Wacke, greywacke and matrix; what approach to immature sandstone classification? SEPM Journal of Sedimentary Research, 34(3), 625–632. https://doi.org/10.1306/74D71109-2B21-11D7-8648-000102C1865D.
5. Flugel, E. (2010). Microfacies of Carbonate Rocks: Analysis, Interpretation and Application (2nd ed.). Springer-Verlag.
6. Folk, R. L., & Lynch, F. L. (2001). Organic matter, putative nanobacteria and the formation of ooids and hardgrounds. Sedimentology, 48, 215–229.
7. Friedman, G. M. (2003). Classification of sediments and sedimentary rocks. In G. V. Middleton (Ed.), Encyclopedia of Sediments & Sedimentary Rocks (pp. 127–135). Kluwer Academic Publishers.
8. Leeder, M. (2011). Sedimentology and Sedimentary Basins: From Turbulence to Tectonics (2nd ed.). Wiley-Blackwell.
9. Neuendorf, K. K. E., Mehl, J. P., & Jackson, J. A. (2005). Glossary of Geology (5th ed.). American Geological Institute.
10. Nichols, G. (2009). Sedimentology and Stratigraphy (2nd ed.). John Wiley & Sons Ltd.
11. Seibold, E., & Berger, W. H. (1996). The Sea Floor: An Introduction to Marine Geology. Springer.
12. Tucker, M. E. (2001). Sedimentary Petrology (3rd ed.). Blackwell Science.
13. Tucker, M. E. (2003). Sedimentary rocks in the field (3rd ed.). John Wiley & Sons Ltd.
14. Walker, R. G. (1979). Facies Models. Geological Association of Canada.
15. Williams, G. D. (1966). Origin of shale-pebble conglomerate. American Association of Petroleum Geologists Bulletin, 50(3), 573–577.

CHAPTER

3

Biogenic, Chemical, and Volcanogenic Sediments

LEARNING OBJECTIVES

After studying this chapter, you will be able to:

- Discuss on limestone;
- Learn about the evaporite minerals;
- Understand about the cherts and volcaniclastics.

KEY TERMS FROM THIS CHAPTER

Biogenic sediments
Cenozoic
Fertile soils
Limestone
Mesozoic
Organic matter
Precambrian
Biological activity
Coquina
Foraminifera
Marine organisms
Metamorphism
Porosity
Recrystallization

INTRODUCTION

Additional processes play a significant role in the accumulation of sediments in areas where clastic detritus is scarce. There are deposits in a variety of environments made up of the hard parts of plants and animals, from microscopic algae to the bones of vertebrates. Most important are the numerous species that during their lives create calcium carbonate shells and structures, and which after they die, leave behind these hard parts as calcareous sediments that eventually form limestone. The production of evaporites, which are precipitated from salt-concentrated waters, is the primary application of chemical processes, though they also contribute to the formation of limestone. Ashes from primary volcanic processes are deposited subaerobically, or beneath the surface of the earth, and this process produces the majority of volcaniclastic sediments. These deposits can completely engulf all other types of sediment in regions where volcanism is occurring. Ironstones are chemical deposits, but most of the other deposits this chapter discusses are primarily of biogenic origin, such as phosphates, carbonaceous deposits, and siliceous sediments.

The three main categories of sediments are biogenic, chemical, and volcanogenic, according to where they came from. Most of the organic materials found in the remains of living things make up biogenic sediments. Among these are the skeletal remains of marine creatures that gather on the ocean floor, such as corals, foraminifera, and diatoms. Sedimentary rocks like limestone and chalk are created when these organic materials gradually compact and lithify. In the geological record, biogenic sediments are important because they serve as indicators of ancient climates and ecosystems and offer insights into past biological activity and environmental conditions. On the other hand, minerals precipitate out of solution to form chemical sediments. Mineral-laden water undergoes this process when its chemical composition, temperature, or pressure changes, causing the minerals to crystallize and separate from the water column. Evaporative basins, where water evaporation exceeds inflow and concentrates and crystallizes salts, gypsum, and other minerals, are common environments for chemical sedimentation. Travertine, a type of limestone deposited by mineral springs, and rock salt, which is made of halite, are two examples of chemical sedimentary rocks. These sediments can contain economically significant mineral deposits and are essential for comprehending historical hydrological and geochemical conditions. Pyroclastic materials, also known as volcanogenic sediments, are products of volcanic activity. Ash, pumice, and other volcanic debris are ejected into the atmosphere during violent volcanic eruptions, which create these sediments. These materials gather and solidify into layers of pyroclastic rocks, volcanic breccia, and tuff as they settle. Volcanogenic sediments can be dispersed over great distances by wind and water and are frequently found close to volcanic regions. They offer insightful knowledge about the effects of volcanic activity on the environment. These sediments can also have an impact on ecosystems, the surrounding landscape, and the development of fertile soils. Reconstructing Earth's past requires understanding the unique formation processes and depositional environments of each type of sediment. Volcanogenic sediments capture the dynamics of volcanic processes and their effects on the environment, while biological sediments document the occurrence and development of life, and chemical sediments provide insight into historical water chemistry and climate. Geologists can interpret past geological events and processes by analyzing these sediments, which provides a thorough understanding of Earth's complicated history.

3.1. LIMESTONE

The primary source of lime is limestone, or calcium carbonate ($CaCO_3$), a kind of carbonate sedimentary rock. The minerals calcite and aragonite, which are distinct crystal forms of $CaCO_3$, make up the majority of its composition. When these minerals separate from water that has dissolved calcium, limestone is created. Both biological and nonbiological processes can cause this, though for the past 540 million years, biological processes—like the buildup of corals and shells in the sea—have probably been more significant. Fossils found in limestone frequently provide scientists insights into past habitats and the evolution of life.

Carbonate rock makes up 20–25% of sedimentary rock, the majority of which is limestone. Dolomite, a closely related rock with a high concentration of the mineral dolomite, $CaMg(CO_3)_2$, makes up the majority of the remaining carbonate rock. Magnesian limestone is an antiquated and imprecise term that can be applied to dolomite, limestone that contains a considerable amount of dolomite (also known as dolomitic limestone), or any other type of limestone that has a substantial percentage of magnesium. Though smaller amounts were formed in many other environments, the majority of limestone was formed in shallow marine environments, such as continental shelves or platforms. Most dolomite is secondary dolomite, which is created when limestone undergoes chemical changes. A significant portion of the Earth's surface is covered in exposed limestone, and because limestone dissolves somewhat in rainwater, these exposures are frequently worn down to create karst landscapes. The bedrock of limestone is home to the majority of cave systems. Limestone serves a variety of purposes, including being a chemical feedstock for the production of lime, which is needed to make cement, an essential component of concrete, aggregate for road bases, a white pigment or filler for paints and toothpaste, a soil conditioner, and a well-liked decorative element for rock gardens. Approximately 30% of the global petroleum reservoirs are found in limestone formations.

3.1.1. Description

The minerals calcite and aragonite, which are distinct crystal forms of calcium carbonate ($CaCO_3$), make up the majority of limestone. In limestone, dolomite, or $CaMg(CO_3)_2$, is a rare mineral; siderite and other carbonate minerals are less common. On the other hand, calcite in limestone frequently has a small amount of magnesium in it. The dividing line between the two types of calcite found in limestone is set at 4% magnesium content. The two types of calcite are low- and high-magnesium. Calcite with a high magnesium content keeps its unique mineral structure, unlike dolomite. Magnesium is not typically present in large amounts in argonite. The majority of limestone exhibits a relatively pure chemical composition, with clastic sediments comprising primarily of fine-grained quartz and clay minerals accounting for less than 5 to 10% of the total composition. The percentage of organic matter in limestone is usually between 0.2 and 1%. In the form of chert or siliceous skeletal fragments (such as sponge spicules, diatoms, or radiolarians), limestone frequently contains varying amounts of silica. In limestone, fossils are also prevalent. The typical color of limestone is white to gray. The color of limestone can vary from off-white to yellow to red depending on the presence of iron or manganese traces. Limestone that is exceptionally rich in organic matter can have an almost black hue. Porosity, which ranges from 0.1% for the densest limestone to 40% for chalk, determines how dense limestone is. Accordingly, the density varies from 1.5 to 2.7 g/cm^3. Dense limestone can have a crushing strength of up to 180 MPa even though it has a Mohs hardness of 2 to 4. In contrast, the average crushing strength of concrete is approximately 40 MPa. Limestones exhibit a wide variety of textures despite having a relatively constant mineral content. However, the majority of limestone is made up of a carbonate mud matrix with grains the size of sand. Classifying limestone typically involves considering its grain type and mud content because limestones are typically of biological origin and are typically made of sediment that is deposited near the place where it formed.

3.1.1.1. Grains

The majority of the grains found in limestone are the skeletal remains of marine creatures like foraminifera or coral. These organisms secrete calcite or aragonite structures, which they leave behind upon their demise. Limeclasts (intraclasts and extraclasts [ca]), ooids, and peloids are additional carbonate grains that makeup limestones. The composition of skeletal grains reflects both the environment and the organisms that produced them. Planktonic (free-floating) foraminifera, coccoliths, and articulate brachiopods are known to have skeletal grains composed of low-magnesium calcite. Coralline algae, echinoderms, and benthic (bottom-dwelling) foraminifera are known to have skeletal grains of high magnesium calcite. Mollusks, calcareous green algae, stromatoporoids, corals, and tube worms are known to have argonite skeletal grains. In addition, the skeletal grains exhibit distinct geological epochs and habitats. For instance, whereas bryozoan grains are more common in low-energy settings (characterized by calm water), coral grains are more common in high-energy environments (characterized by strong currents and turbulence).

Ooids, also known as ooliths, are sand-sized grains with a diameter of less than 2 mm that are surrounded by one or more layers of calcite or aragonite around a central quartz grain or fragment of carbonate mineral. These most likely develop as a result of calcium carbonate precipitating directly onto the ooid. Similar to ooids, pisoliths have a diameter greater than 2 mm and a more asymmetrical shape. An oolite, or occasionally an oolitic limestone, is a limestone that is primarily made up of ooids. Oolites are found in high-energy settings like the Bahama platform, where they are typically characterized by crossbedding and other characteristics of deposition in strong currents. Oncoliths are similar to ooids, but they have an internal structure that is radial rather than layered, suggesting that algae in a typical marine environment formed them. Peloids are amorphous grains of microcrystalline carbonate that are probably the result of several different processes. Numerous are believed to be fecal pellets made by marine life. Others could be created by breaking down mollusk shells, endolithic (boring) algae, or other microbes. They are less common in ancient limestones, maybe because the compaction of carbonate sediments disturbs them, and are difficult to see in limestone samples outside of thin sections. Limeclasts are pieces of partially lithified carbonate sediments or existing limestone. Whereas extraclasts originate from outside the depositional area, intraclasts in limestone are limeclasts that originate near where they are deposited. One type of intraclast is grapestone, which is made up of aggregates of peloids bound together by mineral or organic cement. Extraclasts are rare, typically found with other clastic sediments, and suggest that the sediment was deposited either as part of a turbidity current or in a region that is tectonically active.

3.1.1.2. Mud

Most limestones have carbonate mud embedded in their grain structure. This usually makes up the largest portion of an old carbonate rock. Micrite is mud that has individual crystals smaller than 5 µm (0.20 mils) in length. Micrite is primarily made up of tiny aragonite needles found in fresh carbonate mud. These needles can be produced by abrasion of carbonate grains in a high-energy environment, secreted by algae, or directly precipitated from seawater. After being deposited for a few million years, this turns into calcite. Microspar, whose grains range in diameter from 5 to 15 µm (0.20 to 0.59 mils), is created when micrite undergoes additional recrystallization. Larger calcite crystals, known as sparry calcite or sparite, that range in size from 0.02 to 0.1 mm (0.79 to 3.94 mils), are frequently found in limestone. The grain size of sparite is larger than that of micrite, measuring over 20 µm (0.79 mils). Additionally, sparite can be easily identified under a hand lens or in thin sections as transparent or white crystals. The distinctive crystal shapes of sparite set it apart from carbonate grains, which lack internal structure. Geologists take great care to differentiate between sparite that is formed by the recrystallization of micrite or carbonate grains and sparite that is deposited as cement. The presence of sparite cement in the pore spaces between the grains suggests that the carbonate mud was removed in a high-energy depositional environment. Sparite that has recrystallized is not indicative of the depositional environment.

3.1.1.3. Other Characteristics

In the field, limestone outcrops can be identified by their softness (calcite and aragonite both have Mohs hardness values of less than 4, which is significantly lower than that of common silicate minerals) and by the vigorous bubbling that occurs when diluted hydrochloric acid is dropped on them. Dolomite is likewise soft, but it reacts weakly to diluted hydrochloric acid. Because ferrous iron is present, dolomite typically weathers to a dull yellow-brown color. The dolomite weathers and releases this, which then oxidizes. Limestones will display a variety of colors due to impurities (such as sand, clay, organic remnants, iron oxide, and other materials), especially on weathered surfaces. In the field, one can estimate the composition of a carbonate rock outcrop by etching its surface with diluted hydrochloric acid. This removes the aragonite and calcite, leaving the dolomite or silica grains in place. The rhombohedral shape of the latter allows for their identification. Small rock cavities called "vugs" may be lined with crystals of calcite, quartz, dolomite, or barite. Vugs are a type of secondary porosity that are created in limestone by an environmental shift that makes calcite more soluble. Marble is a term sometimes used to describe massive, dense limestone. For instance, the well-known Italian Portoro marble is actually a dense black limestone. Real marble is created when limestone recrystallizes as a result of the orogeny, a regional metamorphism that occurs concurrently with the formation of mountains. Its coarse crystalline texture and the formation of unique minerals from the silica and clay present in the original limestone set it apart from dense limestone.

3.1.1.4. Classification

Two major classification schemes, the Folk and Dunham, are used for identifying the types of carbonate rocks collectively known as limestone.

3.1.1.4.1. Folk Classification

The categorization scheme that prioritizes the specific makeup of the grains and interstitial material in carbonate rocks. There are three primary components based on composition: cement (sparite), matrix (primarily micrite), and allochems (grains). The first part of the names in the Folk system refers to the grains, and the second part refers to the cement. An oosparite, for instance, would be a limestone with a crystalline matrix and a predominantly ooid composition. Using a petrographic microscope makes it easier to identify the components in each sample when utilizing the Folk scheme.

3.1.1.4.2. Dunham Classification

Robert J. in 1962, Dunham released his limestone system. Its main subject is the carbonate rocks' depositional structure. Dunham categorizes the rocks into four primary groups according to the relative amounts of coarser clastic particles. These groups are determined by factors like the presence of frame builders and algal mats in the rock or whether the grains were initially in mutual contact and thus self-supporting. Dunham addresses the original porosity of the rock, in contrast to the Folk scheme. Because the Dunham scheme is based on texture rather than sample grains, it is more applicable to hand samples. Wright proposed a revised classification in 1992. It enhances the classification scheme with a few diagenetic patterns.

3.1.1.4.3. Other Descriptive Terms

The word "travertine" refers to calcium carbonate deposits that originate in freshwater settings, especially hot springs, waterfalls, and cascades. These deposits are usually banded, massive, and dense. Usually referred to as tufa, the deposits have a sponge-like texture due to their high porosity. Travertine is another term used to describe secondary calcite that is deposited in caves by supersaturated meteoric waters (groundwater). Speleothems like stalactites and stalagmites are created as a result. Coquina is a poorly consolidated limestone made up of fragments of abraded coral, shells, and other detritus from fossils. It is referred to as coquinite when it is more consolidated. While marl is an earthy mixture of silicate sediments and carbonates, chalk is a soft, earthy, fine-textured limestone formed of the tests of planktonic microorganisms like foraminifera.

3.1.2. Formation

Limestone forms when calcite or aragonite precipitates out of water containing dissolved calcium, which can take place through both biological and nonbiological processes. The solubility of calcium carbonate ($CaCO_3$) is controlled largely by the amount of dissolved carbon dioxide (CO_2) in the water. This is summarized in the reaction:

$$CaCO_3 + H_2O + CO_2 \rightarrow Ca^{2+} + 2HCO^{-3}$$

Reduced amounts of dissolved CO_2 and precipitated $CaCO_3$ are typically the result of temperature increases or pressure drops. In fresh water as opposed to seawater, a decrease in salinity also causes a reduction in $CaCO_3$ solubility that is several orders of magnitude greater. The earth's oceans' near-surface water has a $CaCO_3$ oversaturation of greater than six. $CaCO_3$'s inability to precipitate out of these waters quickly is probably caused by dissolved magnesium ions interfering with the nucleation of calcite crystals, which is a prerequisite for precipitation. The presence of naturally occurring organic phosphates in the water may inhibit aragonite precipitation. Even though ooids most likely originate from entirely inorganic processes, biological activity is responsible for the majority of $CaCO_3$ precipitation in the oceans. A large portion of this occurs on carbonate platforms.

Research is still being done on the origins of carbonate mud and the methods used to turn it into micrite. The majority of the needles in modern carbonate mud are aragonite, with a length of about 5 μm (0.20 mils). This is a plausible source of mud because calcareous algae like Penicillus produce needles with this shape and composition. The water's direct precipitation is an additional possibility. Whittings is a phenomenon where white streaks with scattered micrite show up on the water's surface. It happens in shallow waters. There's some evidence that whitings are caused by biological precipitation of aragonite as part of a cyanobacteria or microalgae bloom, though it's unclear if this is newly precipitated aragonite or just material stirred up from the bottom. Abrasion of carbonate grains in high-energy environments has been proposed as a third possibility, but stable isotope ratios in contemporary carbonate mud seem to contradict either of these theories. Throughout the Phanerozoic, the last 540 million years of Earth's history, biological processes most likely played a major role in the formation of limestone. In the Precambrian, before 540 million years ago, microorganisms might have deposited limestone, but inorganic processes were most likely more significant and occurred in an ocean that was more highly oversaturated in calcium carbonate than the ocean we know today.

3.1.2.1. Diagenesis

The process of compacting sediments and transforming them into solid rock is known as diagenesis. Carbonate sediments undergo substantial chemical and textural changes during diagenesis. For instance, low-magnesium calcite is produced from aragonite. Pisoliths, which are concentrically layered particles with a diameter ranging from 1 to 10 mm (0.039 to 0.394 inches) and found in certain limestones, are thought to have originated via diagenesis. Although pisoliths fit together tightly and have no foreign matter nucleus, they differ from ooids in other ways, indicating that they formed after the sediments were originally deposited. Silicification plays a role in the preservation of fossils and happens early in diagenesis at low pH and temperature. Fossils are often preserved in exquisite detail as chert.

Silicification takes place through the following reaction:

$$CaCO_3 + H_2O + CO_2 + H_4SiO_4 \longrightarrow SiO_2 + Ca^{2+} + 2HCO_3^- + 2H_2O$$

In carbonate sediments, cementing occurs quickly; usually, it happens within a million years of deposition. Hardgrounds are created when some cementing takes place while the sediments are still submerged. When rainwater seeps into the sediment beds following the sea's retreat from the depositional environment, cementing speeds up—often in a matter of millennia. Rainwater and high-magnesium calcite are transformed into low-calcium calcite when they combine with groundwater. Because rainwater can seep over 100 km (60 miles) into the sediments beneath the continental shelf, it can begin cementing thick carbonate deposits even before the sea recedes. The chemical and mechanical compaction of carbonate sediments increases with their deeper burial beneath younger sediments. The process of chemical compaction involves forcing the sediments into solution. The porosity of the limestone is decreased from an initial high value of 40% to 80% to less than 10% by this process, which dissolves minerals from grain-to-grain contact points and redeposits them in pore space. Pressure solution creates characteristic stylolites, which are uneven limestone surfaces where silica-rich sediments collect. These could be the result of a sizable portion of the limestone bed dissolving and being lost. The final step in the lithification process is burial cementation, which occurs at depths greater than 1 km (0.62 miles). Styletolites are not produced by burial cementation.

The last stage of diagenesis occurs when limestone is brought closer to the surface due to erosion of the overlying beds. As rainwater seeps into the beds, some of the cement dissolves, creating secondary porosity. This could involve the development of vugs, which are limestone cavities lined with crystals. Magnesium-rich fluids may convert limestone to dolomite during diagenesis. There is ample evidence that dolomite has replaced limestone, including jagged replacement boundaries that sever bedding. Although the mechanisms underlying the dolomitization process are still being investigated, they may involve exposure to concentrated brines in hot environments (evaporative reflux) or diluted seawater in delta or estuary environments (dorag dolomitization). But as a mechanism for dolomitization, dorag dolomitization has lost favor; a 2004 review paper called it outright a myth. If regular seawater flushing—such as that caused by the tides—occurs through the rock, ordinary seawater can change calcite into dolomite. Dolomitization starts quickly and continues until very little carbonate rock with a mixture of calcite and dolomite remains. Either nearly all dolomite or nearly all calcite/aragonite make up carbonate rocks.

3.1.3. Occurrence

Carbonate rock makes up 20–25% of sedimentary rock, the majority of which is limestone. Sedimentary sequences dating back up to 2.7 billion years contain limestone. The geologic record exhibits an uneven distribution of time in the compositions of carbonate rocks. Magnesium-rich calcite and aragonite account for about 95% of the composition of contemporary carbonates. Since low-magnesium calcite is the most stable form of calcium carbonate, it transforms the aragonite needles in carbonate mud into that material in a matter of millions of years. Dolomite is abundant in ancient Precambrian and Paleozoic carbonate formations, but limestone predominates in Mesozoic and Cenozoic carbonate beds. Dolomite nowadays is extremely uncommon. There is evidence that the oceans of the Paleozoic and middle to late Cenozoic favored the precipitation of calcite, whereas the oceans of the modern era favored the precipitation of aragonite. This could suggest that the ocean water during those eras had a lower Mg/Ca ratio. The removal of magnesium from ocean water by faster sea floor spreading could be the cause of this magnesium depletion. The modern ocean and the ocean of the Mesozoic have been described as "aragonite seas."

The majority of limestone was created in platforms or continental shelves that are located in shallow marine environments. Although these environments only make up 5% of ocean basins, deep sea and continental slope environments are rarely found to contain limestone. Warm waters are ideal for deposition because they have higher calcium carbonate saturation and high organic productivity because of their lower dissolved carbon dioxide concentrations. The majority of limestones are relatively pure because modern limestone deposits are almost exclusively found in regions with very little silica-rich sedimentation. Muddy, brackish river water kills reef organisms and much harder silicate grains crush carbonate grains. In contrast to clastic sedimentary rock, nearly all of the sediments that form limestone originate at or close to the place of deposition. Thickness variations in limestone formations are frequently sudden. Big mound-like structures in a limestone formation are thought to be ancient reefs, or bioherms when they are found in the geologic record. While many have abundant fossil resources, most do not have the interconnected organic framework found in contemporary reefs. The fossil remains are scattered throughout the abundant mud matrix. A large portion of the sedimentation appears to be occurring in the supratidal or intertidal zones, indicating that sediments quickly occupy the shelf's or platform's available accommodation space. Additionally, deposition is preferred on the seaward edge of shelves and platforms due to nutrient-rich deep ocean water upwelling that boosts organic productivity. Reefs are common in this area, but in their absence, ooid shoals can be found. Near the coast, finer sediments are deposited.

The rapid subduction of oceanic crust is partially to blame for the lack of deep sea limestones, but the dissolution of calcium carbonate at depth is the main cause. Pressure and higher carbon dioxide concentrations—produced by decomposing organic matter that settles in the deep ocean and isn't eliminated by photosynthesis in the shadowy depths—increase the solubility of calcium carbonate. This causes a fairly abrupt change in water composition, known as the lysocline, from calcium carbonate-saturated to calcium carbonate-unsaturated at the calcite compensation depth of 4,000 to 7,000 m (13,000 to 23,000 feet). Below this depth, foraminifera tests and other skeletal particles dissolve quickly, and the ocean floor's sediments suddenly change from silicic mud devoid of carbonates to carbonate ooze rich in foraminifera and coccolith remains (known as Globogigerina ooze). Benthic (deep ocean) carbonate deposits are occasionally buried and preserved by turbidites or other silica-rich sediments. Because of their tectonic setting, ancient benthic limestones can be distinguished as microcrystalline rocks. Fossils are mostly composed of coccoliths and foraminifera. There are no known pre-Jurassic benthic limestones, most likely due to the lack of evolution of carbonate-shelled plankton. Environments with freshwater can also produce limestones. These limestones are similar to marine limestone, but they contain more silica and clay minerals, which are typical of marls, and less diversity of organisms. One well-known freshwater sedimentary formation with lots of limestone beds is the Green River Formation. Generally speaking, freshwater limestone is micritic. These environments are typified by fossilized charophyte (stonewort) fossils, a type of freshwater green algae that produces and traps carbonates. Additionally, evaporite depositional environments can produce limestones. One of the first minerals in marine evaporites to precipitate is calcite.

3.1.3.1. Limestone and Living Organisms

The majority of limestone is created by living things that are in close proximity to reefs; however, these organisms have changed throughout geologic time. Stromatolites, for instance, are uncommon in younger limestones but are mound-shaped structures found in older limestones that are thought to be the remains of cyanobacteria colonies that accumulated carbonate sediments. In addition to precipitating limestone directly through their skeletons, organisms also precipitate it indirectly through photosynthesis, which lowers the solubility of calcium carbonate in water and removes carbon dioxide from it. The variety of sedimentary structures present in other sedimentary rocks is also present in limestone. However, the burrowing activities of organisms (bioturbation) frequently destroy finer structures, like lamination. The limestone found in playa lakes, which are devoid of organisms that burrow, has fine lamination. Geopetal structures, which are formed when curved shells sink to the bottom with their concave faces downward, are another characteristic found in limestones. This creates a vacuum that sparite can subsequently fill. Because highly deformed limestone formations can make it difficult to tell which way was up at the time of deposition, geologists use geopetal structures to make this determination. The cyanobacterium Hyella balani, the fungus Ostracolaba implexa, and the green alga Eugamantia sacculata are all capable of boring through limestone.

3.1.3.1.1. Micritic mud Mounds

Mud mounds made of micritic calcite with no internal structure are shaped like subcircular domes. Present-day specimens can measure several 100 meters in thickness and one kilometer in diameter, featuring sharp inclinations (with an approximate 50° slope angle). Mangroves or Thalassia grass stabilize them, and they could be made of peloids carried by currents together. By aiding in the retention of sediments, bryozoa may also play a role in the formation of mounds. From the early Ordovician onward, mud mounds dominated both shallow and deep water reef types. They can be found throughout the geologic record. Microbial activity is probably the source of these mud mounds. Mud mounds were primarily limited to deeper water after the emergence of reef organisms that build frames.

3.1.3.1.2. Organic Reefs

In shallow water that is only a few meters deep, organic reefs develop at low latitudes. Throughout the fossil record, these intricate and varied structures can be found. In the early Cambrian, archaeocyathids first appeared; by the late Cambrian, sponges had replaced them; later successions included stromatoporoids, corals, algae, bryozoa, and rudists, a type of bivalve mollusk. These frame-building organisms are indicative of distinct geologic time periods and are responsible for the formation of organic reefs. Over geologic time, the extent of organic reefs has changed, but it is most likely that they were at their most extensive in the middle Devonian when they occupied an estimated 5,000,000 km^2 (1,900,000 sq mi). This is about 10 times larger than the current reefs. Due to their devastation during the late Devonian extinction, stromatoporoids, and tabulate corals played a major role in the construction of the Devonian reefs. The interior structure of organic reefs is usually complex. While interclasts and ooids are uncommon in the reef, whole-body fossils are typically common. A reef's talus, which has a larger volume than the reef's core, encircles the reef's normally massive, undyed core. The talus is composed primarily of large grains with sparse matrix, or rudstone, which is composed of at least 10% of grains larger than 2 mm embedded in abundant matrix. It also contains a large number of intraclasts. Once the talus moves away from the reef, it grades into fine-grained planktonic carbonate mud and then noncarbonate mud.

3.1.4. Limestone Landscape

Due to limestone's partial solubility—particularly in acid—many erosional landforms are created. These consist of gorges, pot holes, cenotes, limestone pavements, and caves. Karsts are landscapes formed by erosion. Compared to most other sedimentary rocks, limestone is more resistant to erosion than most igneous rocks. It therefore typically occurs in areas with other sedimentary rocks, most often clays, and is linked to hills and lowlands. Because surface water readily drains downward through joints in limestone bedrock, karst regions that are overlying limestone bedrock typically have fewer visible above-ground sources (ponds and streams). Water and organic acid from the soil gradually enlarge these cracks during drainage, dissolving the calcium carbonate and removing it in solution over thousands or millions of years. The bedrock of most cave systems is limestone. Cave formation can also be facilitated by cooling groundwater or combining various groundwaters.

Organisms erode coastal limestones by penetrating the rock through a variety of techniques. We call this process bioerosion. It is known from the fossil record and is most prevalent in tropical regions. Limestone bands appear as stunning rocky outcrops and islands that rise from the Earth's surface. Some locations that exemplify this are the Burren in County Clare, Ireland; the Rock of Gibraltar; the Great Orme in Wales; the Niagara Escarpment in Canada and the United States; the Rock of Gibraltar; the Burren in North Yorkshire and the Isle of Wight, England; the Ha Long Bay National Park in Vietnam; and the hills surrounding the Lijiang River and Guilin city in China. The islands off Florida's southern coast are known as the Florida Keys, and they are primarily made up of oolitic limestone in the Lower Keys and the carbonate skeletons of coral reefs in the Upper Keys. These reefs flourished in the region during interglacial periods when sea levels were higher than they are now. Alvars are exceptionally flat limestone stretches with thin soil mantles that are home to unique habitats. Situated on the Swedish island of Öland, the Stora Alvaret is the largest expanse of its kind in Europe. Sweden's Gotland Island is another place that has a lot of limestone. Across more than a 100 kilometers are located massive quarries in northwest Europe, like those of Mount Saint Peter in Belgium/Netherlands (Thornbury, 1969; Bates & Jackson, 1984; Wright, 1992).

3.2. EVAPORITE MINERALS

A sedimentary mineral deposit that is soluble in water and forms as a result of evaporation from an aqueous solution is called an evaporite. Evaporite deposits come in two varieties: marine deposits, which are also known as ocean deposits, and non-marine deposits, which are found in lakes and other bodies of water that are not moving. Chemical sediments are what form evaporites, which are classified as sedimentary rocks. Layered crystalline sedimentary rocks known as evaporites are formed from brines produced in regions where evaporation losses outweigh total water input from rivers and streams and rainfall. Although there are nearly 100 possible varieties in the complex mineralogy of evaporite rocks, less than a dozen species are significant in terms of volume. Evaporite rocks contain a variety of borates, silicates, nitrates, and sulfocarbonates in addition to carbonates (especially calcite, dolomite, magnesite, and aragonite), sulfates (anhydrite and gypsum), and chlorides (especially halite, sylvite, and carnallite). Both marine and nonmarine sedimentary successions contain evaporite deposits.

Even though they are localized, contemporary evaporites add to genetic models that explain prehistoric evaporite deposits. Only arid regions (those with high temperatures and low precipitation rates) are home to modern evaporites. Examples of such regions include the floors of semidry ephemeral playa lakes in the Great Basin of Nevada and California, the Middle East's coastal salt flats (sabkhas), and the salt pans, estuaries, and lagoons surrounding the Gulf of Suez. In the Phanerozoic geologic record, ancient evaporates are common, especially in sedimentary sequences dating from the Cambrian (570–505 million years ago), Permian (286–245 million years ago), and Triassic (245–208 million years ago) periods. However, they are uncommon in Precambrian sedimentary sequences. They are usually closely related to fine (usually iron oxide-rich) mudrocks and carbonates from shallow marine shelves. The presence of evaporite sediments in time and space clearly constrains inferences of paleoclimatology and paleogeography, since evaporite sedimentation requires a particular climate and basin setting. Geologists who study structural geology are particularly interested in evaporite beds because they tend to concentrate and facilitate major thrust fault horizons. In addition, evaporites are valuable economically as a source of fertilizer and salts.

Brine produced by evaporation precipitates into all evaporite deposits. Experiments in the lab can precisely follow the development of brines as different evaporite minerals crystallize. The salinity of normal seawater is 30.5% or 35,000 parts per million, and the two most significant dissolved elements are sodium and chlorine. Evaporite precipitation begins in an orderly manner when the volume of seawater is lowered to one-fifth of its initial volume. The more insoluble components, such as gypsum and anhydrite, form first. More soluble minerals like sylvite and halite form when the solution reaches a tenth of the original volume. Natural evaporite sequences exhibit less systematic vertical mineralogy changes that roughly correlate to the ordered appearance of mineralogy as a function of solubility (Figure 3.1).

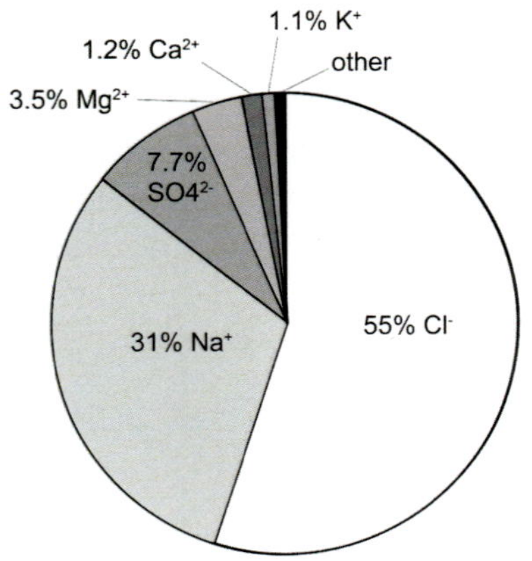

Figure 3.1. The proportions of the principal ions in seawater of normal salinity and 'average' river water.

Source: https://raregeologybooks.wordpress.com/wp-content/uploads/2014/09/sedimentology-and-stratigraphy-by-gary-nichols.pdf.

As water evaporates, the ions in the solution become more concentrated and precipitate out to form these minerals. Seawater typically contains 35 g L^{-1} (35 parts per 1,000) of dissolved ions, primarily sodium, potassium, chloride, sulfate, magnesium, and calcium (Figure 3.1). Lake waters have a varied chemistry, frequently containing the same principal ions in various ratios. When anions and cations become concentrated and the water becomes saturated with regard to a specific compound, they combine to form minerals. When the waters get more concentrated, calcium carbonate, calcium sulfate, and sodium chloride precipitate out of the seawater first because the least soluble compounds precipitate first. Magnesium and potassium chlorides don't precipitate until seawater concentration is very high. Figure 3.2 lists the seawater contents (Thornbury, 1969; Bates & Jackson, 1984; Wright, 1992).

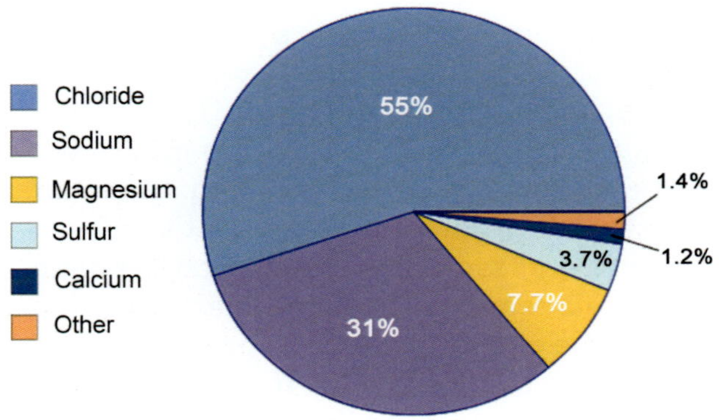

Figure 3.2. Ocean water contents.

Source: https://courses.lumenlearning.com/suny-earthscience/chapter/composition-of-ocean-water/.

3.2.1. Gypsum and Anhydrite

The most frequent evaporite minerals found in sedimentary rocks are anhydrite or gypsum, which are both calcium sulfate forms. When evaporation reduces seawater to 19% of its initial volume, calcium sulfate is precipitated out of it. The mineral's hydrous form is called gypsum ($CaSO_4.2H_2O$). Under all but the most arid conditions, it precipitates at the surface, but upon burial, it may become dehydrated to anhydrite. Anhydrite, which has no water in its crystal structure ($CaSO_4$), is created when gypsum is altered by burial or occurs directly as precipitation in dry shorelines. If water is added, it might hydrate to gypsum. When primary gypsum precipitates out of water, it takes the form of elongate selenite crystals. It takes on a fine crystalline form in alabaster nodules if it originates from the rehydration of anhydrite. Secondary veins can also contain fibrous forms of gypsum.

3.2.2. Halite

When seawater is concentrated to 95% of its initial volume, halite (NaCl) precipitates out of it. Thick crystalline beds or single crystals with a recognizable cubic symmetry may be present, occasionally along with a stepped crystal face (a hopper crystal). Due to its high solubility, sodium chloride can only be found preserved in rocks when there isn't any diluted groundwater present to dissolve it. Some arid regions have surface exposures of halite because rainwater does not carry it away. The easiest way to determine whether a mineral is present is to taste it. Since halite is found in rock salt naturally, the only other mineral that could be mistaken for it is sylvite (see below), a potassium chloride mineral that is far less common and has a more bitter taste than "regular salt." Halite is white or colorless, soft (hardness 2.5, slightly harder than gypsum but still scratchable with a fingernail). Since halite is a cubic mineral, it is isotropic and can exhibit a strong cleavage in thin-section crystals with planes at right angles.

3.2.3. Other Evaporite Minerals

Other minerals, which are uncommon to find in large quantities but may have significant economic implications, can be obtained by evaporating seawater. Specifically, potassium chloride, also known as sylvite (KCl), is a significant industrial potash source that is found in conjunction with halite and is thought to be the result of excessive marine water evaporation. But instead of producing sylvite, the evaporation of modern waters produces a variety of magnesium sulfate ($MgSO_4$) minerals, which has given rise to theories that the chemical makeup of seawater has changed over hundreds of millions of years (Hardie, 1996). These variations in water chemistry, which favor KCl or $MgSO_4$ precipitation at different times, are thought to be caused by differences in the relative importance of meteoric waters (run-off from land) and hydrothermal waters (from mid-ocean ridge vents). Although the amounts of dissolved ions in saline lakes (10.3) are typically different from those in seawater, this leads to a suite of evaporite minerals that are typical of various lake chemistries. The majority of these minerals, like trona (Na_2CO_3), are sodium and magnesium sulfates, carbonates, and bicarbonates. Epsomite ($MgSO_4.7H_2O$), mirabilite ($Na_2SO_4.10H_2O$), and $NaHCO_3.2H_2O$. Naturally, all of these minerals are soluble in water and are rather soft minerals.

3.2.4. Depositional Environments

3.2.4.1. Marine

Research on marine evaporites is typically more in-depth and their deposits tend to be thicker. Minerals are deposited in a specific order when ocean water is evaporated in a lab, as was initially shown by Usiglio in 1884. When roughly half of the initial water depth is still present, the first stage of precipitation starts. Minor carbonates start to form at this point. When the experiment has about 20% of its initial level left, the procedure moves on to the next step. At this stage, the mineral gypsum starts to form, and 10% of the minerals are halite, with the exception of carbonate minerals, which are typically not evaporites. Calcite, gypsum, and anhydrite, halite, sylvite, carnallite, langbeinite, polyhalite, and kainite are the most prevalent marine evaporites. It is also possible to include kieserite ($MgSO_4$), which typically accounts for less than 4% of the total content. Though only a few dozen are common enough to be regarded as significant rock formers, about 80 different minerals have been reported to be found in evaporite deposits.

3.2.4.2. Nonmarine Environment

Marine evaporites are typically the subject of more in-depth study and have thicker deposits. Usiglio (1884) was the first to demonstrate that minerals are deposited in a specific order when ocean water is evaporated in a laboratory. About 50% of the initial water depth is left when the first phase of precipitation starts. This is when the formation of minor carbonates starts. When the experiment is down to roughly 20% of its starting level, the next step in the procedure begins. At this point, the mineral gypsum starts to form, followed by 10% halite, omitting carbonate minerals, which are generally not evaporites. The most prevalent marine evaporites are anhydrite and gypsum, kainite, halite, sylvite, carnallite, langbeinite, and polyhalite. Kieserite ($MgSO_4$) is another possible addition, though it usually accounts for less than 4% of the total content. Still, only about a dozen of the roughly 80 different minerals that have been reported to be found in evaporite deposits are common enough to be regarded as significant rock formers. Since the water from which non-marine evaporites precipitate typically has different proportions of chemical elements than those found in marine environments, non-marine evaporites are typically composed of minerals that are uncommon in marine environments. Minerals like blödite, borax, epsomite, gaylussite, glauberite, mirabilite, thenardite, and trona are frequently found in these deposits. Although these minerals did not originate from ocean deposits, they can still be found in non-marine deposits and in some cases even dominate them. These minerals include halite, gypsum, and anhydrite. Nevertheless, this does not lessen the significance of non-marine deposits, which frequently contribute to the understanding of past Earth climates. Some specific deposits even exhibit significant climatic and tectonic shifts. These deposits might also include valuable minerals that are beneficial to the modern economy. Where there is enough soluble supply and where evaporation rates will surpass inflow rates, thick non-marine deposits typically form. For the sediment to accumulate and form in a lake or other body of water, the inflow must also take place in a closed basin or one with limited outflow. Salinity lake deposits are prime instances of this. Saline lakes can be any of the terms used to describe locations that hold standing bodies of water periodically or year-round, such as perennial lakes, which are lakes that are present year-round, or playa lakes, which are lakes that only appear during specific seasons. The Dead Sea, which is located between Jordan and Israel, and Utah's Great Salt Lake are two examples of contemporary non-marine depositional environments.

Evaporite depositional environments that meet the above conditions include:

- Graben areas and half-grabens within continental rift environments fed by limited riverine drainage, usually in subtropical or tropical environments.
- Example environments at the present that match this is the Denakil Depression, Ethiopia; Death Valley, California.
- Graben environments in oceanic rift environments fed by limited oceanic input, leading to eventual isolation and evaporation.

- Examples include the Red Sea, and the Dead Sea in Jordan and Israel.
- Internal drainage basins in arid to semi-arid temperate to tropical environments fed by ephemeral drainage.
- Example environments at the present include the Simpson Desert, Western Australia, the Great Salt Lake in Utah.
- Non-basin areas fed exclusively by groundwater seepage from artesian waters.
- Example environments include the seep-mounds of the Victoria Desert, fed by the Great Artesian Basin, Australia.
- Restricted coastal plains in regressive sea environments.
- Examples include the sabkha deposits of Iran, Saudi Arabia, and the Red Sea; the Garabogazköl of the Caspian Sea.
- Drainage basins feeding into extremely arid environments.
- Examples include the Chilean deserts, certain parts of the Sahara, and the Namib.

The most significant known evaporite depositions happened during the Messinian salinity crisis in the basin of the Mediterranean (Thornbury, 1969; Bates & Jackson, 1984; Wright, 1992).

3.2.4.3. Shallow Marine Environment

In desert coastal regions, evaporite deposition takes place in the shallow marine environment, also known as the salina. This occurs especially near the edges of semi-restricted bodies of water like the Red Sea, Persian Gulf, and Gulf of California. Since free and unlimited mixing with the open sea would make it easy for the bodies of water to overcome the high evaporation rates of arid areas and dilute these waters to near-normal salinity, the restriction is generally one of the critical requirements for evaporite deposition. In actuality, the main factor influencing brine production is coastal physiography, so this semi-restriction cannot stop a significant amount of dilution through mixing. Fine-grained mudrock and tidal flat limestone and dolomite are commonly interfinger with shallow-water evaporites, which are nearly exclusively composed of gypsum, anhydrite, and halite.

3.2.4.4. Deep-Basin Environment

The majority of the dense, widely spaced evaporite deposits seem to have formed in secluded, deep basins that emerged during periods of extreme global dryness. Aridity is the most important condition for evaporite production; water needs to evaporate more quickly than it can be refilled by precipitation and inflow. Furthermore, to stop brines created by evaporation from returning to the open ocean, the evaporite basin needs to be isolated, if only partially. Brine concentration can be achieved by limiting it to a solitary basin and allowing it to remain there until evaporite mineral precipitation takes place. The basin is periodically refilled by the periodic breaching of the barrier, which is caused by either crustal downwarping or changes in the global sea level. This replenishment of seawater allows for the exceptionally thick and regionally extensive evaporite sequences that are visible in the geologic record (Figure 3.3).

Figure 3.3. Marine evaporites the dead sea.

Source: By Wilson44691 – Own work, Public Domain, https://commons.wikimedia.org/w/index.php?curid=4549752.

The precise mechanisms responsible for producing thick evaporite deposits are still up for debate. There are three potential models for limiting "barred" evaporite basins. They vary in specifics, and not one has gained widespread acceptance. According to the deep-water, deep-basin model, brine escapes seaward and maintains a constant brine concentration, allowing for the gradual accumulation of thick evaporites and replenishment of the basin across the barrier or sill. The shallow-basin, shallow-water model causes thick evaporites because the basin floor continuously sinks. According to the shallow-water, deep-basin model, evaporation has caused the brine level in the basin to drop below sea level; brines are refilled by groundwater recharge from the open ocean (Thornbury, 1969; Bates & Jackson, 1984; Wright, 1992).

3.3. CHERTS

A type of silica composed of radiating fibers a few tens to hundreds of microns long, chalcedony, and silt-sized interlocking quartz crystals (microquartz) make up the fine-grained siliceous sedimentary rocks known as cherts. Chert beds can arise from diagenetic processes or from primary sediments. Siliceous ooze is created when microscopic organisms' skeletal remains accumulate on the bottoms of lakes and seas. These are lake-dwelling diatoms, which can also accumulate in marine environments, though marine siliceous oozes typically consist primarily of Radiolaria. Radiolarians belong to the class of microscopic planktonic animals known as zooplankton, while diatoms are classified as free-floating algae known as phytoplankton. As these oozes solidify, chert beds are formed. The opaline silica found in radiolaria and diatoms is metastable and recrystallizes into microquartz or chalcedony. Opal is a type of cryptocrystalline silica with water in its mineral structure. Since the amount of clay-sized material varies, ooze-formed cherts frequently have thin bedded layers. Deep-ocean habitats are where they are most common. The formation of dialectic cherts occurs when silica-rich waters permeate the rock, replacing other materials like calcium carbonate. The opaline silica of diatoms, radiolarians, and siliceous sponges is primarily produced by biogenesis, thus it is primarily sourced from these sources. Chert that is formed in this manner often appears as irregular layers and nodules within other limestones and mudstones, as well as dark flint nodules that are frequently found within Cretaceous Chalk.

Chert is the hardest sedimentary rock due to its dense internal structure of interlocking microquartz grains and fibers. This rock's ability to break into tiny shards and have a conchoidal fracture made it crucial for early humans to use in the making of tools. The hue varies based on the amounts of impurities present: haematite gives jasper its intense red color, while organic material residue produces grey or black chert. Thin sections through chert show distinctive patterns of microquartz grains that completely interlock or radiating fibers in chalcedony. The petrified remnants of siliceous ooze, a biogenic sediment that covers a significant portion of the deep ocean floor and contains the silicon skeletons of radiolarians, silicoflagellates, and diatoms, are the usual components of chert. Fossil cyanobacteria are a notable feature of Precambrian cherts. Chert occasionally contains macrofossils in addition to microfossils. But some chert has no fossils in it at all. Chert can be white or black, but its most common colors are gray, brown, grayish brown, light green, rusty red, and sometimes dark green. The rock's trace elements are expressed through its color. Red and green are typically associated with iron traces in their reduced and oxidized forms, respectively.

3.3.1. Description

The word "chert" in petrology refers to all sedimentary rocks that have precipitated chemically and are mainly made of microcrystalline, cryptocrystalline, and microfibrous silica. With less than 5% of other minerals (mostly calcite, dolomite, clay minerals, hematite, and organic matter) present, the majority of cherts are almost pure silica. Cherts, however, vary in purity, ranging from impure nodular cherts with less than 65% silica content to extremely pure cherts with over 99% silica content. After iron and manganese, or potassium, sodium, and calcium, aluminum is the most prevalent minor element. The majority of cherts contain less than 1% extracrystalline water, which is defined as microscopic water inclusions inside and surrounding the quartz grains. Chert is categorized into three textural groups according to the Folk classification. The part of chert that is made up of roughly equidimensional quartz grains, granular microquartz can range in size from a fraction of a micron to 20 microns but is usually 8 to 10 microns. Chalcedony is a type of quartz that is microfibrous and is made up of radiating bundles of incredibly thin crystals that are only a few microns long. The grains that make up megaquartz are equidimensional and larger than 20 microns. Cherts vary in purity from almost pure opal to almost pure quartz chert, with the majority being microcrystalline quartz with trace amounts of opal and chalcedony. On the other hand, little opal is older than 60 million years. Glass sponge spicules, radiolarians, and diatoms are frequently found in visible fossil form in opaline chert.

Chert can be found in a variety of environments, such as alkaline lakes, banded iron formations, hot spring deposits (siliceous sinter), and so on. But the majority of chert is found either nodule- or bedded-like. Precambrian beds are more likely to contain bedded chert, but as the amount of chert in the rock record decreased overall during the Phanerozoic, nodular chert became more prevalent. After the early Mesozoic, bedded chert is uncommon. During the Devonian and Carboniferous periods, chert experienced a moderate abundance, which persisted until the Jurassic period (Thornbury, 1969; Bates & Jackson, 1984; Wright, 1992).

3.3.1.1. Bedded Chert

Bedded chert, also called ribbon chert, is composed of extremely thin layers of silica-rich shale sandwiched between thinly bedded, nearly pure chert layers that range in thickness from a few centimeters to a meter. The color of it ranges from black to green, and the entire row of beds may be several 100 meters thick. Usually black in color, the shale occasionally contains pyrite, a sign that it was deposited in an anoxic environment. On active tectonic plate margins, bedded chert is typically found in conjunction with turbidites, deep water limestone, submarine volcanic rock, ophiolites, and mélanges. In bedded cherts, sedimentary structures are uncommon. Similar to other chemically precipitated rocks, bedded chert is typically highly pure, indicating that it was deposited in an area with minimal inflow of detrital sediments (such as silty river water containing clay particles). Authigenic pyrite and hematite, which formed in the sediments after they were deposited, are among the impurities that are present, along with minute amounts of detrital minerals.

Seawater normally has a silica content of 0.01 to 11 parts per million (ppm), with 1 ppm being the average. This is well below saturation, suggesting that inorganic processes are not typically able to precipitate silica from seawater. Instead, silica is extracted from seawater by living things like glass sponges, diatoms, and radiolarians. These organisms are estimated to currently produce 12 cubic kilometers (2.9 cu mi) of opal annually in the world's oceans. They are also capable of efficiently extracting silica even from highly unsaturated water. Under perfect circumstances, diatoms can multiply eight times a day (although in typical seawater, doubling once a day is more common). They can also extract silica from water containing as little as 0 ppm silica. The organisms arm their skeletons with metal ions to prevent them from dissolving. If the organisms die and settle on the ocean floor, their skeletons will dissolve rapidly, forming a siliceous ooze that contains between 30 and 60% silica. As a result, the majority of fossil remains found in bedded cherts are typically those of organisms secreting silica skeletons, which are typically changed by solution and recrystallization.

These creatures' skeletons are made of opal-A, an amorphous silica that lacks a long-range crystal structure. This eventually changes into opal-CT, a microcrystalline silica that is primarily made up of bladed crystals of tridymite and cristobalite. A large portion of opal-CT is found as lepispheres, which are groups of bladed crystals with a diameter of roughly 10 microns. Afterward, opal-CT changes into microquartz. The transition from deep ocean water to opal-CT happens at roughly 45°C (113°F), and the transition from microquartz to opal-CT happens at roughly 80°C (176°F). However, the transition temperature varies a lot, and the presence of magnesium hydroxide, which acts as a nucleus for recrystallization, speeds up the transition. Megaquartz forms at high temperatures that are characteristic of metamorphism. There's evidence that porcelainite, a chert variety named for its high opal-CT content, recrystallizes at very shallow depths. Additionally, the Texas Caballos Novaculite exhibits evidence of very shallow water deposition in the form of shallow water sedimentary structures and evaporite pseudomorphs, which are casts of soluble mineral crystals that could only have formed in conditions close to the surface. This novaculate seems to have developed as a result of chert replacing carbonate fecal pellets.

3.3.1.1.1. Subvarieties

The types of organisms that produced the silica skeletons allow for further division of bedded cherts. Diatomaceous chert is made up of diatomite beds and lenses that were transformed into dense, hard chert during the diagenesis process. Seabeds of several 100-meter-thick marine diatomaceous chert strata have been documented from sedimentary sequences as old as the Cretaceous, including the Miocene Monterey Formation in California. The main siliceous organism that extracted silica from seawater during the Jurassic Period and beyond was diatoms. Radiolarian remnants make up the majority of radiolarite. Radiolarian chert is the term for remains that have been thoroughly cemented with silica. While many exhibit evidence of a deep-water origin, some seem to have formed in as shallow as 200 meters (660 feet), possibly in shelf seas where high organic productivity is supported by upwelling of nutrient-rich deep ocean water. Before the Jurassic, radiolarians were the main organisms that extracted silica from seawater. Chert made of glass sponge and other invertebrate spicules is called spilicarite. It is referred to as spicular chert when it is densely cemented. They are associated with nonvolcanic rocks typical of water a few 100 meters deep, such as phosphorites, black shale, clay-rich limestone, and glauconite-rich sandstone. Even under a microscope, certain bedded cherts seem to be devoid of fossils. They may originate from completely dissolved fossil remains in fluids that migrate and precipitate their silica load in a nearby bed, though their exact origin is unknown. Another silica source that has been proposed for chert beds is eolian quartz. Bedded cherts from the Precambrian period are widespread, accounting for 15% of middle Precambrian sedimentary rock. These rocks may have been deposited in environments with higher silica saturation than the current ocean through nonbiologic means. The high level of silica saturation was caused by either a lack of modern organisms that remove silica from seawater or by intense volcanic activity (Thornbury, 1969; Bates & Jackson, 1984; Wright, 1992).

3.3.1.2. Nodular Chert

Although it can also be found in shales and sandstones, nodular chert is most frequently found in limestone. It is not as prevalent in dolomite. In carbonate rocks, nodular chert is present as irregular to oval nodules. The size of these ranges from tiny quartz particles to nodules that are several meters in diameter. Though they are occasionally found cutting across bedding surfaces, where the chert fills fossil burrows, fluid escape structures, or fractures, the nodules are most commonly found along bedding planes or stylolite (dissolution) surfaces, where fossil organisms tended to accumulate and provided a source of dissolved silica. Smaller nodules typically have an egg-shaped shape, while larger nodules form irregular bodies with surfaces that are knobby. Large nodules may have desiccation cracks with secondary chert visible in the outer few centimeters. These cracks most likely developed concurrently with the nodule. There are sometimes fossils of calcareous materials that have been fully silicified. When found in chalk or marl, chert is commonly referred to as flint. The rind of nodular chert weathers to a white color and is typically dark in hue.

With silica deposited in place of clay minerals or calcium carbonate, the majority of chert nodules have textures that indicate they were created by diagenetic replacement. This might have happened when seawater from the sediment beds mixed with meteoric water (water from snow or rain), trapping carbon dioxide and creating an environment that was undersaturated with calcium carbonate and supersaturated with silica. Ecosystems on continental shelves are especially prone to nodular chert. The carbonate buildup zone of the Permian Basin (North America) contains relatively few chert nodules and chertified fossils, but they are abundant in the basin limestones. This might be the result of strong currents removing siliceous skeletons and redepositing them in the deep basin, the dissolution of opal where carbonate is actively being deposited, or a lack of siliceous organisms in these environments. Depending on internal banding in nodules, the silica in nodular chert most likely precipitates as opal-A and may recrystallize straight to microquartz without first recrystallizing to opal-CT. Because silica is not highly supersaturated, some nodular chert may precipitate as microquartz directly.

3.3.1.3. Other Occurrences

Iron oxide and chert layers alternate to form the Precambrian banded iron formations. In saline-alkaline lakes, nonmarine cherts can form as thin lenses or nodules with sedimentary structures that point to an evaporite origin. In the East African Rift Valley's alkaline lakes, these cherts are currently forming. These lakes are typified by extremely high pH sodium carbonate brines that have a silica content of up to 2,700 parts per million. Freshwater runoff episodes into lakes cause the pH to drop and precipitate the unusual sodium silicate minerals, magadiite or kenyaite. These minerals undergo burial and diagenesis, which transforms them into Magadi-type chert. Magadi-type chert, which may have originated in the alkaline Lake T'oo'dichi, is found in the Morrison Formation. Chert can also occur when silica dissolved from the beds of volcanic ash above replaces calcrete in paleosols, or fossil soils.

3.3.2. Fossils

The cryptocrystalline nature of chert, combined with its above average ability to resist weathering, recrystallization, and metamorphism has made it an ideal rock for the preservation of early life forms.

For example:

- The 3.2 Ga chert of the Fig Tree Formation in the Barberton Mountains between Eswatini and South Africa preserved non-colonial unicellular bacteria-like fossils.
- In addition to bacteria and cyanobacteria, the Gunflint Chert in western Ontario (1.9 to 2.3 Ga) contains organisms thought to consume ammonia as well as some that resemble green algae and fungus-like organisms.
- The Apex Chert (3.4 Ga) of the Pilbara craton, Australia preserved 11 taxa of prokaryotes. However, these findings have been disputed.
- The Bitter Springs Formation of the Amadeus Basin, Central Australia, preserves 850 Ma cyanobacteria and algae.
- The Rhynie chert (410 Ma) of Scotland has remains of a Devonian land flora and fauna with preservation so perfect that it allows cellular studies of the fossils.

3.3.3. Prehistoric and Historic Uses

As a source of silica, chert is currently only marginally significant economically—quartz sand is far more valuable. Nonetheless, valuable deposits of iron, phosphorite, manganese, uranium, and petroleum may be linked to chert deposits.

1. Tools: Chert was a common raw material used in prehistoric times to make stone tools. Chert fractures in a Hertzian cone when struck with enough force, just like obsidian and some rhyolites, felsites, quartzites, and other tool stones used in lithic reduction. Conchoidal fractures are the result, and all minerals lacking cleavage planes have these. Similar to what happens when a small object, like an air gun projectile, strikes a plate-glass window, this type of fracture occurs when a cone of force propagates through the material from the point of impact, eventually removing a full or partial cone. Flakes, which are partial Hertzian cones formed during lithic reduction, have striking platforms, bulbs of force, and occasionally eraillures, which are tiny secondary flakes that break off from the flake's bulb of force. These characteristics are typical of this type of breakage. Sparks fly when a

chert stone strikes something that contains iron. This makes chert a great fire starter, and throughout history, tinderboxes, and other fire starters have been made from both flint and common chert. The main historical application for common chert and flint was for flintlock weapons, in which a spark from the chert striking a metal plate ignites a small reservoir of black powder, causing the weapon to discharge.

2. Construction: When used as aggregate in concrete, cherries can lead to a number of issues. Because deeply weathered chert has a high porosity, it develops surface pop-outs when used in concrete that is subjected to freezing and thawing. The other issue is that high-alkali cements cause an alkali-silica reaction with some cherts. Concrete expands and cracks as a result of this reaction, finally failing the material (Thornbury, 1969; Bates & Jackson, 1984; Wright, 1992).

3.3.4. Varieties

There are numerous varieties of chert, classified based on their visible, microscopic, and physical characteristics. Some of the more common varieties are:

- A compact microcrystalline quartz is called flint. When silica was substituted for calcium carbonate in chalk or marly limestone formations, chert was first identified by this name. This variety, which is frequently found as nodules, was once widely employed to create bladed tools. Some geologists still refer to any chert that ranges from dark gray to black as flint. Organic matter inclusions are the source of the dark color. When comparing chert and flint, non-geologists frequently assume that chert is of lower quality than flint. In Great Britain, where most true flint (found in chalk formations) was of higher quality than common chert (found in limestone formations), this terminology is especially common.
- "Common chert" is a variety of chert that forms in limestone formations by replacement of calcium carbonate with silica. This is the most abundantly found variety of chert. It is generally considered to be less attractive for producing gemstones and bladed tools than flint.
- Hematite inclusions give jasper, a type of chert that was formed as primary deposits and is found in or near magmatic formations, its characteristic red hue. Jasper often comes in green, yellow, and black hues as well, depending on the kind of iron it contains. The typical Jasper is opaque to almost opaque. In banded iron formation, jasper is also found and is referred to as jaspilite there.
- Radiolarite is a variety of chert formed as primary deposits and containing radiolarian microfossils. Many show evidence of a deep-water origin, but some appear to have formed in water as shallow as 200 meters (660 ft),
- Chalcedony is a microfibrous quartz.
- Agate is distinctly banded chalcedony with successive layers differing in color or value.
- Onyx is a banded agate with layers in parallel lines, often black and white (sardonyx).

- Magadi-type chert is a variety that forms from a sodium silicate precursor in highly alkaline lakes such as Lake Magadi in Kenya.
- Novaculite is a very dense, fine-grained, and uniform form of very pure white chert with a high content of extracrystalline water. It is most common in the mid-Paleozoic rocks of Texas, Oklahoma, and Arkansas in the south-central United States, where it has undergone some metamorphism.
- Porcelanite is a term used for fine-grained siliceous rocks with a texture and a fracture resembling those of unglazed porcelain. It likely forms in shallow water and is composed mostly of opal-CT.
- Tripolitic chert (or tripoli) is a light-colored porous friable siliceous (largely chalcedonic) sedimentary rock, which results from the weathering (decalcification) of chert or siliceous limestone.
- Siliceous sinter is a porous, low-density, light-colored siliceous rock deposited by waters of hot springs and geysers.
- Mozarkite a varicolored, easily polished Ordovician chert that takes a high polish. It is the state rock of Missouri.

Other lesser-used archaic terms for chert are firestone and silex.

3.4. VOLCANICLASTICS

Clasts, or fractured pieces of volcanic rock, make up volcaniclastics, which are rocks found in the earth. No matter how the rock was broken up, how it was moved afterwards, what kind of environment it was deposited in, or whether nonvolcanic material mixed with the volcanic clasts, these apply to all clastic volcanic materials. Volcaniclastic rocks are defined by the US Geological Survey to be composed solely of volcanic rock fragments that have been transported from their original location, which is a somewhat more restrictive definition. Volcanic rocks, such as the Bandelier Tuff, cinder cones and other tephra deposits, the basal and capping breccia that characterize 'a'ā lava flows, and lahars and debris flows of volcanic origin are all considered to be part of volcaniclastics in the broad semantic sense. Many volcanoes contain more volcaniclastics than lava flows in their volume. A third or more of the sedimentation recorded in the geologic record could have been attributed to volcaniclastics.

On the surface of the Earth, igneous and sedimentary rocks are formed primarily by volcanic eruptions, which are also the most visually striking examples of this process. Molten lava flowing from the volcano's fissures and particulate matter ejected from the vent to form volcaniclastic deposits are among the materials produced by volcanoes during eruption (Cas & Wright, 1987). Volcano locations are influenced by plate tectonics and are typically found near plate borders and other regions of intense crustal heat flow. An important tectonic setting indicator for the formation of the sedimentary succession can be found in the presence of beds created by volcanic processes. Although ash may be dispersed tens, hundreds, or even thousands of kilometers away from the eruption site, lava can be found nearby. Hence, volcaniclastic material can be found in a wide range of other sedimentary rocks and can occur in any depositional environment. Because volcanic rocks can frequently be radiometrically dated, which places an exact time limit on the sedimentary succession, they are also very valuable in stratigraphy (Thornbury, 1969; Bates & Jackson, 1984; Wright, 1992).

3.4.1. Classification of Volcaniclastics

Variable depositional environments are responsible for the formation of volcaniclastics, which are made up of a mixture of epiclastic sediments and pyroclastic debris. Volcaniclastic materials comprise autoclastic, alloclastic, and epiclastic rocks; pyroclastic rock and tephra; and fault gouges, which are formed when faults shift volcanic rock. Each is described below. Primary volcaniclastics and secondary volcaniclastics, also known as epivolcaniclastics, are two categories for these (Table 3.1).

Table 3.1. Primary versus Secondary Volcaniclastic Rocks

Primary Volcaniclastic Deposits (Characteristic Clasts must be More Than 75% of Volume)	Secondary Volcanic Deposits (Epivolcaniclastic Deposits)
Pyroclastic tuff or tephra (pyroclastic deposit). Autoclastic rock or loose deposit. Hyaloclastic rock or loose deposit (Hyaloclastite). Redeposited pyroclastic tephra. Redeposited autoclastic loose deposit. Redeposited hyaloclastic loose deposit.	Epiclastic deposits have more than 75% epiclasts in the general case. They are epivolcanicastic deposits if they have recognizable volcanic fragments in any proportion.

Source: https://en.wikipedia.org/wiki/Volcaniclastics.

3.4.1.1. Pyroclastic

Without regard to the origin of the particles or the type of eruption, pyroclastic material is made up of rock fragments that released as individual particles from the vent as a result of explosive volcanism. A portion of the country rock that may be entrained in the vent could be among these. Pyroclast material that has accumulated but not yet undergone significant consolidation is referred to as pyroclastic rock, whereas tephra is the term used to describe such accumulations. When pyroclastic material is created at the interface between magma and water, it can take many different forms, one of which is hydroclastic material.

3.4.1.2. Autoclastic

Processes that are active when solid or semisolid lava moves produce autoclastic volcanic material. These comprise fragments of rock formed by motion or gas explosions within volcanic flows, fragments of rock produced by the gravitational collapse of lava domes or spines, and fragments of rock produced within volcanic vents but not extruded. 'a'ā lava flows are characterized by autoclastic volcaniclastics that form the basal and capping breccia.

3.4.1.3. Other Kinds of Volcaniclastic Material

Alloclastic volcanic material is formed by fragmentation of existing igneous rock by subsurface igneous activity that may or may not involve magma intrusions. Fault gouge produced by motion along a fault in volcanic rock is also a type of volcaniclastic material.

3.4.1.4. Epivolcaniclastics

Volcanic epiclastic material (epivolcaniclastics) contains a substantial fraction of epiclasts (rock fragments produced by weathering and erosion) derived from volcanic rock.

3.4.1.5. Mixed Pyroclastic-Epiclastic Rocks

One of the materials that is most helpfully characterized as simply volcaniclastic is deposits containing pyroclastic material that has been reworked in stream or lake environments or mixed with epiclastic material (volcanic or nonvolcanic). These deposits present a unique challenge. In these situations, a more precise classification presents challenges. A rock unit that is simply referred to as volcaniclastic is the Espinaso Formation in New Mexico. It is made up of a complex mixture of pyroclastic and volcanic epiclastic material. Another is the Washburn Group in the Yellowstone region, which consists of volcanic epiclastic rock and debris flows of reworked volcanic ash. The percentage of pyroclastic material and average clast size can be used to categorize mixed pyroclastic-epiclastic deposits (Table 3.2).

Table 3.2. Terms for Mixed Pyroclastic-Epiclastic Rocks

Pyroclastic	Tuffites (Mixed Pyroclastic-Epiclastic)	Epiclastic (VOLCANIC or nonvolcanic)	Average Clast Size (mm)
Agglomerate, agglutinate pyroclastic breccias	Tuffaceous conglomerate, tuffaceous breccia	Conglomerate, breccia	> 64
Lapillistone			2–64
Coarse ash tuff	Tuffaceous sandstone	Sandstone	0.0625–2
Fine ash tuff	Tuffaceous siltstone	Siltstone	0.004–0.0625
	Tuffaceous mudstone, tuffaceous shale	Mudstone, shale	< 0.004
75–100% pyroclasts	25–75% pyroclasts	0–25% pyroclasts	–

Source: https://en.wikipedia.org/wiki/Volcaniclastics.

3.4.2. Types of Volcaniclastic Rocks

The type of eruption depends on the makeup of the magma. Volcanoes formed by basaltic magmas typically spew out copious amounts of lava and relatively little volcanic ash. Greater silicic magma content in volcanoes makes them far more explosive, ejecting large amounts of molten rock as particulate matter. Pyroclastic material, or tephra as a collective term, is the name given to the expelled particles. Keep in mind that any deposit that is primarily made up of volcanic detritus is referred to as volcaniclastic, while pyroclastic refers to material that is ejected from the volcano as particles. Individual crystals, volcanic rock fragments (lithic fragments), or pumice—the extremely vesicular, chilled rock "froth"—can all be considered examples of pyroclastic material. The fragments of the pyroclastic debris can be several meters across, or they can be as small as fine dust a few microns across (Thornbury, 1969; Bates & Jackson, 1984; Wright, 1992).

3.4.3. Nomenclature of Volcaniclastic Rocks

Volcaniclastic deposits are categorized texturally (Figure 3.4) The Wentworth scheme has been modified. Volcanic breccia and agglomerate are the terms used to describe the two types of coarse material (over 64 mm) that are separated into volcanic blocks, which were solid during eruption, and volcanic bombs, which were partially molten and cooled in the air before congealing into a rock.

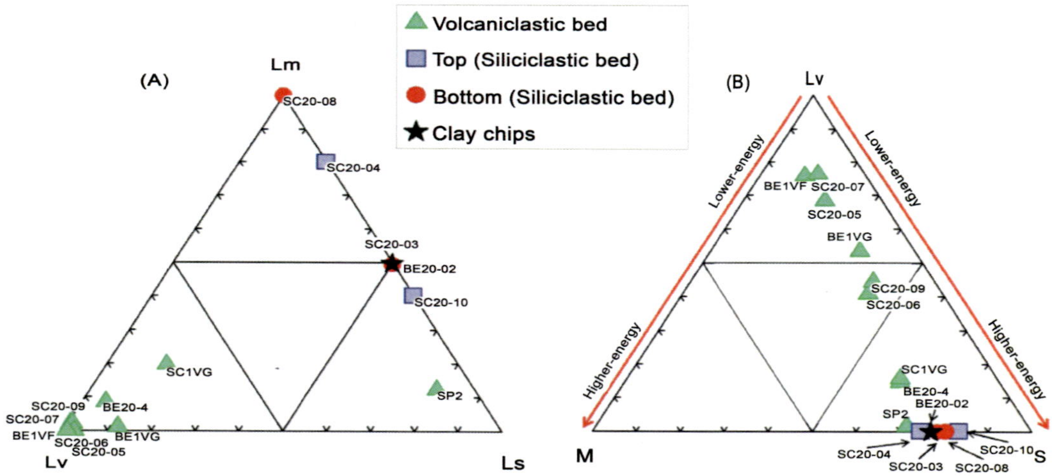

Figure 3.4. (A) Ternary plot of the relative proportions of Lm:Lv:Ls. Lm: metamorphic lithics; Lv: volcanic lithics; Ls: sedimentary lithics). (B) Relative proportions of Lv:M:S. Lv: All volcanic lithic textures (Lvl + ALvl + Algl + Lvf + Pm + Apm + Shard); M: mafic single crystal grains (pyroxene + amphibole + garnet + biotite + opaques); S: sialic single crystal grains (plagioclase + K-feldspar + monocrystalline and polycrystalline quartz without tectonic fabric).

Source: Chiara, B. C., Rosanna, D. R., Paola, D., Consuele, M., & Francesco, M., (2021). High preservation potential volcaniclastic sedimentation in the Serravallian sequence of the Amantea basin (Coastal Chain, North-Western Calabria). Geosciences, 11(9), 360. https://doi.org/10.3390/geosciences11090360.

Lapillistones are made up of granule to pebble-sized particles (2–64 mm) known as lapilli. Air fall produces spherical aggregates of fine ash, which are called accretionary lapilli. Tephra of the sand, silt, and clay grades is ash when it is unconsolidated and turns to tuff when it lithifies. Fine ash/tuff is a material of silt and clay grade, whereas coarse ash/tuff is sand-sized. The relative amounts of crystals, lithic fragments, and vitric material—fragments of volcanic glass created when the molten rock cools extremely quickly—are what determine compositional descriptions. Pumice is occasionally formed from these fragments.

3.4.4. Recognition of Volcaniclastic Material

If the lithology of the larger clasts can be identified as an igneous rock, such as basalt, then it is usually easy to determine the origin of coarse-grained volcaniclastic sediments. Except for rounded volcanic bombs, well-rounded accretionary lapilli found in some air fall ashes, and the unusual shape of fiamme, glassy pumice fragments that, when compacted, may resemble a tuning fork, the tephra particles are primarily angular during formation. Since tephra can only be mixed with other types of sediment through further reworking, the material's uniformity is another helpful indicator. More rhyolitic deposits have a lighter color, while volcaniclastic sediments with a basaltic composition tend to be darker in color. Certain field identification of fine ash and tuff can be more challenging, particularly if the material has weathered. Ash beds can occasionally change, resulting in the formation of brightly colored green and orange strata. Verification of the composition of volcaniclastic sediments usually requires petrographic analysis. If a high magnification is utilized to identify the minerals that comprise the rock fragments, the composition of lithic fragments can be ascertained through thin-section analysis. If the deposit is a crystal tuff, feldspar crystals are typically prevalent. Other silicate minerals may also be present in the form of euhedral to subhedral crystal grains. Fiamme are transparent, isotropic grains with distinct shapes. Volcanic glass is unstable and can change into clay minerals in older tuffs or have a very finely crystalline structure (Thornbury, 1969; Bates & Jackson, 1984; Wright, 1992).

CASE STUDY

APPLICATION OF EVAPORITE MINERALS IN IDENTIFICATION OF MARINE CARBONATE SEDIMENTARY ENVIRONMENT: A CASE STUDY OF MA 54 IN THE CENTRAL-EAST ORDOS BASIN:

The Central-East Ordos Basin in China is renowned for its complex geological history and rich hydrocarbon resources. In this case study, we explore the significance of evaporite minerals in identifying marine carbonate sedimentary environments within the Ma 54 section of this basin. The Ma 54 section, a part of the Middle Ordovician Majiagou Formation, offers a unique opportunity to study the interplay between marine carbonate deposition and evaporite mineral precipitation.

The geological context of the Central-East Ordos Basin is crucial for understanding the depositional environments within the Ma 54 section. Over millions of years, this basin has witnessed dynamic changes, including marine transgressions and regressions, which have left behind a diverse stratigraphic record. The Ma 54 section, situated within this basin, contains a sequence of carbonate rocks interbedded with evaporite minerals, providing valuable insights into past environmental conditions. Evaporite minerals, such as halite, gypsum, and anhydrite, play a pivotal role in unraveling the depositional history of sedimentary sequences. Their presence within carbonate rocks often indicates periods of high evaporation rates in shallow marine settings. In the Ma 54 section, the identification of evaporite minerals serves as a key diagnostic feature for interpreting the paleoenvironment. These minerals precipitate from concentrated brines in restricted marine environments, such as lagoons or sabkhas, where evaporation exceeds replenishment. To elucidate the depositional environment and understand the formation of evaporite minerals in the Ma 54 section, a multidisciplinary approach was employed. Petrographic analysis of core samples revealed the textural features and mineral compositions, while X-ray diffraction (XRD) and scanning electron microscopy (SEM) provided detailed mineralogical data. Additionally, stable isotope ($\delta^{18}O$ and $\delta^{13}C$) analyses were conducted to infer the origin of carbonates and the paleoenvironmental conditions.

The findings from the Ma 54 section yielded compelling evidence of a marine carbonate sedimentary environment punctuated by episodes of evaporite mineral precipitation. Halite casts and nodules of gypsum and anhydrite were observed within certain intervals, indicating periods of high salinity and evaporation. These evaporite minerals often exhibit characteristic textures and crystal morphologies, further supporting their interpretation as primary precipitates in shallow marine settings. The stable isotope signatures of carbonates provided additional insights into the depositional history of the Ma 54 section. The $\delta^{18}O$ and $\delta^{13}C$ values reflected variations in seawater chemistry and paleotemperature, corroborating the presence of evaporite minerals and the cyclical nature of marine transgressions and regressions. Sedimentary structures such as mud cracks and stromatolites within the carbonates further reinforced the interpretation of shallow, periodically exposed marine environments conducive to evaporite formation. By integrating petrographic, mineralogical, and geochemical data, a comprehensive reconstruction of the Middle Ordovician paleoenvironment in the Central-East Ordos Basin was achieved. The Ma 54 section represents a dynamic interplay between marine carbonate deposition and evaporite mineral precipitation, offering valuable insights into past climatic and sea-level changes. This understanding has broader implications for regional stratigraphy, resource exploration, and paleoenvironmental reconstructions.

In conclusion, the application of evaporite minerals in identifying marine carbonate sedimentary environments, as exemplified by the Ma 54 section of the Central-East Ordos Basin, underscores their significance as proxies for paleoenvironmental studies. This case study highlights the intricate relationship between sedimentary processes, depositional environments, and the formation of mineral deposits, providing a deeper understanding of Earth's geological history and its implications for resource exploration and environmental management.

CLASS ACTIVITY

STUDENTS WILL EXAMINE LIMESTONE ROCK AND PERFORM A TEST:

Tell the students that the largest and greatest number of caves are found in areas of extensive deposits of limestone. Explain that limestone and other carbonate rocks are somewhat unique because their mineral grains can be dissolved in nature. Ask the students why they think this would be important in the process of forming a cave. Explain to the students that a geologist can often tell what kinds of minerals are found in rocks by the physical properties of the rock. Geologists look at the appearance of the rock and its characteristics such as its grain size, hardness, color, texture, and whether it can be dissolved.

Ask the students if they know what it means to dissolve something. Show the students a clear plastic cup filled with room-temperature water and a small amount of sugar. Ask the students what they think will happen when the sugar, a solid, is added to the water, a liquid. Add the sugar to the water and stir. Allow the students to observe the results. Ask the students to explain what happened to the sugar. Is the sugar still present even though they cannot see it? Explain to the students that the main mineral that makes up limestone is called calcite. Tell the students that calcite will dissolve in most forms of acid. Show the students several rock samples. Tell the students that one of the samples of rock is limestone. Ask the students if they can suggest a way to determine which rock is the limestone. Explain to the students that the rock samples can be tested for calcite by checking to see which ones dissolve when they come into contact with acid.

SUMMARY

- In areas where there is not a large supply of clastic detritus other processes are important in the accumulation of sediments. The hard parts of plants and animals ranging from microscopic algae to vertebrate bones make up deposits in many different environments.
- Chemical sediments, on the other hand, form through the precipitation of minerals from solution. This process occurs when mineral-laden water changes temperature, pressure, or chemical composition, causing minerals to crystallize and settle out of the water column
- Limestone (calcium carbonate $CaCO_3$) is a type of carbonate sedimentary rock that is the main source of the material lime. It is composed mostly of the minerals calcite and aragonite, which are different crystal forms of $CaCO_3$.
- Limestone is composed mostly of the minerals calcite and aragonite, which are different crystal forms of calcium carbonate ($CaCO_3$). Dolomite, $CaMg(CO_3)_2$, is an uncommon mineral in limestone, and siderite or other carbonate minerals are rare.
- Most grains in limestone are skeletal fragments of marine organisms such as coral or foraminifera. These organisms secrete structures made of aragonite or calcite and leave these structures behind when they die.
- The grains of most limestones are embedded in a matrix of carbonate mud. This is typically the largest fraction of an ancient carbonate rock. Mud consisting of individual crystals less than 5 µm (0.20 mils) in length is described as micrite.
- Travertine is a term applied to calcium carbonate deposits formed in freshwater environments, particularly waterfalls, cascades, and hot springs. Such deposits are typically massive, dense, and banded.
- Limestone forms when calcite or aragonite precipitates out of water containing dissolved calcium, which can take place through both biological and nonbiological processes.
- Diagenesis is the process in which sediments are compacted and turned into solid rock. During diagenesis of carbonate sediments, significant chemical and textural changes take place. For example, aragonite is converted to low-magnesium calcite.
- The cyanobacterium Hyella balani can bore through limestone; as can the green alga Eugamantia sacculata and the fungus Ostracolaba implexa.
- Micricitic mud mounds are subcircular domes of micritic calcite that lacks internal structure. Modern examples are up to several 100 meters thick and a kilometer across, and have steep slopes (with slope angles of around 50°).

REVIEW QUESTIONS

1. How is limestone formed, and what are the key processes involved in its formation?
2. What role do living organisms play in the formation of limestone, and how do their biological activities contribute to limestone deposits?
3. What are micritic mud mounds, and what processes lead to their formation in sedimentary environments?
4. What are the characteristics of nonmarine environments that are conducive to evaporite deposition, and how do these environments differ from marine settings?
5. How has chert been used historically and prehistorically, and what are some significant examples of its applications in ancient cultures?

REFERENCES

1. Bates, R. L., & Jackson, J. (1984). Dictionary of Geological Terms (3rd ed.). American Geological Institute/Doubleday.
2. Benison, K. C., & Goldstein, R. H. (2002). Recognizing acid lakes and groundwaters in the rock record. Sedimentary Geology, 151, 177–185.
3. Blatt, H., & Tracy, R. J. (1996). Petrology: Igneous, Sedimentary, and Metamorphic (2nd ed.). W. H. Freeman.
4. Boggs, S. (2006). Principles of Sedimentology and Stratigraphy (4th ed.). Pearson Prentice Hall.
5. Cas, R. A. F., & Wright, J. V. (1987). Volcanic Successions: Modern and Ancient. Unwin-Hyman.
6. Hardie, L. (1984). Evaporites: Marine or nonmarine? American Journal of Science, 284, 193–240. https://doi.org/10.2475/ajs.284.3.193.
7. Hardie, L. A., & Eugster, H. P. (1971). The depositional environment of marine evaporites: A case for shallow, clastic accumulation. Sedimentology, 16, 187–220.
8. Huwald, E. (2001). Calcium carbonate—Pigment and filler. In F. W. Tegethoff (Ed.), Calcium Carbonate (pp. 160–170). Birkhäuser. https://doi.org/10.1007/978-3-0348-8245-3_7.
9. McBride, E. F., & Folk, R. L. (1977). The Caballos Novaculite revisited: Part II: Chert and shale members and synthesis. SEPM Journal of Sedimentary Research, 47. https://doi.org/10.1306/212F731A-2B24-11D7-8648-000102C1865D.
10. McGhee, G. R. (2013). When the Invasion of Land Failed: The Legacy of the Devonian Extinctions. Columbia University Press.
11. Pratt, B. R. (1995). The origin, biota, and evolution of deep-water mud-mounds. Special Publications International Association of Sedimentologists, 23, 49–123.
12. Thornbury, W. D. (1969). Principles of Geomorphology (2nd ed.). Wiley.
13. Wright, V. P. (1992). A revised classification of limestones. Sedimentary Geology, 76(3–4), 177–185. https://doi.org/10.1016/0037-0738(92)90082-3.

CHAPTER

4

Processes of Transport and Sedimentary Structures

LEARNING OBJECTIVES

After studying this chapter, you will be able to:

- Discuss about sediment transport and deposition;
- Focus on transport media;
- Learn about sedimentary structures and textures.

KEY TERMS FROM THIS CHAPTER

Bedload transport
Heterogeneous fluids
Laminar flows
Sediment
Fluids
Hydraulic engineering
Landslides
Sediment transport

INTRODUCTION

Physical characteristics created by the deposition and subsequent modification of sediments are known as sedimentary structures. They are essential for comprehending sedimentary processes and interpreting historical environments. A multitude of transport mechanisms, such as water, wind, ice, and gravity, combine to create these structures. Different sedimentary structures are left behind by each transport process; these structures are divided into primary and secondary structures.

Among the primary sedimentary structures that emerge during or soon after sediment deposition are graded bedding, ripple marks, cross-bedding, and stratification. The most basic type of sedimentary structure is called stratification, or bedding, and it is defined by horizontal sediment layers with clear boundaries. These layers show the various depositions that occurred over time. Cross-bedding is characterized by slanted layers inside horizontal beds and happens when sediments are deposited by flowing fluids, such as water or wind. This type of structure is typical of deserts, coastal dunes, and river environments. Little, wave-like structures called ripple marks are created when water or wind passes over loose sediment. There are two types of ripples: asymmetrical ripples, which are produced by unidirectional currents like river flows, and symmetrical ripples, which are produced by oscillating water in shallow environments. Turbidity currents under water frequently cause graded bedding, a layer in which particle sizes decrease from bottom to top. Mixed-size sediments are carried by these currents and settle in a size-sorted fashion.

Compaction, bioturbation, and chemical modification are some of the processes that lead to the development of secondary sedimentary structures after deposition. Because of the weight of the sediments above, compaction can cause grains to reorient and reduce pore spaces. When organisms disturb sediments, they produce structures like trails, burrows, and tracks, a process known as bioturbation. Chemical alteration involves processes such as dissolution, which removes soluble minerals and changes the texture of the rock, and diagenesis, which involves minerals precipitating from groundwater and binding sediments.

These structures are formed in large part by the transport mechanisms. Structures like point bars and channel deposits, where sediments are deposited and sorted based on flow velocity, are produced by the flow of water through rivers and streams. Driven by wind, aeolian processes produce deposits of loess and dunes, and their well-sorted, rounded grains are indicative of wind transport. The movement of ice during glacial processes has the potential to carry and deposit sorted sediments, resulting in the formation of structures such as drumlins and moraines. Mass wasting deposits are composed of erratic and poorly sorted sediments that are produced by gravity-driven processes like landslides and debris flows.

Understanding past environmental conditions is possible through the interpretation of sedimentary structures. For example, symmetrical ripple marks may indicate a shallow marine environment, while cross-bedding in sandstones may indicate ancient desert dunes or river channels. Graded bedding frequently indicates past seismic activity or slope instability by pointing to underwater landslides or turbidity currents. Reconstructing the geological past and depositional settings of sedimentary rocks depends heavily on the sedimentary structures. Geologists can infer historical sea levels, tectonic events, and climates thanks to the variety of structures and the transport processes that give rise to them. Comprehending these procedures is crucial for multiple uses, such as investigating natural resources, reconstructing the surroundings, and forecasting upcoming geological alterations.

4.1. SEDIMENT TRANSPORT AND DEPOSITION

The two main processes in sedimentology that control the creation of sedimentary rocks are sediment transport and deposition. There are several different ways that sediment can move, such as wind, water, ice, and gravity. The energy of the transporting medium, which can vary from mild river currents to the ferocious flow of a pyroclastic flow from a volcanic eruption, greatly influences the type of sediment transport.

The movement of solid particles, or sediment, usually results from the combined action of gravity acting on the sediment and the fluid in which the sediment is entrained. This process is known as sediment transport. In natural systems, sediment is transported by clastic rocks, which include boulders, gravel, sand, and other similar particles), mud, or clay; the fluid is either ice, water, or air; and the particles are moved by gravity along the sloping surface they are resting on. Currents and tides cause sediment to be transported by fluid motion in rivers, lakes, seas, and other bodies of water.

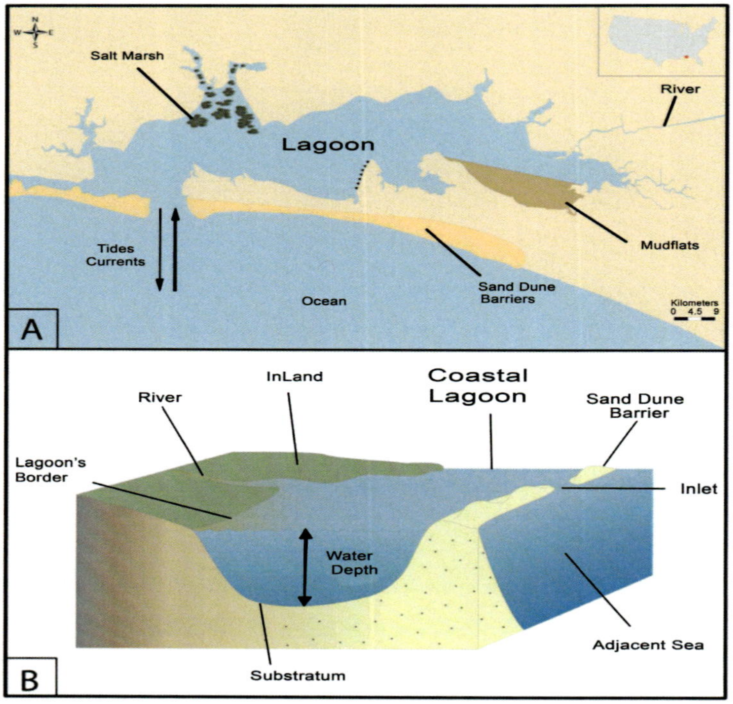

Figure 4.1. Sediment transport in coastal lagoons.

Source: El Mahrad, B., Newton, A., & Murray, N., (2022). Coastal lagoons: Important ecosystems. Front. Young Minds, 10, 637578. doi: 10.3389/frym.2022.637578.

Additionally, transport is a result of glaciers moving and wind-driven movement on land surfaces. On sloping surfaces in general, such as hillslopes, scarps, cliffs, and the continental shelf—continental slope boundary, sediment can be transported solely by gravity. In the domains of sedimentary geology, geomorphology, hydraulic engineering, environmental engineering, and civil engineering, sediment transport is crucial. The most common methods for predicting whether or not erosion or deposition will occur, how much it will be, when it will occur, and over what distance are all dependent on our understanding of sediment transport (Figure 4.1).

After being moved, sediment can be left behind in a variety of locations, such as deep marine basins, lakes, river channels, deltas, and beaches. Grain size, composition, and sorting of the deposited sediment are all influenced by the depositional environment.

Understanding the formation of sedimentary rocks and the environments in which they were deposited requires a thorough understanding of sediment transport and deposition. The management of water resources, the building of dams and levees, and the anticipation and mitigation of geological hazards like landslides and debris flows are just a few of the practical uses for this knowledge.

4.1.1. Agents of Sediment Transport

The agents of sediment transport are the natural forces that move sediments from one location to another. These agents can be classified into several categories:

1. Water: Sediment can be transported by rivers, streams, ocean currents, and waves. The amount of sediment transported by water depends on the velocity and turbulence of the water, the size and shape of the sediment particles, and the availability of sediment.
2. Wind: Sediment can be transported by wind, especially in arid regions where there is little vegetation to hold the soil in place. Wind can move both fine-grained and coarse-grained sediment, and the amount of sediment transported depends on the wind speed and the size of the sediment particles.
3. Ice: Sediment can be transported by glaciers, which move slowly downhill under their own weight. Glaciers can transport large amounts of sediment, including boulders, cobbles, and sand.
4. Gravity: Sediment can be transported by gravity, especially on steep slopes and in mountainous regions. Gravity can cause landslides, rockfalls, and debris flows that transport sediment downslope.
5. Biological Agents: Sediment can also be transported by living organisms, such as burrowing animals, root systems of plants, and microbes.

The type and amount of sediment transported by each agent depends on a variety of factors, including the strength and duration of the agent, the type and size of the sediment, and the characteristics of the terrain (Figure 4.2).

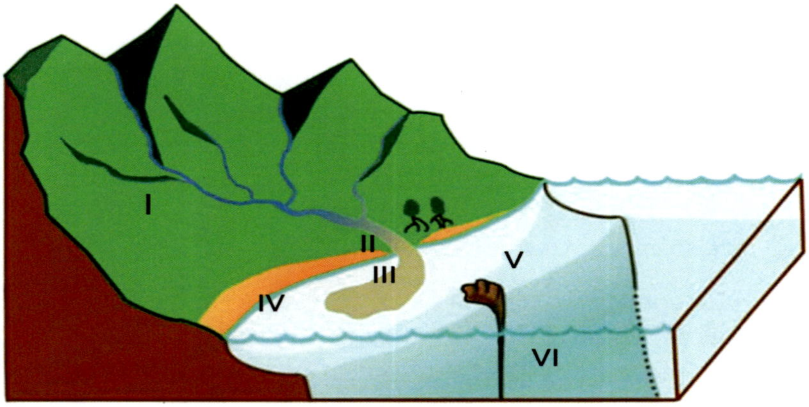

Figure 4.2. Six main domains/geomorphological units of sediment transport studies. (I) the watershed; (II) estuary/delta; (III) the estuary-littoral transition, including the river plume, the bottom nepheloid layer and the ROFI zone; (IV) the littoral zone; (V) the continental shelf; and (VI) canyon and deep sea.

Source: Ouillon, S., (2018). Why and how do we study sediment transport? Focus on coastal zones and ongoing methods. *Water*, 10(4), 390. https://doi.org/10.3390/w10040390.

4.1.2. Modes of Sediment Transport

There are several modes of sediment transport, including:

1. Bedload Transport: This is the movement of sediment particles along the bottom of a river, stream, or ocean by rolling, sliding, or bouncing.
2. Suspended Load Transport: This is the movement of small sediment particles that are suspended in the water column by the force of the flowing water.
3. Saltation: This is a type of bedload transport in which sediment particles bounce along the bottom in a series of short hops.
4. Traction: This is a type of bedload transport in which sediment particles are rolled or dragged along the bottom by the force of the flowing water.

The mode of sediment transport is influenced by a number of factors, including the velocity of the water, the size and shape of the sediment particles, and the density of the particles.

4.1.2.1. Bedload Transport

The movement of sediment particles on a river or stream bed caused by the flow of water is known as "bedload transport." Larger sediment particles, like sand and gravel, are carried along the bottom of the stream channel by the force of the water and roll, slide, or bounce during this process. Intermittent, high-energy flows that happen during times of increased stream discharge, like storms or snowmelt events, are typical characteristics of bedload transport. The size and shape of the sediment particles, the water's velocity, the slope and roughness of the stream channel, and other factors all affect the rate of bedload transport. The physical and ecological features of river and stream ecosystems, such as the creation of riffles and pools, the erosion of stream banks, and the modification of aquatic habitat, can all be significantly impacted by bedload transport.

4.1.2.2. Suspended Load Transport

One method of moving sediment is called suspended load transport, in which the moving fluid—usually air or water—carries fine-grained particles like silt and clay in suspension. The fluid flow creates turbulent eddies, which have upward components that suspend the particles. There are many different environments where suspended load transport can happen, including rivers, estuaries, oceans, and the atmosphere. The formation of sedimentary deposits, such as floodplains and deltas, as well as erosion and sedimentation downstream, can be attributed to the suspended load, which is generally regarded as the most important component of sediment transport in rivers.

The velocity and turbulence of the fluid, the size and density of the particles, and the concentration of the sediment in the fluid are some of the variables that determine how much sediment can be transported in suspension. Larger particles and higher concentrations of transported sediment are generally possible when the fluid's velocity and turbulence levels are higher.

4.1.2.3. Saltation

Saltation is a type of sediment transport where air or water movement bounces particles, typically sand-sized, along a streambed or the top of a dune. The particles are propelled into the air during saltation and then return to the surface, hopping or bouncing in short bursts. In rivers, streams, and other waterways where water is moving sand and other small particles along the bottom, this kind of sediment transport is crucial. It is also significant in deserts where wind carries sand and on the surface of Mars, where a thin atmosphere permits similar wind-borne sand transportation.

4.1.2.4. Traction

Particles of sediment are moved by traction, which involves rolling or sliding them along the bottom of a fluid body like a body of water. Coarser sediment particles, like sand or gravel, that are too heavy to be carried in suspension or saltation are frequently linked to traction. The particles are pulled along the bottom by the force of the flowing fluid as it passes over them, be it water or another fluid. The strength of the current or flow, the size and shape of the sediment particles, and the properties of the sediment bed, such as the surface roughness and slope, all affect how much sediment is transported by traction. In addition to forming sedimentary deposits, traction plays a significant role in the morphology of rivers, streams, and other bodies of water (Bovis, 2003; Chanson, 2004).

4.1.3. Depositional Environments

The places and circumstances where sediment accumulates are known as depositional environments, and they are defined by the physical, chemical, and biological processes that are at work. The features of the sediment and the main processes that form the deposit determine how they are categorized (Figure 4.3).

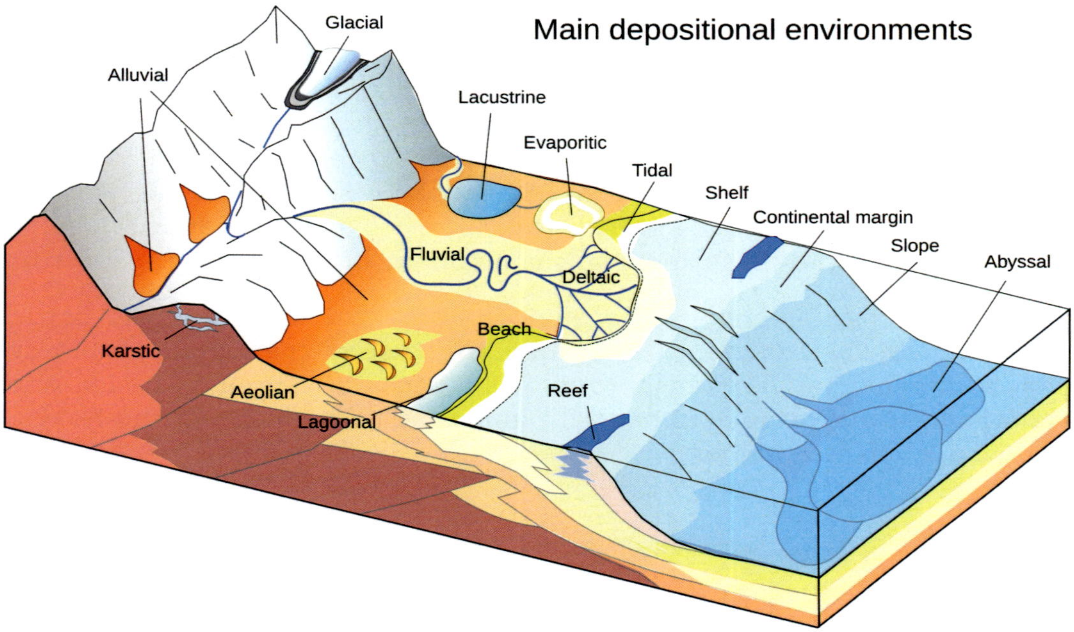

Figure 4.3. Depositional environments.

Source: https://upload.wikimedia.org/wikipedia/commons/6/66/Main_depositional_environments.svg.

Some common depositional environments include:

1. Alluvial: These deposits made by water moving through alluvial fans, floodplains, and river channels. The term "alluvial" describes sedimentary materials that are deposited by flowing water in rivers, streams, or floods, such as sand, gravel, clay, or silt. The Latin word "alluvium," which means "wash against," is where the word "alluvial" originates. River channels, floodplains, deltas, and alluvial fans are home to alluvial deposits, which are frequently abundant in minerals and organic matter. They are frequently employed in the mining of precious metals and other resources, as well as in the production of building materials.
2. Deltaic: Accumulation of sediment at the mouth of a river as it enters a standing body of water, often forming branching patterns of channels and distributaries.

At the mouths of rivers, where they enter still bodies of water like lakes, seas, or oceans, deltaic deposition takes place. When a river's velocity drops because of a standing body of water, sediment that is carried by the river and deposited at its mouth accumulates to form deltas. The sediment is usually fine-grained and composed of layers of sand, silt, and clay. The type of sediment being carried and the circumstances at the river's mouth, such as the direction and intensity of the tidal currents, determine the precise features of the deltaic deposit. Because deltaic deposits are often very rich in organic matter, they are significant hydrocarbon sources.

3. Coastal: These deposits formed along the coast by the action of waves, currents, and tides, including beaches, barrier islands, and lagoons.

The presence of marine or coastal processes that affect sediment transport and deposition, such as tides, waves, and currents, characterizes coastal depositional environments. Estuaries, tidal flats, barrier islands, beaches, and dunes are a few examples of these habitats. With sediment that is the size of sand, beaches are the most prevalent coastal depositional environment.

They are created by wave action. They can be categorized as high-energy dissipative beaches or low-energy reflective beaches based on how much wave energy they receive. Wind action shapes dunes that can develop behind beaches. They can act as a natural barrier against coastal erosion and be significant habitats for flora and fauna. Long and narrow offshore landforms known as barrier islands run parallel to the coast and are divided from it by estuaries or lagoons.

They are created when sediment is deposited by waves and currents, and they have the ability to move over time in response to variations in the availability or level of sediment. Low-lying coastal areas that are exposed at low tide and submerged at high tide are known as tidal flats. They are distinguished by fine-grained silt and can serve as vital habitats for marine life, both as breeding and feeding grounds.

Estuaries are coastal water bodies that are partially enclosed and connected to the sea. They are subject to the influence of both freshwater and marine processes. They can serve as crucial areas for shipping and recreation in addition to serving as fish and other aquatic organisms' nidification habitats.

4. Marine: These deposits formed in the ocean by currents, waves, and other processes, including continental shelves, slopes, and deep-sea basins.

Marine depositional environments are those in which sedimentation occurs in the ocean or sea. These environments are characterized by the presence of marine organisms and can range from shallow water to deep sea environments. Examples of marine depositional environments include:

i. Continental Shelf: This is a shallow, gently sloping area of the ocean floor that extends from the shoreline to the shelf break, where the slope steepens. Sediments here are typically fine-grained, including sand, mud, and silt.

ii. Continental Slope and Rise: Beyond the continental shelf lies the continental slope, which is a steeply sloping area that extends down to the deep ocean floor. Sediments here are transported downslope by gravity and can include turbidites, which are deposits of sand and mud that result from underwater avalanches.

iii. Abyssal Plain: This is a flat, featureless area of the deep ocean floor that lies beyond the continental rise. Sediments here are typically very fine-grained and can include clay and silt.

iv. Mid-Ocean Ridge: This is an underwater mountain range that runs through the middle of the world's oceans. Sediments here are typically volcanic in origin and can include pillow lavas and basalt.

v. Hydrothermal Vents: These are deep sea environments where hot, mineral-rich water is discharged from the seafloor. Sediments here can include metal sulfides and other minerals.

vi. Coral Reefs: These are areas of shallow, warm water where coral animals build reefs. Sediments here can include sand and shell fragments, as well as coral debris.

vii. Estuaries: These are semi-enclosed coastal environments where freshwater and saltwater mix. Sediments here are typically fine-grained and can include sand, mud, and silt.

viii. Mangrove Swamps: These are coastal environments where mangrove trees grow in shallow water. Sediments here can include mud and organic debris.

5. Glacial: These deposits formed by glaciers, including moraines, till, and glacial outwash plains.

The process by which glaciers carry and erode sediment and deposit it somewhere else when the ice melts is known as "glacial deposition." As they travel, glaciers erode rocks and other materials, taking the debris with them. The material can vary in size depending on the strength and size of the glacier, from large boulders to fine silt.

Generally speaking, glacial sediments fall into two groups: till and stratified drift. Stratified drift is the sediment that has been sorted and deposited by meltwater streams that flow out of the glacier, whereas till is the unsorted sediment that is directly deposited by the glacier as it melts.

Glacial sediments can be found in a variety of environments, including mountain valleys, coastal regions, and continental interiors. These deposits are important indicators of past glacial activity and can provide valuable insights into the climate and landscape of the past. They also have economic importance, as glacial deposits often contain valuable minerals such as gold, silver, and copper.

6. Aeolian: These deposits formed by wind, including sand dunes and loess.

The term "aeolian deposition" describes how wind carries sediment and deposits it. Since the wind tends to carry only the tiniest particles over great distances, aeolian sediment is frequently distinguished by its small size and well-sorted nature. Aeolian sediment is frequently deposited in dunes, which, depending on the direction and intensity of the wind, can take on a variety of shapes. Understanding current and historical climates, as well as the discovery and exploitation of natural resources like oil and gas, depend on the study of aeolian sediment and landforms.

7. Lacustrine: These deposits formed in lakes, including deltas, and deep lake basins.

Areas connected to or associated with lakes are referred to as lacustrine environments. Lake sedimentation, which may eventually be buried and preserved as sedimentary rocks, is a defining feature of these environments. The size, shape, depth, and type of sediment input from rivers or other sources, as well as the surrounding climate, all influence the depositional processes in lacustrine environments.

The depositional features of lacustrine environments can vary greatly and are subject to various influences. Fine-grained silt and clay to coarser-grained sand and gravel can be found deposited in a lacustrine environment. Additionally, the sedimentary structures can range from thin layers of laminated silt and clay to thick-bedded sandstone and conglomerate.

The depositional environment can also vary from shallow-water to deep-water. Mud cracks, varves (thin laminations of sediment), and rhythmites (repeated layering of sediments) are a few of the common sedimentary characteristics linked to lacustrine environments. Lacustrine sedimentary rocks contain fossils that can reveal crucial details about the development of fish, insects, and plant life in freshwater environments. Moreover, lacustrine sedimentary rocks can shed light on past climatic conditions, such as variations in precipitation and temperature as well as tectonic or volcanic activity (Bovis, 2003; Chanson, 2004).

Every one of these habitats has distinct features, including the kinds of sediment, sedimentary structures, and fossils that are found there, and they can all offer valuable insights into the region's environmental and geological past. Sedimentary rocks that originated in these various settings are utilized as markers for tectonic activity, past climates, changes in sea level, and the evolution of life on Earth.

4.1.4. Applications

Numerous geological, geotechnical, and environmental issues are resolved by sediment transport. This makes measuring or quantifying erosion or sediment transport crucial for coastal engineering. To quantify sediment erosion, a number of devices have been developed (e.g., Simulator for Particle Erosion (PES)). To measure the rates of sediment erosion, one such tool—also known as the BEAST (Benthic Environmental Assessment Sediment Tool)—has been calibrated. Sediment movement in rivers is crucial for giving fish and other organisms a place to live. Thus, it is frequently recommended to managers of heavily regulated rivers—which are frequently dam-starved of sediment—to organize brief floods to replenish the bed material and restore bars. This is also necessary to restore shoreline habitats that are used as campgrounds, as in the Colorado River's Grand Canyon.

A reservoir delta is created by the discharge of sediment into a reservoir created by a dam. As a result of this delta filling the basin, the dam or the reservoir will eventually need to be removed by dredging. Planning to prolong the life of a dam can be done effectively with an understanding of sediment transport. Inverse solutions of transport relationships can be used by geologists to comprehend the direction, depth, and velocity of flow from sedimentary rocks and young alluvial deposits.

4.2. TRANSPORT MEDIA

The substances or procedures in charge of transferring sediments from their source to their deposition sites are known as transport media. These means of transportation consist of gravity, ice, wind, and water. The characteristics and mechanisms of each medium vary, affecting the kind, volume, and arrangement of the sediments they carry.

Here's an overview of each type of transport medium:

1. Water: Of all the transport mechanisms, material transport in water is by far the most important. Water moves overland and through channels on the surface of the land. Ocean circulation, wind, and tides all influence sea currents. These flows might be powerful enough to carry finer material in suspension and coarser material along the flow's base.

Types of water transport:

- Rivers and Streams: They carry sediments ranging from clay to boulders. The competence (largest particle size a river can carry) and capacity (total sediment load) of a river depend on its velocity and discharge.
- Marine Currents: These can transport fine sediments like silt and clay over large distances. Tidal currents and longshore drift are examples.
- Floods: High-energy events that can transport large amounts of sediments, including coarse materials.
- Glaciers: Glacial meltwater streams can transport sediments far from their original location.

- Characteristics:
  - Sediments transported by water tend to be well-sorted and rounded due to prolonged abrasion and selective transport.
  - Depositional environments include alluvial fans, deltas, beaches, and deep-sea basins.

2. Wind (Aeolian Processes):

- Characteristics:
  - Wind primarily transports fine sediments like sand, silt, and clay.
  - Sediments are generally well-sorted and rounded due to wind abrasion.
  - Common depositional features include dunes, loess deposits, and desert pavements.
- Depositional Environments:

- Deserts, coastal dunes, and loess plains.

3. Ice (Glacial Processes):
   - Characteristics:
     - Glaciers can transport a wide range of sediment sizes from clay to large boulders.
     - Sediments transported by ice are often poorly sorted and angular.
     - As glaciers move, they erode the underlying rock and carry debris along with them.
   - Depositional Features:
     - Moraines, drumlins, eskers, and glacial till deposits.
4. Gravity (Mass Wasting): The movement of particles down a slope due to gravity is the most basic mechanism of sediment transport. At the foot of slopes, rock falls produce sediment piles that are primarily made up of coarse debris that hasn't been further processed by other processes. In mountainous regions, these accumulations are viewed as scree along valley sides. With a surface angle equal to the rest of the gravel, they accumulate into talus cones, the maximum angle at which the material is stable and prevents clasts from falling farther down the slope. The distribution of clast sizes and the shape of the clasts affect the slope angle for loose debris, which varies from slightly over 308 for well-sorted sand to roughly 368 for angular gravel (Carson, 1977; Bovis, 2003). Seldom do they survive in the stratigraphic record.

Types of gravity-driven transport:

- Rockfalls: Large blocks or fragments fall freely under gravity.
- Landslides: Movement of a mass of rock, earth, or debris down a slope.
- Mudflows and Debris Flows: Flow of water-saturated debris down a slope.

- Characteristics:
  - Sediments are typically unsorted and angular.
  - Rapid transport and deposition can occur during these events.
- Depositional Features:
  - Talus slopes, debris fans, and landslide deposits (Bovis, 2003; Chanson, 2004).

4.2.1. The Behavior of Fluids and Particles in Fluids

A brief introduction to some aspects of fluid dynamics, the behavior of moving fluids, to give some physical basis to the discussion of sediment transport and the formation of sedimentary structures.

4.2.1.1. Laminar and Turbulent Flow

Two different kinds of fluid flow exist. All of the molecules in a fluid move parallel to one another in the direction of transport during laminar flows; in heterogeneous fluids, virtually no mixing happens during laminar flow. Heterogeneous fluids are thoroughly mixed in turbulent flows, where molecules in the fluid move in all directions with a net movement in the direction of transport. Experiments with dye threads in tubes demonstrate that at low flow rates, the lines of flow are parallel; however, as the flow becomes turbulent at higher flow velocities, the dye thread breaks up (Figure 4.4).

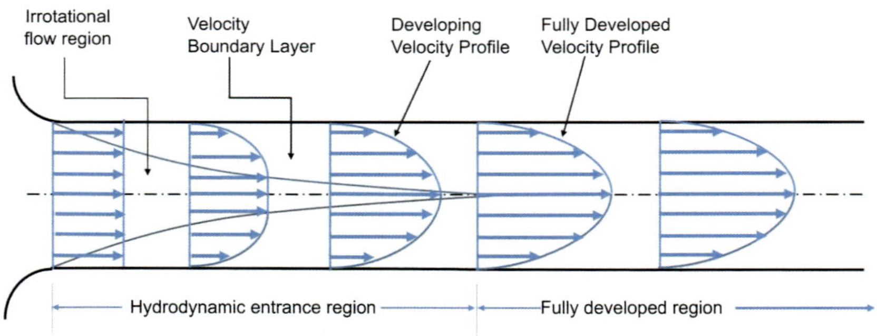

Figure 4.4. Laminar and turbulent flow of fluids through a tube.

Source: By DEVENDER KUMAR5908 – Own work, CC BY-SA 3.0, https://commons.wikimedia.org/w/index.php?curid=33253407.

A Reynolds number (Re), named for Osborne Reynolds, who distinguished between turbulent and laminar motion in the late 1800s, can be applied to flows. This number, which lacks dimensions, describes how laminar or turbulent a flow is. The following variables are related to determine the Reynolds number: the flow velocity (y), the fluid's viscosity (n), the fluid's density, and a "characteristic length" (l), which is the pipe's diameter or the depth of flow in an open channel. The equation to define the Reynolds number is:

$$Re = v \cdot l / \nu$$

Fluid flow in pipes and channels is found to be laminar when the Reynolds value is low (<500) and turbulent at higher values (>2,000). With increased velocity the flow is more likely to be turbulent and a transition from laminar to turbulent flow in the fluid occurs. With their high kinematic viscosities, debris flows, moving ice, and lava flows all exhibit laminar flow. All naturally occurring flows in air that have the capacity to carry particles are turbulent because low-kinematic-viscosity fluids, like air, are turbulent at low velocities. In aqueous sediment transport and deposition processes, turbulent flows are far more common because laminar water flows only occur at very low velocities or very shallow water depths. The majority of air and water flows that are likely to transport sizable amounts of sediment are turbulent (Figure 4.5).

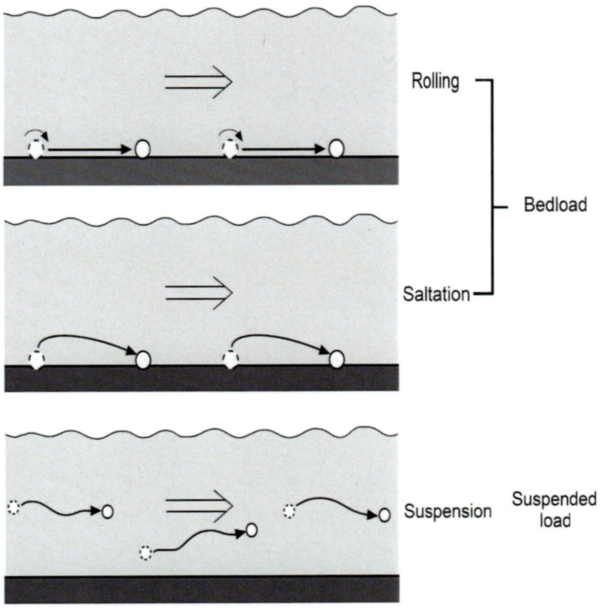

Figure 4.5. Particles move in a flow by rolling and saltating (bedload) and in suspension (suspended load).

Source: https://raregeologybooks.wordpress.com/wp-content/uploads/2014/09/sedimentology-and-stratigraphy-by-gary-nichols.pdf.

4.2.2. Transport of Particles in a Fluid

Three mechanisms can be used to move particles in a fluid, regardless of their size (Figure 4.2). Rolling: The clasts follow the water's or air's bottom while maintaining contact with the bed surface. Saltation is the process by which the particles jump repeatedly, occasionally departing from the bed surface and traveling brief distances within the fluid's body before coming back to the bed. Suspension: Enough upward motion is produced by turbulence in the flow to maintain particles in the moving fluid almost continuously. The material in suspension is known as the suspended load, and the particles being carried by rolling and saltation are referred to as the bedload. Sand-size particles move through rolling and some saltation in water at low current velocities, while only fine (fine silt and clay) and low density particles are maintained in suspension. Higher flow rates can cause coarser material to roll and granules and fine pebbles to saltate, keeping all silt and part sand in suspension. These processes work essentially the same in both water and air, but because air is less dense and viscous than water, higher velocities are needed in air to move particles of a given size.

4.2.3. Entraining Particles in a Flow

Frictional drag between the clasts and the flow moves the rolling grains. But more force is needed to cause grains to saltate and, as a result, move momentarily upward from the flow's base. The Bernoulli effect, which gives birds and airplanes the ability to fly and yachts the ability to sail "close to the wind," provides this force. The best way to understand the Bernoulli effect is to think about how a fluid would flow through a tube that is narrower at one end than the other—it could be water, air, or any other fluid medium (Figure 4.5). The tube's cross-sectional area is smaller at one end than the other, but the same amount of fluid must enter one end and exit the other in a predetermined amount of time to ensure continuous fluid transport along the tube. The fluid must travel through the narrow end at a faster speed to pass through a smaller gap with the same amount of fluid. Anyone who has squeezed or constricted the end of a garden hose is familiar with this effect: when the end of the hose is partially closed off, the water shoots out in a faster jet (Figure 4.6).

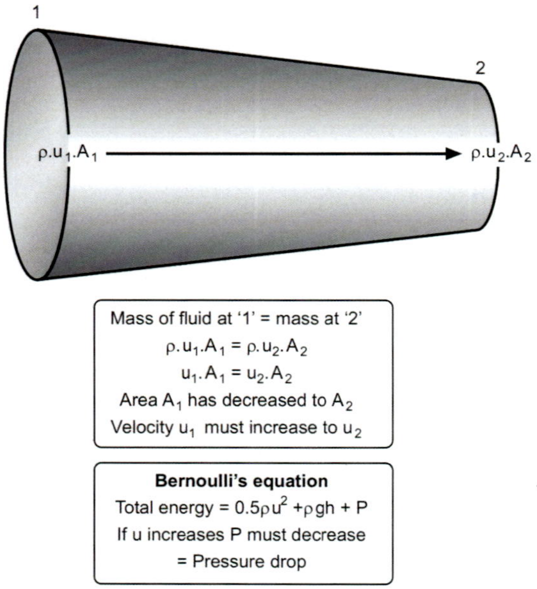

Figure 4.6. Flow of a fluid through a tapered tube results in an increase in velocity at the narrow end where a pressure drop results.

Source: https://raregeologybooks.wordpress.com/wp-content/uploads/2014/09/sedimentology-and-stratigraphy-by-gary-nichols.pdf.

The next thing to consider is the conservation of mass and energy along the length of the tube. The variables involved can be presented in the form of the Bernoulli equation:

$$\text{total energy} = \rho g h + \rho v^2/2 + p$$

where; ρ is the density of the fluid; y the velocity; g the acceleration due to gravity; h is the height difference; and p the pressure. The three terms in this equation are potential energy (ρgh), kinetic energy ($\rho v^2 = 2$) and pressure energy (p). This equation assumes no loss of energy due to frictional effects, so in reality the relationship is:

$$\rho g h + \rho v^2/2 + p + E_{\text{loss}} = \text{constant}$$

Since there is an equal difference in level between the fluid's beginning and ending points, the potential energy (ρgh) remains constant. With a change in flow velocity, kinetic energy ($\rho v^2=2$) changes as well. The last term, the pressure energy (p), needs to change to preserve the system's overall energy. The energy that is retained when a fluid is compressed, such as in a compressed gas canister, is known as pressure energy. Compressed fluids have more energy than uncompressed ones. Referring back to the flow in the tapered tube, the constriction of the flow at the end of the tube causes an increase in kinetic energy ($\rho v^2=2$), which must be offset by a reduction in pressure energy (p) to balance the Bernoulli equation. This indicates that the tube's narrower end is experiencing a drop in pressure.

Now if we apply these ideas to a flow through a channel (Figure 4.4) The cross-section of the flow over the channel will be smaller due to a clast in the bottom. The Bernoulli equation requires a decrease in pressure over the clast for the velocity over it to equal that of the upstream and downstream directions. The clast is lifted momentarily off the bottom of the flow by the pressure drop that occurs. In a single saltation event, the clast is then momentarily entrained in the flowing fluid before returning to the channel base under the force of gravity (Bovis, 2003; Chanson, 2004) (Figure 4.7).

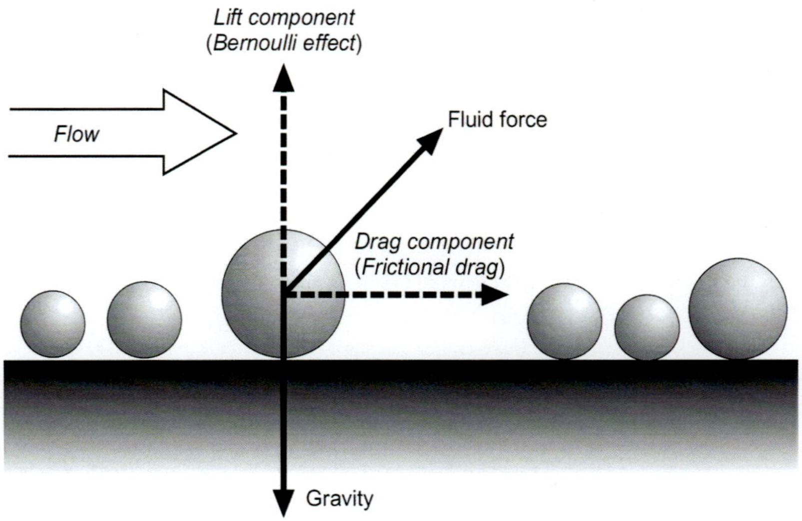

Figure 4.7. The lift force resulting from the Bernoulli effect causes grains to be moved up from the base of the flow.

Source: https://raregeologybooks.wordpress.com/wp-content/uploads/2014/09/sedimentology-and-stratigraphy-by-gary-nichols.pdf.

4.2.4. Grain size and Flow Velocity

The critical velocity is the fluid velocity at which a particle entangles itself in the flow. A straightforward relationship between the critical velocity and the particle's mass would be anticipated if the forces acting on the particle in the flow are taken into account. Both the lift force needed to raise a particle into the flow and the drag force needed to move it along in a flow will increase with mass. For sand and gravel, there is a straightforward linear relationship between the flow velocity and the lift and drag forces; however, things get more complex when fine grain sizes are involved.

The Hjülstrom diagram (Figure 4.8) shows the relationship between water flow velocity and grain size and although this diagram has largely been superseded by the Shields diagram (Miller et al., 1977) it nevertheless demonstrates some important features of sediment movement in currents. The relationship between flow velocity and particles in motion is shown on the graph by the lower line. This demonstrates that a clay particle will settle at practically zero flow velocity, a pebble will settle at about 20 to 30 cms^{-1}, and a medium sand grain at 2 to 3 cm s^{-1}. Therefore, if the sediment is deposited as isolated particles, the grain size of the particles in the flow can be used as an indicator of the velocity at that time. The flow velocity needed to move a particle from rest is indicated by the upper, curved line. This line parallels the first on the right half of the graph, but for any given grain size, the velocity needed to start a particle moving is greater than the velocity needed to keep it moving. The lines in the diagram sharply diverge on the left, indicating that, paradoxically, smaller particles need to travel faster to get below coarse silt size. This is because clay minerals have certain characteristics that cause them to predominate in a sediment's fine fraction. It is challenging to entice them into a flow once they are deposited because of their propensity to adhere to one another. Take note that cohesive material has two lines. "Unconsolidated" mud has solidified but still has a plastic, sticky consistency. "Consolidated" mud is stiff and has lost a significant amount of water. The way that small particles behave in a flow, as shown by the Hjülstrom diagram, has significant impacts on deposition in natural environments. Without this behavior, clay would only be eroded in certain conditions, such as in standing water. However, mud can build up in any environment where the flow stops long enough for clay particles to settle. Once deposited, clay is not easily re-entrained by flowing water unless the velocity is high.

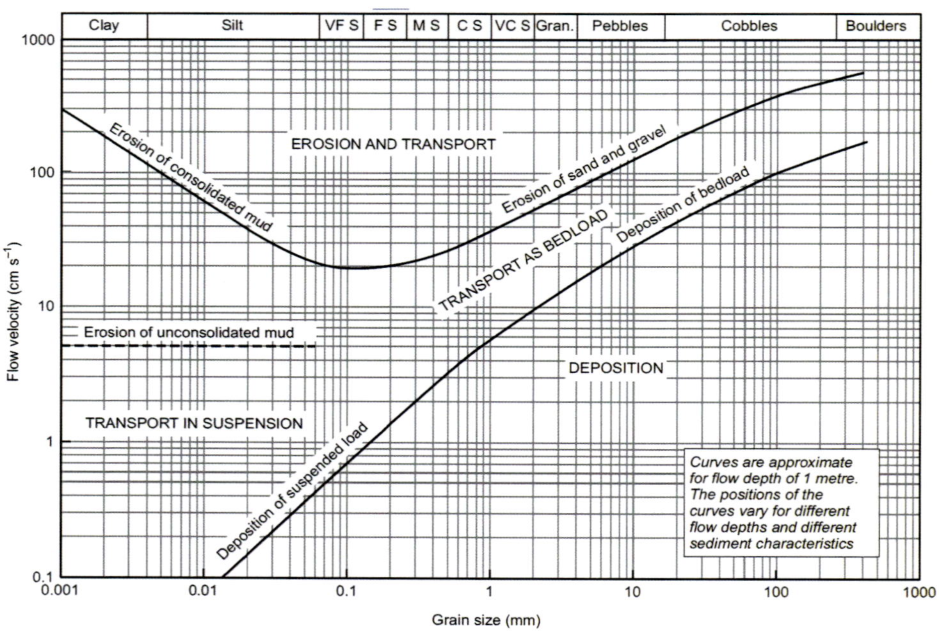

Figure 4.8. The Hjülstrom diagram shows the relationship between the velocity of a water flow and the transport of loose grains. Once a grain has settled it requires more energy to start it moving than a grain that is already in motion. The cohesive properties of clay particles mean that fine-grained sediments require relatively high velocities to re-erode them once they are deposited, especially once they are compacted.

Source: https://raregeologybooks.wordpress.com/wp-content/uploads/2014/09/sedimentology-and-stratigraphy-by-gary-nichols.pdf.

4.2.5. Clast-Size Variations: Graded Bedding

Reverse grading, which is a pattern of increasing average size from base to top, is also known as normal grading (Figure 4.9). The more frequently seen pattern, known as normal grading, can be caused by particles settling out of suspension or by a gradual decline in flow strength. Particle size, fluid viscosity, and the difference in density between the particle and the fluid all affect the settling velocity of particles in a fluid. The relationship, known as Stokes Law, can be expressed in an equation:

$$V = g \cdot D^2 \cdot (\rho_s - \rho_f)/18\mu$$

where; V is the terminal settling velocity; D is the grain diameter; (ρs-ρf) is the difference between the density of the particle (ρs); and the density of the fluid (ρf) and m is the fluid viscosity; g is the acceleration due to gravity.

Larger diameter clasts can travel at higher velocities, which means that grading of particles occurs when sediment falls out of suspension in standing water. This has implications for sedimentary processes. Stokes Law only provides an accurate estimate of the settling velocity of small grains (fine sand or smaller), as the velocity is decreased by turbulence caused by larger grains falling through the fluid and creating drag. Because plate-like clasts have a greater drag effect and fall more slowly, the shape of the particle also plays a role. Because mica settles more slowly than quartz and other grains of equivalent mass, it is common to find mica grains concentrated at the tops of beds.

Sand will be deposited coarsely at first, but as the flow velocity decreases from 20 cm s^{-1} to 1 cm s^{-1}, medium and fine sand will be gradually deposited. This slowing flow will produce a sand bed that is typically graded, with coarse grains at the bottom and finer ones at the top. On the other hand, flows that progressively increase in strength over time to produce reverse grading are less common. Instead, an increase in flow velocity over time may lead to an increase in grain size up through a bed, or reverse grading.

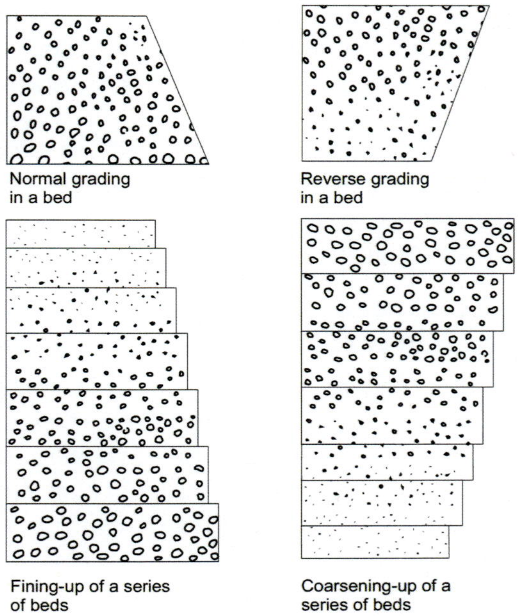

Figure 4.9. Normal and reverse grading within individual beds and fining-up and coarsening-up patterns in a series of beds.

Source: https://raregeologybooks.wordpress.com/wp-content/uploads/2014/09/sedimentology-and-stratigraphy-by-gary-nichols.pdf.

It is helpful to distinguish between grading, which is a grain size trend within a single bed, and grain size trends that span several beds. Fining-upward refers to a pattern of multiple beds where the highest bed contains finer material and the lowest bed contains coarser clast size. The opposite pattern is an upward coarsening succession with the coarsest bed at the top (Figure 4.6). Keep in mind that there may be situations where a bed is normally graded but is part of a succession of beds that are coarsening up.

4.2.6. Fluid Density and Particle Size

The fact that the forces acting on a grain depend on the mass of the particle as well as the viscosity and density of the fluid medium is a second significant implication of Stokes Law. Because of the larger density contrast between the particle and the fluid, as well as the lower fluid viscosity, a clast falling through air will move more quickly than one falling through water. Moreover, for a given flow velocity, fluids with a higher viscosity produce more lift and drag forces. Even at very high wind strengths during storms, the largest rock and mineral particles carried are probably about a millimeter in size. Water flows can carry boulder-sized clasts at the speeds observed in rivers. One of the factors that can be used to differentiate material deposited by water from that transported and deposited by wind is the restriction on the particle size that can be carried by air. Boulders can be transported across distances of meters or tens of meters by higher viscosity fluids like ice and debris flows, which are dense mixtures of sediment and water.

4.2.7. Flows, Sediment, and Bedforms

A morphological feature called a bedform is created on a bed by the interaction of cohesionless sediment and flow. Examples of bedforms are sand dunes in deserts and ripples in the sand in a flowing stream; the former is caused by airflow, the latter by water flow. The flow produces the patterns of ripples and dunes, and the creation of bedforms gives the sediment unique layering and structures that can be preserved in strata. The identification of sedimentary formations caused by bedforms yields data regarding the direction, depth, and strength of the current as well as the flow of sediment.

More thought must be given to fluid dynamics to explain how bedforms are created (a thorough explanation can be found in Leeder, 1999). Three distinct regions can be identified in a fluid flowing over a surface: a boundary layer, where friction with the bed causes the velocity to start decreasing, a viscous sublayer, which is a region of reduced turbulence usually less than a millimeter thick, and the free stream, which is the portion of the flow unaffected by boundary effects (Figure 4.10). Although it is unaffected by flow depth, the viscous sublayer's thickness decreases as flow velocity increases. An essential aspect of the flow is defined by the correlation between the size of the grains on the flow bed and the thickness of the viscous sublayer. Hydraulically smooth surfaces are those in which every particle is contained within the viscous sublayer; hydraulically rough surfaces are those in which some particles protrude through this layer. The development of distinct bedforms is fundamentally influenced by the viscous sublayer and the effects of smooth and uneven surfaces.

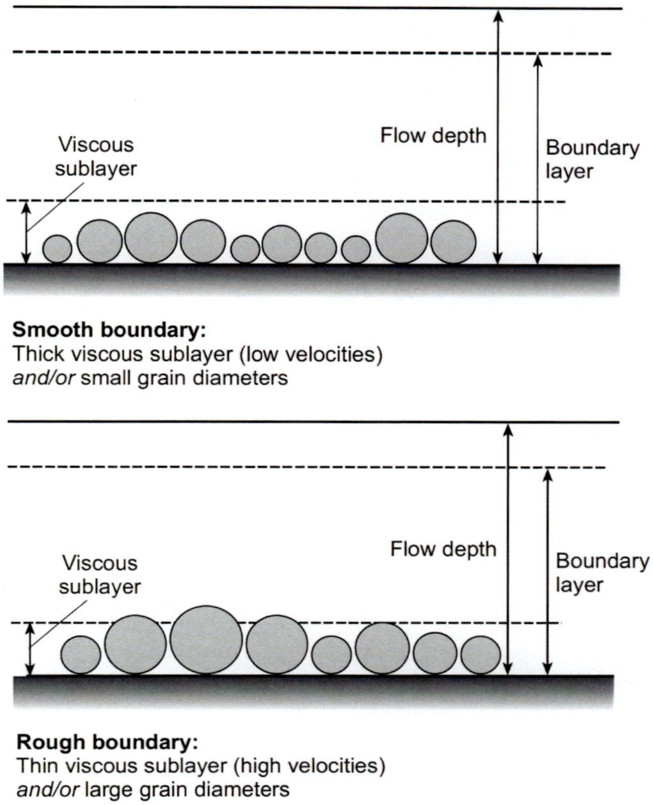

Figure 4.10. Layers within a flow and flow surface roughness: The viscous sublayer, the boundary layer within the flow and the flow depth.

Source: https://raregeologybooks.wordpress.com/wp-content/uploads/2014/09/sedimentology-and-stratigraphy-by-gary-nichols.pdf.

4.3. SEDIMENTARY STRUCTURES AND TEXTURES

The physical features and arrangements of sedimentary rocks, known as sedimentary structures and textures, offer valuable information about the conditions in which they were formed. These structures and textures provide insights into the depositional environment, sediment transport processes, and the history of the sedimentary basin. Geologists use them to interpret the evolution of sedimentary deposits. From bedding planes to ripple marks to fossils, these structures can range in scale and include various features that give clues about the sediment grains' size, shape, sorting, and arrangement within the rock (Figure 4.11).

Figure 4.11. The physical features and arrangements of sedimentary rocks.

Source: By Michael C. Rygel – Own work, CC BY-SA 3.0, https://commons.wikimedia.org/w/index.php?curid=12622448.

4.3.1. Importance of Sedimentary Structures and Textures in Sedimentology

In sedimentology, sedimentary structures and textures are significant because they reveal details about the conditions and processes that existed during the deposition of sediment and the ensuing diagenesis. Sedimentologists can reconstruct the environment of deposition and infer the transport mechanisms, sources of sediment, and depositional energy by examining sedimentary structures and textures. Understanding the evolution of the Earth's surface over time and the geological history of sedimentary basins depend on this knowledge. Furthermore, the identification and correlation of sedimentary units can be facilitated by the use of sedimentary structures and textures, which can help with the exploration and exploitation of natural resources like groundwater, oil, and gas. All things considered, understanding sedimentary structures and textures is crucial to expanding our understanding of Earth's past and its natural resources (Bovis, 2003; Chanson, 2004).

4.3.2. Classification of Sedimentary Structures and Textures

Sedimentary structures and textures can be classified into different types based on their origin, characteristics, and geological setting. Here are some of the common classifications:

1. Primary Structures: These are formed during sediment deposition and include bedding, lamination, cross-bedding, graded bedding, ripple marks, mud cracks, and fossils.
2. Secondary Structures: These are formed after deposition and include deformation structures such as folds and faults, as well as diagenetic structures such as concretions, nodules, and stylolites.
3. Physical Structures: These are formed by physical processes and include erosional structures such as channels, scour marks, and flutes, as well as depositional structures such as dunes, bars, and deltas.
4. Chemical Structures: These are formed by chemical processes and include precipitates, evaporites, and replacement textures.
5. Biological Structures: These are formed by the activity of living organisms and include bioturbation, stromatolites, and trace fossils.

These classifications can be further refined and modified based on specific sedimentary environments, depositional settings, and depositional processes (Figure 4.12).

Figure 4.12. Sedimentary structures and textures.

Source: By Wilson44691 at English Wikipedia – Photograph taken by Mark A. Wilson (Department of Geology, The College of Wooster). Public Domain, https://commons.wikimedia.org/w/index.php?curid=3796685.

4.3.3. Bedding Structures and Textures

The different characteristics present in sedimentary beds are referred to as bedding structures and textures. These characteristics can reveal important details about the sedimentary rock's history and depositional environment. Some common bedding structures and textures include (Figure 4.13):

1. Bedding Planes: The surfaces that divide the layers of sedimentary rocks are known as bedding planes. These strata, also known as beds, are distinguished by differences in sedimentary properties like grain size, composition, sorting, and color.

Figure 4.13. The surfaces that divide the layers of sedimentary rocks are known as bedding planes.

Source: By Weyf – Own work, CC0, https://commons.wikimedia.org/w/index.php?curid=32721922.

2. Graded Bedding: This is a type of bedding where the grain size of sediment within a bed gradually changes from coarse at the bottom to fine at the top. Graded bedding can be caused by changes in the energy of the depositional environment over time (Figure 4.14).

Figure 4.14. Graded bedding.

Source: https://i0.wp.com/opengeology.org/historicalgeology/wp-content/uploads/2020/09/4969229267_ce297ba37d_b-anno.jpg?ssl=1.

3. Cross-Bedding: This refers to sets of inclined beds within a larger sedimentary bed. Cross-beds are formed when sediment is transported by a current of water or wind, and can be used to determine the direction and strength of the current (Figure 4.15).

Figure 4.15. Cross bedding sedimentary structure.

Source: https://i0.wp.com/geologyistheway.com/wp-content/uploads/2021/10/C-02a-IMG_1328.jpg?ssl=1.

4. Ripple Marks: These are small, wavy ridges found on bedding surfaces that are caused by the action of water or wind (Figure 4.16).

Figure 4.16. Ripple marks.

Source: https://i0.wp.com/geologyistheway.com/wp-content/uploads/2021/10/0cc627f30bd87a7ed0364ef3aea266d2.jpg?ssl=1.

5. Mud Cracks: These are polygonal cracks that form when mud dries out and contracts. Mud cracks can be used to determine whether a sedimentary rock was deposited in a wet or dry environment (Figure 4.17).

Figure 4.17. Mud cracks.

Source: https://i0.wp.com/thumbs.dreamstime.com/z/mud-cracks-also-known-as-desiccation-sedimentary-structures-formed-as-muddy-sediment-dries-contracts-78896047.jpg?ssl=1.

6. Biogenic Structures: These are structures formed by living organisms, such as burrows, tracks, and trails. Biogenic structures can provide information about the types of organisms that were present in the depositional environment (Figure 4.18).

Figure 4.18. Biogenic structures.

Source: https://i0.wp.com/d1u1p2xjjiahg3.cloudfront.net/68db8013-db07-45cf-9918-b77cfe7bde4d_l.jpg?ssl=1.

4.3.4. Interpretation of Bedding Structures and Textures in Depositional Environments

Bedding structures and textures can provide important clues about the depositional environment in which sediments were deposited. For example:

- Regular, parallel bedding can indicate deposition in a relatively calm, low-energy environment such as a lake or deep ocean basin.
- Cross-bedding (inclined layers within a larger bed) can indicate deposition in environments such as sand dunes or rivers with flowing water.
- Ripple marks can indicate the direction and strength of water flow in ancient rivers, lakes, and oceans.
- Mud cracks can indicate alternating wet and dry conditions in a depositional environment such as a tidal flat or lake shore.

Sedimentologists can deduce the paleoenvironmental circumstances that prevailed during the deposition of sedimentary rocks by analyzing bedding structures and textures. In addition to being useful in reassembling the geologic past of a region, this data can be used to identify and describe putative petroleum reservoirs.

4.3.5. Interpretation of Sedimentary Structures and Textures in Alluvial Systems

Alluvial systems are areas where sediments are deposited by rivers and other fluvial processes, including river channels, floodplains, and deltas. By studying the sedimentary structures and textures in alluvial systems, researchers can gain valuable insight into the flow dynamics, sediment transport mechanisms, and depositional environments of the sediments.

Some common sedimentary structures and textures found in alluvial systems include:

1. Bedding: The most common sedimentary structure found in alluvial systems is bedding or stratification. Bedding can be used to determine the orientation, thickness, and lateral extent of sedimentary layers, as well as the depositional history of the sediment.

2. Cross-Bedding: It is formed when sediment is deposited at an angle to the underlying bedding plane, resulting in inclined layers within the sediment. Cross-beds can provide information about the direction and intensity of sediment transport, as well as the shape and size of the sediment particles.
3. Ripple Marks: These are small-scale sedimentary structures that are formed by the action of water or wind. They can provide information about the direction and intensity of sediment transport, as well as the flow conditions of the water or wind.
4. Mud Cracks: These are formed when fine-grained sediments, such as mud or silt, dry out and contract. They can provide information about the duration and frequency of dry periods in the depositional environment.
5. Channel Structures: These, such as channels, levees, and point bars, are formed by the action of rivers and streams. They can provide information about the size and shape of the channel, as well as the flow dynamics and sediment transport mechanisms of the water.

By analyzing the sedimentary structures and textures in alluvial systems, sedimentologists can reconstruct the depositional history of the sediments, including the source, transport, and depositional mechanisms of the sediment, as well as the environmental conditions in which they were deposited. This information can be used to understand the evolution of the landscape over time, as well as to identify potential natural resources, such as oil and gas reservoirs (Bovis, 2003; Chanson, 2004).

CASE STUDY

SEDIMENTARY STRUCTURES: INTERPRETING PAST ENVIRONMENTS

Background: Sedimentary structures provide valuable information about past depositional environments, sediment transport processes, and paleoclimates. This case study focuses on a hypothetical sequence of sedimentary rocks observed in a field study. The goal is to interpret the depositional environment and the processes that led to the formation of these structures.

- **Study Area:** The study area is a stratigraphic section exposed in a river valley. The sequence includes various sedimentary structures such as cross-bedding, ripples, graded bedding, mudcracks, and flute casts.

Observations:

1. Layer 1: Cross-Bedded Sandstone:
   - **Description:** Thick beds of well-sorted, medium-grained sandstone with prominent cross-bedding.
   - **Interpretation:** The cross-bedding indicates deposition by migrating dunes in a high-energy environment such as a river channel or desert.
2. Layer 2: Ripple Marks:
   - **Description:** Fine-grained sandstone with symmetrical and asymmetrical ripple marks.
   - **Interpretation:** Symmetrical ripples suggest wave action in a shallow marine environment, while asymmetrical ripples indicate unidirectional flow, possibly in a fluvial or shallow marine setting with tidal influences.
3. Layer 3: Graded Bedding:
   - **Description:** Alternating layers of coarse-to-fine sandstone and siltstone showing normal grading.
   - **Interpretation:** Graded bedding typically forms from turbidity currents in a submarine fan or deep-water setting.
4. Layer 4: Mudcracks:
   - **Description:** Fine-grained mudstone with polygonal mudcracks.
   - **Interpretation:** Mudcracks form in fine sediments exposed to air, indicating periodic subaerial exposure in a tidal flat, floodplain, or evaporative basin.
5. Layer 5: Flute Casts:
   - **Description:** Base of a sandstone bed with flute casts and sole marks.

- **Interpretation:** Flute casts are formed by erosive currents and indicate a high-energy environment with directional flow, such as a turbidity current in a deep marine setting.

Analysis:

1. Sedimentary Structures and Depositional Environments:
   - The cross-bedded sandstone suggests high-energy deposition by wind or water currents, possibly in a river or desert environment.
   - Ripple marks indicate alternating energy conditions, with wave action or tidal currents.
   - Graded bedding points to episodic sediment gravity flows, typical of submarine fan systems.
   - Mudcracks reveal periodic drying and exposure, consistent with tidal flats or ephemeral lakes.
   - Flute casts signify strong erosive currents, supporting the idea of a deep-water turbidity current setting.
2. **Paleocurrent Directions:** Cross-bedding and flute casts can be used to infer paleocurrent directions. Cross-bedding dip directions and flute cast orientations provide clues about the prevailing current directions during deposition.
3. **Environmental Interpretation:**
   - The sequence of sedimentary structures suggests a transition from a high-energy fluvial or desert environment (cross-bedding) to a shallow marine or tidal environment (ripple marks), followed by deep marine conditions (graded bedding and flute casts), and intermittent subaerial exposure (mudcracks).
   - This progression indicates a transgressive sequence where sea levels rise, transitioning from continental to marine depositional environments.

Conclusion: The stratigraphic section in the river valley reveals a complex depositional history with varied energy conditions and environments. By analyzing the sedimentary structures, we infer a transition from terrestrial to shallow marine, and finally to deep marine environments, punctuated by periods of exposure. This case study highlights the importance of sedimentary structures in reconstructing past environments and understanding sediment transport processes.

CLASS ACTIVITY

INVESTIGATING SEDIMENTARY STRUCTURES

Students will explore sedimentary structures through hands-on experiments, including creating cross-bedding, ripple marks, graded bedding, mudcracks, and flute casts. Using a transparent container, sand, silt, water, and simple tools, they will simulate and observe these structures. Each station will allow students to document and analyze how these features form and what they reveal about past depositional environments. Through this activity, students will gain practical understanding and insights into sediment transport processes and environmental interpretations, enhancing their observational and analytical skills. The activity concludes with a discussion and a reflective writing assignment.

SUMMARY

- Sedimentary structures are physical features formed by the deposition and subsequent modification of sediments. They play a crucial role in interpreting past environments and understanding sedimentary processes.
- Sediment transport and deposition are key processes in sedimentology that govern the formation of sedimentary rocks. The movement of sediment can occur through a variety of mechanisms, including wind, water, ice, and gravity.
- Bedload transport is a process by which sediment particles on the bed of a river or stream are moved by the flow of water.
- Depositional environments are the locations and conditions where sediment accumulates and are determined by the physical, chemical, and biological processes at work.
- Sediment transport is applied to solve many environmental, geotechnical, and geological problems. Measuring or quantifying sediment transport or erosion is therefore important for coastal engineering.
- Rolling grains are moved as a result of frictional drag between the flow and the clasts. However, to make grains saltate and therefore temporarily move upwards from the base of the flow a further force is required.
- Alluvial systems are depositional environments that are dominated by fluvial processes, such as river channels, floodplains, and deltas.

REVIEW QUESTIONS

1. What are Sedimentary Structures?
2. What is the Difference between Bedload, Suspended Load, and Dissolved Load?
3. What is Cross-Bedding?
4. How do Ripples Form?
5. What is Graded Bedding?
6. What are Mudcracks and how do They Form?
7. What is a Turbidite?
8. What are Imbricated Clasts?
9. What is a Flute Cast?
10. What are Sole Marks?

REFERENCES

1. Ashton, A., Murray, A. B., & Arnault, O. (2001). Formation of coastline features by large-scale instabilities induced by high-angle waves. Nature, 414(6861), 296–300. https://doi.org/10.1038/35104500.
2. Bovis, M. J. (2003). Avalanche and rock fall. In G. V. Middleton (Ed.), Encyclopedia of Sediments and Sedimentary Rocks (pp. 31–34). Kluwer Academic Publishers.
3. Carson, M. A. (1977). Angles of repose, angles of shearing resistance, and angles of talus slopes. Earth Surface Processes and Landforms, 2, 363–380.
4. Chanson, H. (2004). The Hydraulics of Open Channel Flow: An Introduction (2nd ed.). Butterworth-Heinemann.
5. Cheng, N. S. (2002). Exponential formula for bedload transport. Journal of Hydraulic Engineering, 128(10), 942. https://doi.org/10.1061/(ASCE)0733-9429(2002)128:10(942).
6. Grant, J., Walker, T. R., Hill, P. S., & Lintern, D. G. (2013). BEAST-A portable device for quantification of erosion in intact sediment cores. Methods in Oceanography, 5, 39–55. https://doi.org/10.1016/j.mio.2013.08.001.
7. Harris, C. K. (2003, March 18). Lecture 9: Suspended sediment transport II [PDF]. In Sediment Transport Processes in Coastal Environments. Virginia Institute of Marine Science.
8. Kuhnle, R. A., Wren, D. G., Langendoen, E. J., & Rigby, J. R. (2013). Sand transport over an immobile gravel substrate. Journal of Hydraulic Engineering, 139(2), 167–176. https://doi.org/10.1061/(ASCE)HY.1943-7900.0000688.
9. Leeder, M. R. (1999). Sedimentology and Sedimentary Basins: From Turbulence to Tectonics. Blackwell Science.
10. Miller, R. M. C., McCave, I. N., & Komar, P. D. (1977). Threshold of sediment motion under unidirectional currents. Sedimentology, 24, 507–527. https://doi.org/10.1111/j.1365-3091.1977.tb00165.x.
11. Moore, A. (n.d.). Lecture 20—Some loose ends [PDF]. Lecture Notes: Fluvial Sediment Transport. Kent State.
12. Moore, A. (n.d.). Lecture 21—Suspended sediment transport [PDF]. Lecture Notes: Fluvial Sediment Transport. Kent State.
13. Sharmeen, S., & Willgoose, G. R. (2006). The interaction between armoring and particle weathering for eroding landscapes. Earth Surface Processes and Landforms, 31(10), 1195–1210. https://doi.org/10.1002/esp.1332.
14. Walker, T. R., & Grant, J. (2009). Quantifying erosion rates and stability of bottom sediments at mussel aquaculture sites in Prince Edward Island, Canada. Journal of Marine Systems, 75(1–2), 46–55. https://doi.org/10.1016/j.jmarsys.2008.08.004.
15. Whipple, K. (2004). Hydraulic roughness [PDF]. 12.163: Surface Processes and Landscape Evolution. MIT OCW.

CHAPTER 5

Field Sedimentology, Facies, and Environments

LEARNING OBJECTIVES

After studying this chapter, you will be able to:

- Understand the field sedimentology;
- Discuss graphic sedimentary logs;
- Explain the paleocurrent;
- Define collection of rock samples.

KEY TERMS FROM THIS CHAPTER

Alluvial fans
Carbonate facies
Cross-bedding
Depositional environments
Grain size distribution
Ripple marks
Sedimentary petrology
Bioturbation
Clastic deposits
Deltaic systems
Facies models
Lithofacies analysis
Sedimentary environments
Sedimentary structures

INTRODUCTION

Field sedimentology is a branch of geology that focuses on the study of sediments and sedimentary rocks in their natural environment. This discipline involves the examination of sedimentary processes, including the transportation, deposition, and lithification of sediments. Field sedimentologists analyze various sedimentary structures and features such as bedding, ripple marks, and cross-bedding, which provide insights into past environmental conditions and sedimentary processes. The study often requires extensive fieldwork, where geologists collect samples and make detailed observations to understand the spatial distribution and characteristics of sedimentary deposits.

Facies analysis is a critical component of sedimentology. A facies is a body of rock with specific characteristics that distinguish it from adjacent rock units. These characteristics include lithology, grain size, sedimentary structures, fossil content, and color. Facies analysis involves identifying and interpreting these attributes to reconstruct the depositional environment and its changes over time. By studying facies, geologists can infer whether a sedimentary rock formed in a river, delta, beach, deep-sea, or other environments. This information is vital for understanding Earth's history, including changes in climate, sea level, and tectonics.

Sedimentary environments refer to specific settings where sediment deposition occurs. These environments are classified into three main categories: continental, transitional, and marine. Continental environments include rivers, lakes, and deserts, where sediments are primarily transported by wind, water, and ice. Transitional environments, such as deltas, estuaries, and lagoons, are influenced by both terrestrial and marine processes. Marine environments encompass shallow to deep ocean settings, where sediments are deposited by waves, currents, and biological activity. Each sedimentary environment produces distinct sedimentary deposits, which are characterized by their facies. Understanding these environments helps geologists to interpret the past depositional settings and to predict the distribution of natural resources such as oil, gas, and groundwater.

Overall, field sedimentology, facies analysis, and the study of sedimentary environments are interconnected disciplines that provide a comprehensive understanding of sedimentary processes and the geological history of sedimentary basins. By examining sedimentary rocks and their characteristics in the field, geologists can reconstruct past environments, understand the dynamics of sedimentary processes, and make informed predictions about future geological developments.

5.1. FIELD SEDIMENTOLOGY

Interpreting sediments and sedimentary rocks in terms of transport and deposition processes and their spatial and temporal distribution within sedimentary environments constitutes a significant portion of contemporary sedimentology. This kind of sedimentological analysis requires certain data, which are primarily gathered from rock exposures, though subsurface data are becoming more and more crucial (Wüst & Bustin, 2003; Collinson & Thompson, 2006).

5.1.1. Field Equipment

Fieldwork in sedimentology and stratigraphy requires a minimal number of tools. To capture data, you will need a notebook, and it is highly advised to use a sturdy, hardcover book with weatherproof paper. Geological hammer, compass-clinometer, and hand lens (10 × magnification) are also must-haves. A clipboard and measuring tape or meter stick are required if a sedimentary log is to be recorded, as well as proforma log sheets. Small, sturdy plastic bags and a marker pen are required for the collection of samples. To check if there is calcium carbonate in the field, a tiny bottle with diluted hydrochloric acid is a great tool. Grain-size comparators of any kind are beneficial. One way that sedimentologists create a comparator is by adhering sand with varying sizes to sections of a small piece of card or Perspex. Alternatively, small cards with a printed visual chart of grain sizes are available. Using real grains to make these comparators has the advantage of allowing for both visual and tactile comparison.

A GPS (Global Positioning Satellite) receiver and a camera are probably the most "hi-tech" items that are taken into the field. Images can serve as a great documentation of the features observed in the field, but only if the location of each photo is noted. Additional sketches should also be made. Field geologists now carry GPS receivers as standard equipment, and they can be a convenient and quick way to find locations. They are not a substitute for a compass-clinometer; a compass functions without batteries, and a GPS unit typically does not have a clinometer on it. Instead, they are used in conjunction with one another.

5.1.2. Field Studies: Mapping and Logging

The goals of the project will determine how a field program of sedimentary studies is organized. For limestones, the Dunham classification can be used to characterize the lithology, and for sandstones, the Pettijohn classification. Bed thicknesses, sedimentary structures, fossils, rock color, and any other qualities like weathering, degree of consolidation, and so forth are additional features to be aware of. Field guides with a checklist of features to notice are Tucker (2003); and Stow (2005). It is customary to measure and record a graphic sedimentary log from an appropriate location within each formation as soon as the various formations have been identified. While it is sufficient to refer to a rock unit as simply "red sandstone" when creating a geological map, any report that goes with the map should make an effort to piece together the area's geological past. At this point, it will be necessary to understand the specific characteristics of the sandstone in some detail and to gather enough data to be able to interpret the sandstone in terms of its deposition environment.

Several data points from sedimentary rocks may need to be collected for a comprehensive study to determine the spatial variability of a given unit or the vertical variation in the sedimentary character of a unit (i.e., through time)—or the two. The primary method of recording data for these paleoenvironmental or stratigraphic studies is the graphic sedimentary log, which is used to collect data in an efficient and systematic manner. Even though a sedimentologist may need to invest a lot of time in documenting and sketching these logs under varying circumstances—from wind-swept mountainsides to sunny beaches—the process is essentially the same each time. Samples for petrographic and paleontological analyses will be gathered, along with additional data such as paleocurrent data, in addition to the information entered on logs.

5.2. GRAPHIC SEDIMENTARY LOGS

A sedimentary log is a graphical method for representing a series of beds of sediments or sedimentary rocks. There are many different schemes in use, but they are all variants on a theme. The format presented here (Figure 5.1) closely follows that of Tucker (1996); other commonly used formats are illustrated in Collinson et al. (2006). The objective of any graphic sedimentary log should be to present the data in a way which is easy to recognize and interpret using simple symbols and abbreviations that should be understandable without reference to a key (although a key should always be included to avoid ambiguity)

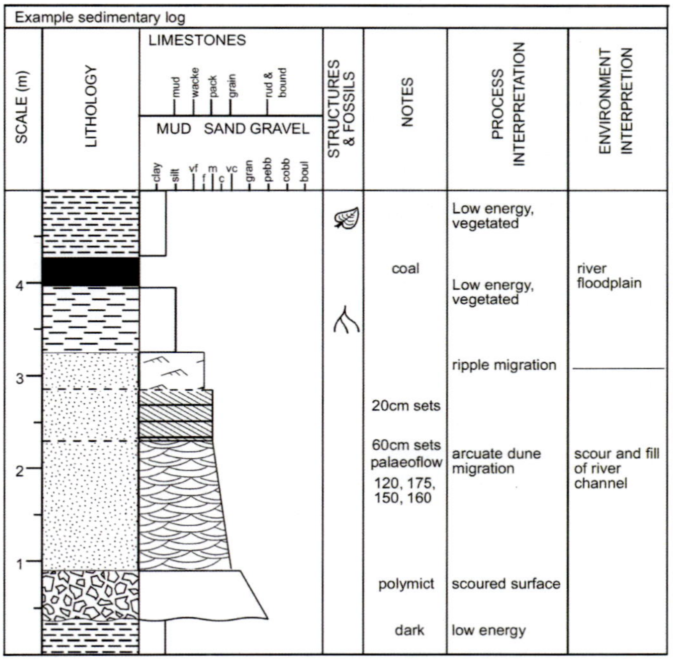

Figure 5.1. An example of a graphic sedimentary log: This form of presentation is widely used to summarize features in successions of sediments and sedimentary rocks.

Source: Gary Nichols; Sedimentology and Stratigraphy; ISBN: 978-1-4051-9379-5.

5.2.1. Drawing a Graphic Sedimentary Log

The required level of detail dictates the vertical scale to be used. A ratio of 1:10 is suitable if data on beds that are one centimeter thick is required. If individual bed records are not required for beds that are less than 10 cm thick, a log drawn across tens or hundreds of meters may be drawn at a ratio of 1:100. Although other scales are also utilized, 1:20 and 1:50 are typically chosen to facilitate scale conversion. Summary logs can be drawn at a scale of 1:500 or 1:1000, providing only an outline of a series of strata.

The majority of widely used symbols for lithologies are essentially standardized: bricks are used to represent limestones to represent sands and sandstone, and so forth (Figure 5.2). The design can be changed to better fit the succession described. For instance, the letter "G" can be superimposed to represent glauconitic sandstone or can be added to the brickwork to represent sandy limestone, and so on. The lithology is displayed in a single column in many schemes. There is space for further details about the type of sediment and for noting sedimentary structures next to the lithology column (to the right). In clastic sediments, the grain size is represented by a horizontal scale. With this kind of scale, the Dunham classification for limestones can also be shown. This scheme provides a quick visual impression of any trends in grain size in successions of beds that are coarsening upward or fining upward, as well as in normal or reverse graded beds.

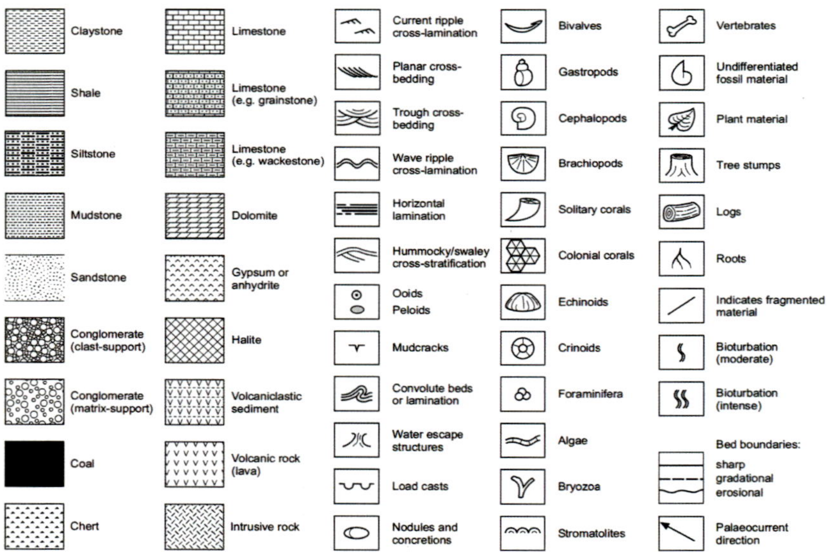

Figure 5.2. Examples of patterns and symbols used on graphic sedimentary logs.

Source: Gary Nichols; Sedimentology and Stratigraphy; ISBN: 978-1-4051-9379-5.

Sedimentary structures are typically represented by symbols that closely resemble how the feature appears in the field or in the core (Figure 5.2). For the purpose of simplicity, this representation has been somewhat stylized, but once more, the symbols can be changed to fit specific situations. Symbols can be drawn to the right of the bed or placed inside if there is enough room. Bed boundaries can be transitional or gradational, meaning that there is a gradual change from one lithology to another, or sharp/erosional, where the upper bed cuts down into the lower one. Additional information regarding the order of beds can also be noted on the graphic log (Figure 5.1). Paleocurrent data can be displayed alongside the log as a rose diagram with a series of arrows pointing in the direction of the measured or summarized paleoflow for a unit. Color is typically noted using words or acronyms; any additional comments or observations can simply be written in the appropriate spot next to the graphic log.

5.2.2. Presentation of Graphic Sedimentary Logs

It is common practice to draw a log in the field and then redraft it at a later stage. The field log can be drawn straight onto a proforma log sheet (such as Figure 5.3), but it can be more convenient to draw a sketch log in a field notebook.

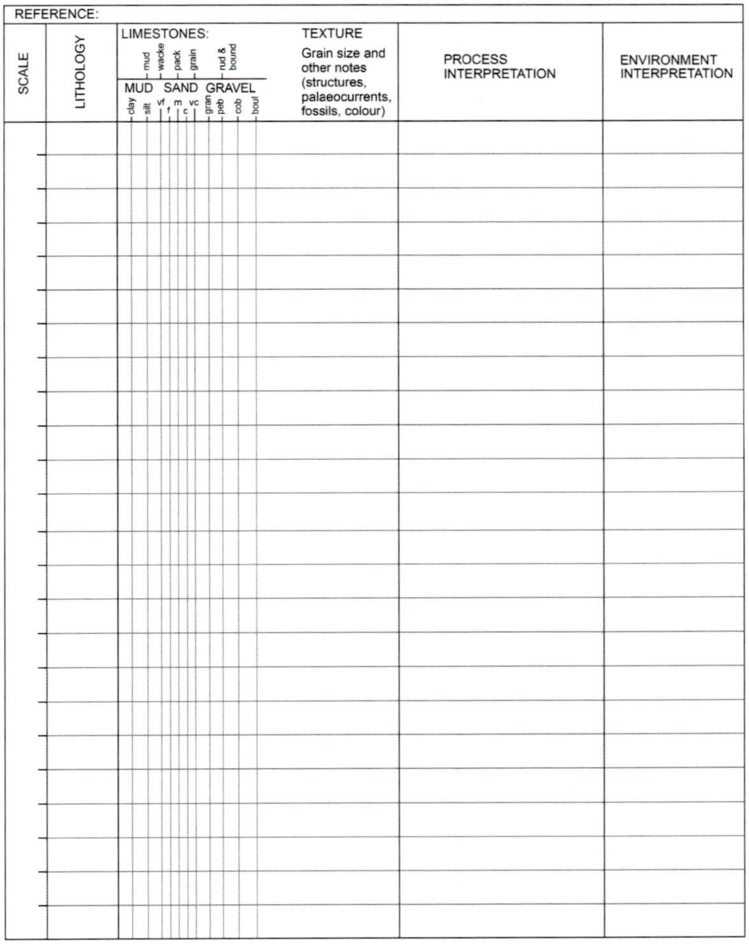

Figure 5.3. A proforma sheet for constructing graphic sedimentary logs.

Source: Gary Nichols; Sedimentology and Stratigraphy; ISBN: 978-1-4051-9379-5.

The field log does not have to be drawn to scale, but the thickness of every bed must be recorded so that a properly scaled version of the log can be drawn later. Sketch logs in the field notebook are also a quick and convenient way of recording sedimentological data, even when there are no plans to present graphic logs in a report. As is usually the case in geology, sketches and graphical representations of data, especially field data, are a quicker and more effective way of recording information than words. Interpretation of the information in terms of processes and environment is normally carried out back in the laboratory.

In recent years, computer-aided graphic log presentation has proliferated, utilizing both standard drawing packages and specialized log drawing software. These are used in publications and can give clear images for presentation purposes, but they must be used carefully to prevent the loss of detailed information from oversimplifying logs. The graphic log created in the field using a pen and pencil is indispensable for fieldwork, and it must always be regarded as the primary source of raw data.

It is possible to present several sedimentary logs on a single sheet and connect them using surfaces of correlation by applying sequence stratigraphic or lithostratigraphic principles. If every log location falls on a straight line, these fence diagrams can be used as straightforward correlation panels. However, they can also be used to display relationships and correlations in three dimensions.

5.2.3. Other Graphical Presentations: Sketches and Photographs

The only way to present sedimentary rock beds with drill core is through a graphic log, which is a one-dimensional depiction that works well for the simplest "layer-cake" strata—beds with consistent thickness and characteristics throughout. In situations where bed exposure indicates notable lateral variation, such as in river channels and overbank deposits within a fluvial setting, a single vertical log is insufficient to accurately depict the characteristics of the deposits. A section of a natural or artificial exposure in a cliff or cutting must be drawn in two dimensions as a two-dimensional representation (Figure 5.4).

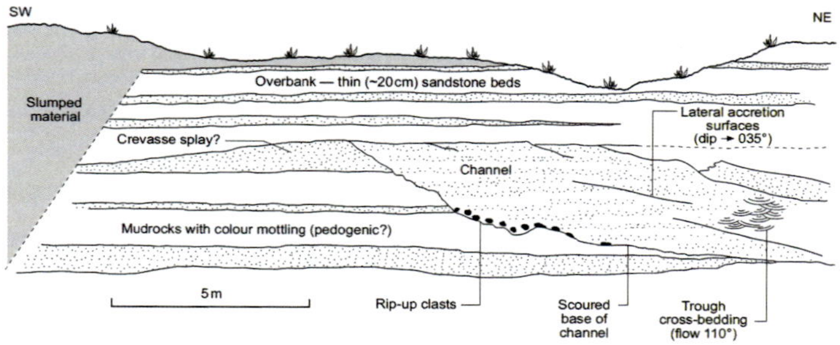

Figure 5.4. An example of an annotated sketch illustrating sedimentary features observed in the field.

Source: Gary Nichols; Sedimentology and Stratigraphy; ISBN: 978-1-4051-9379-5.

Typically, a meticulously sketched and annotated section illustrating all the primary sedimentological features (bedding, cross-stratification, etc.) is adequate and can be enhanced with a photo. Photographs (Figure 5.5) can serve as a model for a field sketch, and with the advent of digital cameras, laptops, and portable printers, the sketch can now be created on location. But a photograph is never a good replacement for a field sketch because sedimentological features are rarely as clear in a photograph as they are in the field, and if significant features and relationships are not sketched at the time, a lot of information may be lost. It's not necessary for a good geological sketch to be artistic. While incidental items like trees and bushes can frequently be overlooked, geological features should be represented clearly and prominently. Every sketch or photo needs to have a scale of some kind, and the view's orientation needs to be noted and annotated to emphasize important geological features (Wüst & Bustin, 2003; Collinson & Thompson, 2006).

Figure 5.5. A field photograph of sedimentary rocks: An irregular lower surface of the thick sandstone unit in the upper part of the cliff marks the base of a river channel.

Source: Gary Nichols; Sedimentology and Stratigraphy; ISBN: 978-1-4051-9379-5.

5.3. PALEOCURRENT

Paleocurrent measurements are essential to the study of sedimentary rocks because they offer insights into the paleogeography, paleoslope, current and wind directions, and facies interpretation. Since the direction of a paleocurrent is a crucial characteristic of a lithofacies and is required for a thorough description, measuring paleocurrents in the field ought to be standard practice.

Paleocurrent indicators can be derived from a wide range of sedimentary rock characteristics. While some structures only record the current's line of movement (trend), others record its direction of movement (azimuth). The most beneficial sedimentary structures are sole structures (flute and groove casts) and cross-bedding, although other structures also produce accurate results.

5.3.1. Paleocurrent Measurements

The more measurements you are able to collect from a bed or set of beds, the more precise the paleocurrent direction you will get; however, the variability (spread) of the measurements must be taken into account.

Examine the outcrop first. In case there is just one lithofacies, measurements can be taken from any or all of the beds. When data from a single bed (or several beds at an outcrop) show a consistent pattern (a unimodal paleocurrent pattern, Figure 5.6), taking a lot of readings is not really necessary. To get an accurate vector mean, about 20 to 30 measurements from an exposure would be enough. The paleocurrent pattern over the region can then be inferred by looking for additional outcrops of the same lithofacies nearby and further away.

To determine the mean direction of a bed, you will need to gather a significant number of readings (more than 20 or more than 50, depending on the variability). At least initially, measurements taken from various sedimentary structures should be kept apart. They can be merged if they are extremely similar. Additionally, record distinct measurements from various lithofacies at an exposure; these may have been deposited by various current kinds or currents originating from various directions. In your field notebook, tabulate the readings (refer to Figure 5.7).

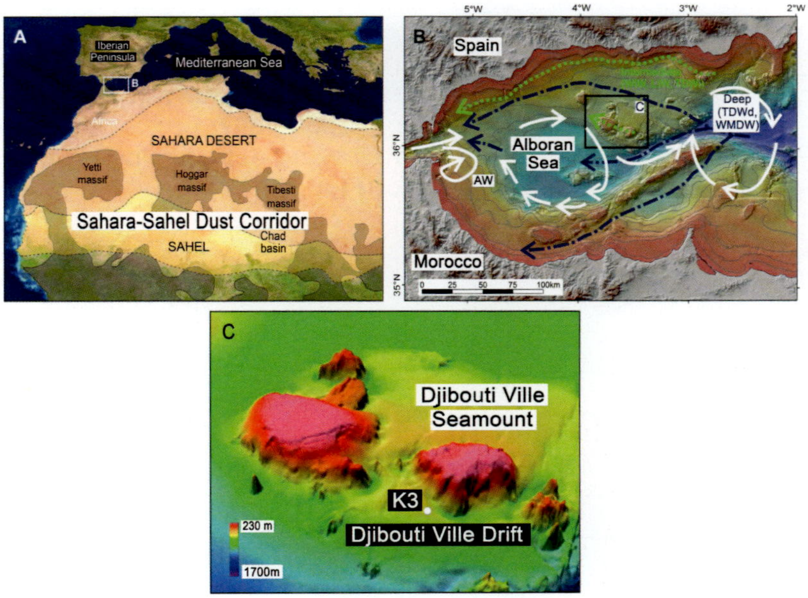

Figure 5.6. (A) Local setting of the study area in the Western Mediterranean Sea and the Sahara–Sahel Dust Corridor. (B) Main trajectories of the water masses circulation in the Alboran Sea.

Legend: AW: Atlantic water; WIW: western intermediate water; LIW: Levantine intermediate water; TDW: Tyrrhenian dense water (i: intermediate and d: deep); WMDW: Western Mediterranean deep water. (C) 3D bathymetric view of the Djibouti Ville Seamount showing the location of the K3 sediment core at the Djibouti Ville drift.

Source: López-González, N., Alonso, B., Juan, C., Ercilla, G., Bozzano, G., Cacho, I., Casas, D., Palomino, D., Vázquez, J.-T., Estrada, F., et al., (2019). 1,33,000 Years of Sedimentary Record in a Contourite Drift in the Western Alboran Sea: Sediment Sources and Paleocurrent Reconstruction. *Geosciences*, 9(8), 345. https://doi.org/10.3390/geosciences9080345.

If tectonism has altered the shape, orientation, or both of the sedimentary structures, your measurements might not accurately reflect the current direction. It is crucial to understand that there are two possible changes: deformation and tilt. The tilt is defined as a straightforward variation in the plane's inclination, of which the sedimentary structure is a part. A sedimentary structure's shape is unaffected by tilt.

Locality: Grid/GPS ref:		
Formation:		
Date:	Bedding strike:	Bedding dip:
Facies code	Sedimentary structure	Palaeocurrent measurement
	Average:	
	Dispersion:	

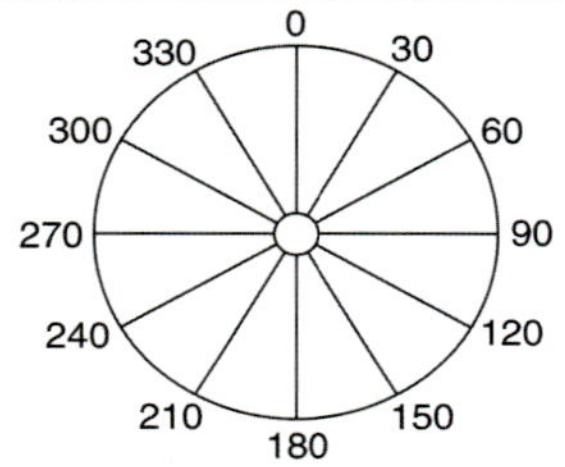

Figure 5.7. Proforma for paleocurrent data. If dip of cross-beds is required, add another column. Corrections for tectonic tilt are required for linear structures such as groove marks if the strata are dipping more than 25° and for planar structures such as cross-bedding if the strata are dipping more than 10°. If this is the case, then add columns for the pitch and down-dip trend of linear structures or the dip and strike of cross-beds.

Source: https://jiwaji.edu/pdf/ecourse/earth_sci/SNM_geology_202_Paleo_Current_Analysis.pdf.

5.3.1.1. Correction of Measurements for Tectonic Tilt

To ascertain the direction of paleocurrents from structures in dipping beds it is necessary to remove the effects of tilting, a straightforward process that is described below.

5.3.1.1.1. Linear Structures

If a tilt of more than 25° has altered the trend of a linear structure, like a flute or groove cast or parting lineation, the following actions should be taken (also refer to Figure 5.8).

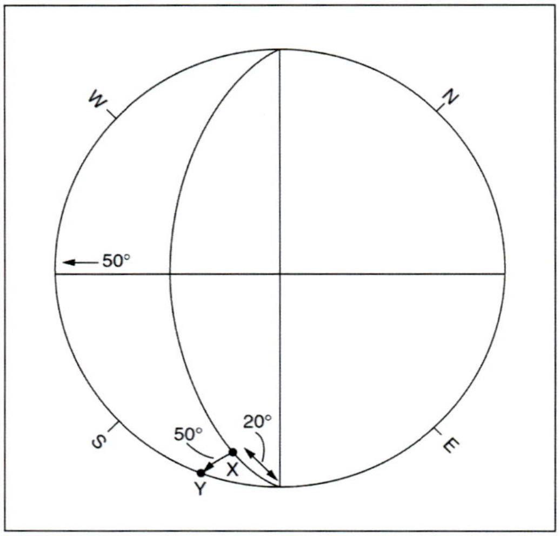

Figure 5.8. An example of correcting the orientation of a linear structure for tectonic tilt. A bed dips at 50° to 225° and has a linear structure, such as a groove mark or parting lineation, with a pitch of 20° to the south-east. The bed is plotted as a great circle on a stereonet and the pitch of the structure is marked on that circle (X). The bed is restored to the horizontal and with it the structure along a small circle to give its original orientation at point Y. On rotating back, the azimuth of Y is given: 155°.

Source: https://jiwaji.edu/pdf/ecourse/earth_sci/SNM_geology_202_Paleo_Current_Analysis.pdf.

1. First measure the direction of dip (or strike) and the angle of dip of the bedding surface (or undersurface); then measure the acute angle between the elongation of the structure and the strike of the bedding (this is the pitch or rake of the structure) and note the down-dip direction of the structure.

2. Place tracing paper on the net, draw the circle and mark north, south, east and west and the central point; mark the direction of dip on the circle (circumference of the net) and then rotate the paper so this mark is on the east–west (equatorial) line of the net; count in the angle of dip of the bed from the mark on the circumference and draw the great circle using the line already there on the net.
3. On the great circle, from the appropriate end, mark the acute angle between the trend of the sedimentary structure and the strike of the bedding surface (i.e., the pitch). To do this, have the paper in the same position as when you drew the great circle (dip direction on the net equatorial line) and then count the pitch angle from the end of the great circle to which the structure is directed down-dip.
4. From this point of the pitch angle on the great circle, move along the small circle to the nearest point on the circumference of the net and mark this place.
5. Finally, rotate the tracing paper back to its original position with regard to the stereonet (north on north) and read off the new direction for the structure. For asymmetric linear structures such as flute marks the azimuth is obtained, whereas with structures giving a trend such as groove marks and current lineation, the current could have been in the opposite direction too.

5.3.1.1.2. Planar Structures, Especially Cross-Bedding

The following steps should be taken to determine the original orientation and angle of dip of a planar structure (principally cross-bedding) that has been changed by a tilt of more than 10°.

1. First measure the direction of dip (or strike) and the angle of dip of the bedding surface, and then the direction of dip (or strike) and angle of dip of the cross-bedding.
2. Place tracing paper on the net, draw the circle and mark north, south, east and west and the central point. Mark the point of the direction of dip on the circle (circumference of the net) and then rotate the paper so this point is on the east–west/equatorial line of the net; count the angle of dip of the bed from the net central point along the equatorial line away from the point of the dip direction on the circumference. This point is the pole to the bedding on the lower hemisphere.
3. Plot the pole for the cross-bed surface in the same way as the pole to the bedding (so start with the tracing paper back at the north, coinciding with the stereonet).
4. Restore the bedding to the horizontal by rotating the tracing paper to bring the pole of the bedding to the equatorial line of the net and (notionally) moving the pole to the center of the net; then move the pole of the cross-bedding along the small circle it is now situated upon, in a similar direction, by the same number of degrees as the angle of dip of the bedding, to give a new point.
5. Rotate the tracing paper back to its original position relative to the net (north on north) and read off the new direction by drawing a line from the central

point, through the new pole position for the cross-bedding, to the circumference. This azimuth is the original orientation (direction of dip) of the cross-bedding, i.e., the direction to which the current was going.

6. The original angle of dip of the cross-bedding can be obtained by rotating the tracing paper so that the new pole position for the cross-bed is on the equatorial line, and then noting the number of degrees from the net center to the new pole position. See Figure 5.9 for a worked example.

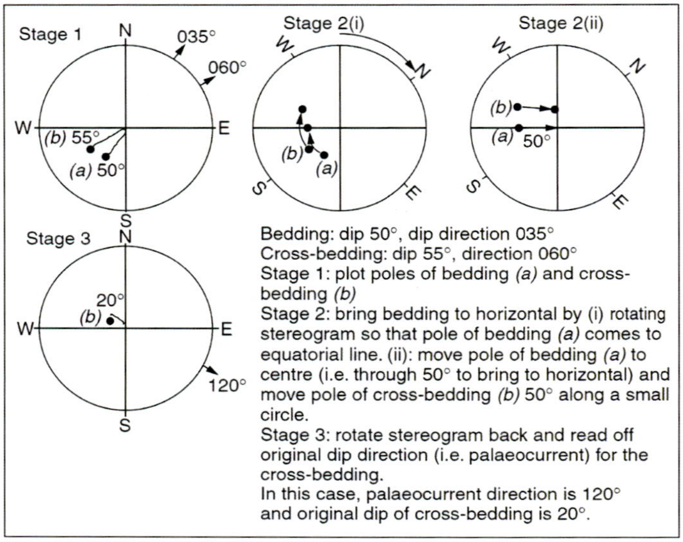

Figure 5.9. Correction of cross-bedding for tectonic tilt using stereographic projection.

Source: https://jiwaji.edu/pdf/ecourse/earth_sci/SNM_geology_202_Paleo_Current_Analysis.pdf.

Finally, a note of warning: if tilted strata are part of a fold whose axis is inclined (i.e., plunging), the fold axis must be brought to the horizontal, so changing the orientation of the tilt, before the tilt is removed. Having completed these corrections, can you be certain that the paleocurrent direction is that at its time of formation? Unfortunately not, for there may also have been rotations of the strata about a vertical axis; rarely can these be determined accurately.

5.3.2. Structures for Paleocurrent Measurement

5.3.2.1. Cross-Bedding

One of the best structures to use is this one, but you should first ascertain what kind of cross-bedding is present. It is ideal for measuring paleocurrents or paleowinds if it was created by the migration of subaqueous dunes, sand waves (many of which are of this type), or aeolian dunes. Verify whether the cross-bedding is trough or planar (tabular).

The direction of the maximum angle of dip in planar cross-bedding provides the paleocurrent direction. Direct measurement of the exposure is unproblematic if it is three-dimensional or two-dimensional with a bedding-plane surface. Since you are only measuring the face's orientation—which is unlikely to be precisely in the paleocurrent direction—taking readings when there is only one vertical face displaying the cross-bedding is less satisfying. You might be able to see a small portion of the cross-bed surface by closely examining the rock face, which will allow you to ascertain the true dip direction. Measuring the rock surface's orientation is the only option if even this is not feasible.

To precisely measure the direction down the trough axis and observe the crossstrata's shape, trough cross-bedding requires a three-dimensional exposure or one with a bedding-plane section. Vertical sections may exhibit cross-bedding dipping up to 90° from the true current direction due to the geometry of the trough cross-beds. Thus, vertical trough cross-bedding sections should only be used as a last resort because they are unreliable for paleocurrent measurements.

5.3.2.2. Ripples and Cross-Lamination

Current ripples and cross-lamination provide easy means of determining paleocurrent directions. It is simple to measure the asymmetry of the ripples (steeper lee-side downstream) and the cross-laminae's dip direction. On the other hand, ripples and the resulting cross-lamination are typically caused by local flow directions that do not correspond to the regional paleoslope. For instance, in turbidite beds, the orientation of cross-lamination within the bed can vary greatly and deviate significantly from the paleocurrent direction noted by the single structures. When the turbidity current slows down and starts to meander across the seafloor, cross-lamination occurs. Despite their drawbacks, it is always important to note the orientation of the ripples and cross-lamination if no other, better directional structure (cross-bedding or sole structures) is present.

Small-scale structures known as wave-formed ripples serve as indicators of local shoreline trends and wind directions. Their crest orientation and, if visible, the direction of internal cross-lamination dip should be measured (Wüst & Bustin, 2003; Collinson & Thompson, 2006).

5.3.2.3. Sole Structures

A current azimuth is provided by flute casts; a line of movement is provided by groove casts; some tool marks also provide a current azimuth, if not a current line of movement. Since flute casts are often all oriented in the same direction, multiple measurements taken from one bed along with measurements from additional beds during the exposure will be adequate. Since there may be a significant amount of orientation variation with groove casts, a greater number should be measured so that the vector mean for each bed can be determined.

5.3.2.4. Preferred Orientations of Clasts and Fossils, and Imbrication

It is possible to align fossils and pebbles parallel or normal to the prevailing currents if they have an elongation ratio of at least 3:1. Look for this preferred orientation, and if you see or suspect it, measure the elongation of enough objects (20 or more readings should suffice) and plot it. You will frequently get a bimodal distribution with one dominant mode that is parallel to the current. The majority of the time, stones, grains, and fossils provide a line or trend of movement. However, some fossils, such as high-spired gastropods, belemnites, and orthoconic cephalopods, can also provide a direction of movement because of their preferentially upstream-directed pointed ends. Stromatolite columns and domes may be asymmetrical, with growth occurring more readily on the side that faces the incoming waves or currents.

5.3.2.5. Other Directional Structures

Parting lineation captures the current's trend and poses no measurement issues. The paleocurrent's trend can also be preserved by scour structures and channels. The direction of the paleoslope down which slumping occurred is recorded by slump folds (fold axes parallel to paleoslope strike, anticlinal overturning downslope). The bedrock's glacial striations indicate the direction of the ice movement (Wüst & Bustin, 2003; Collinson & Thompson, 2006).

5.3.3. Presentation of Results and Calculation of Vector Means

Based on the number of readings and variability, paleocurrent measurements are grouped into classes of 10°, 15°, 20°, or 30° intervals. These classes are then plotted on a rose diagram, with the appropriate scale selected along the radius for the number of readings. In contrast to wind roses, the rose diagram is typically constructed to show the current-to-sense for data from structures providing current azimuths. In the case of trend-giving structures' data, the rose diagram will be symmetrical. Little rose diagrams can be positioned next to a graphic log at the reading location for presentational purposes.

A rose diagram will typically show the dominant paleocurrent (or paleowind) direction, but for precise work, the mean paleocurrent direction—that is, the vector mean—must be determined. It is also worthwhile to compute the data's dispersion, or variance. It is only possible to compute vector means and dispersions for a unimodal paleocurrent pattern.

Each observation is taken to have both direction and magnitude (the magnitude is usually taken to be unity but can be weighted) to determine the vector means from structures that provide azimuths. The magnitude of each vector is then multiplied by the cosine and sine of the azimuth, respectively, to determine its north–south and east–west components. To find the tangent of the resulting vector, simply add the values for each column in the table and divide the summed E–W components (sine values) by the N–S components (cosine values). Alternatively, you can use a calculator to find the sine and cosine of each paleocurrent reading. Using a calculator, convert the tangent to an angle to find the vector mean, or average paleocurrent direction.

$$\text{E–W component} = \Sigma(n \sin \sigma)$$

$$\text{N–S component} = \Sigma(n \cos \sigma)$$

$$\tan \overline{\sigma} = \frac{\Sigma(n \sin \sigma)}{\Sigma(n \cos \sigma)}$$

where;

σ = azimuth of each observation from 0° to 360°
n = observation vector magnitude, generally 1, but if data are grouped into classes (0–15, 16–30, 31–45, etc.), then it is the number of observations in each group
$\overline{\sigma}$ = azimuth of resultant vector (i.e. vector mean)

If trends only can be measured then each observation, measured in the range 0° to 180°, is doubled before the components are calculated:

$$\text{E–W component} = \Sigma(n \sin 2\sigma)$$

$$\text{N–S component} = \Sigma(n \cos 2\sigma)$$

$$\tan 2\overline{\sigma} = \frac{\Sigma(n \sin 2\sigma)}{\Sigma(n \cos 2\sigma)}$$

The magnitude (r) of the vector mean gives an indication of the dispersion of the data, comparable to the standard deviation or variance of linear data:

$$r = \sqrt{(\Sigma n \sin \sigma)^2 + (\Sigma n \cos \sigma)^2} \text{ for azimuthal data}$$

$$r = \sqrt{(\Sigma n \sin 2\sigma)^2 - (\Sigma n \cos 2\sigma)^2} \text{ for non-azimuthal data}$$

To calculate the magnitude of the vector mean in percentage terms (L):

$$L = \frac{r}{\Sigma n} \times 100$$

A vector magnitude of 100% means that all the observations either have the same azimuth or lie within the same azimuth group. In a vector magnitude of 0% the distribution is completely random. There would be no vector mean in this case

5.3.4. Interpretation of the Paleocurrent Pattern

The paleocurrent pattern can be classified as unimodal, which has a single dominant current direction, bimodal bipolar, which has two opposing directions, bimodal oblique, which has two current directions at an angle of less than 180°, and polymodal, which has multiple dominant directions.

To maximize the amount of information, lithofacies research must be done in conjunction with analysis of the paleocurrent pattern. Table 5.1 displays the characteristics of the paleocurrent (or paleowind) pattern of the main depositional environments, which include the river, deltaic, aeolian sand, shoreline–shallow shelf, and turbidite basin.

Table 5.1. Paleocurrent Patterns of Principal Depositional Environments, Together with Best and Other Directional Structures

Environment	Directional structures	Typical dispersal patterns
Aeolian	large-scale cross-beddin	unimodal common, also bimodal and polymodal; depend on wind directions/dune type
Fluvial	cross-bedding, also parting lineation, ripples, cross-lamination, channels	unimodal down palaeoslope, dispersion reflects river sinuosity
Deltaic	cross-bedding, also parting lineation, ripples, channels	unimodal directed offshore, but bimodal or polymodal if marine processes important
Marine shelf	cross-bedding, also ripples, fossil orientations, flutes/grooves on bases of storm beds	bimodal common through tidal current reversals but can be normal or parallel to shoreline; unimodal and polymodal patterns
Turbidite basin	flutes, also grooves, parting lineation, ripples	unimodal common, either downslope or along basin axis if turbidites, parallel to slope if contourites

Source: https://jiwaji.edu/pdf/ecourse/earth_sci/SNM_geology_202_Paleo_Current_Analysis.pdf.

To determine the regional paleoslope in fluvial facies, paleocurrents are best measured from the largest crossbeds. Directions derived from smaller features, such as ripples and parting lineation, typically depict the river's minor currents during low tide and do not accurately represent the paleogeography on a larger scale. Measure any lateral accretion surfaces that may exist to determine the meandering's direction. Whereas meandering-stream facies will display a greater range of directions, braided-stream deposits typically yield paleocurrent directions with a lower dispersion.

Different paleocurrent patterns can be produced by deltaic facies, depending on the kind of delta (lobate versus elongate) and the functions of coastal and fluvial processes. Depending on the type of delta, unimodal patterns with dispersion will be obtained in a system that is dominated by rivers. Wave and tidal effects can yield polymodal patterns where the reworking of delta-front sands by marine processes is significant.

Because they are subject to storm, tidal, and wave action, shelf, shoreface, and coastline sandstones can exhibit intricate paleocurrent patterns. Bipolar patterns can occur where tidal currents predominate, but they are typically asymmetrical because one tidal current direction is stronger than the other. There are three types of tidal currents: parallel, oblique, and normal to the coast. Cross-beds are created by storm waves and currents and are typically directed offshore, though they can spread widely. Once more, measure larger structures (cross-beds, sole structures) whenever feasible rather than ripples, which might indicate small reworking.

In basinal successions, deep-sea currents may flow along the basin axis after reaching the basin center, although paleocurrents from turbidites are frequently linked to regional paleoslopes that flow basin-ward. To interpret the paleocurrent pattern, it helps to know the orientation of the basin and the larger-scale tectonic context of the succession. The best structures to measure are grooves and flutes because, once a turbidity current slows down, it starts to wander and structures like cross-lamination and ripples on the bed's upper surface might not accurately reflect the direction of the main current. For paleoslope orientation, slump folds are useful. Certain deepwater currents, such as those found in contourite deposits, flow parallel to the slope contours.

Based on large-scale cross-bedding, paleowind patterns in desert sandstones can be extremely simple or complex. It is dependent upon the wind pattern, the local topography, and the type of sand deposit—big ridges (seif draas) or sand seas (ergs). The effects of recurring trade winds have caused unidirectional cross-beds in certain desert sandstones. Paleowind directions are not connected to local paleoslopes, so caution must be used when interpreting the paleogeography (Wüst & Bustin, 2003; Collinson & Thompson, 2006).

5.4. COLLECTION OF ROCK SAMPLES

It is common practice to gather samples for additional analysis because field studies only reveal a fraction of the information that can be found in sedimentary rocks. Materials may be needed for mineralogical and geochemical analyses, paleontological research, or determining the biostratigraphic age of the strata. The rock's texture and composition can be thoroughly examined using thin-sections; alternatively, the sample can be dissolved to perform chemical analyses or broken down to determine the amount of heavy minerals present. The provenance of a product is determined using several of these processes.

The material's intended use will determine the sample's size and condition, but generally speaking, pieces that are roughly 50 mm across will work well. It is best practice to gather "fresh" samples, i.e., after the worn exterior has been removed. An arrow pointing in the direction of younging and perpendicular to the bedding planes should typically be marked on the sample to indicate its orientation with respect to the bedding. When a sample is gathered in the field, it should be assigned a special identification number, and its location should be noted in the field notebook. The sample's location within the logged succession should be noted if it was taken as part of the sedimentary log recording process.

Samples must always be stored separately in suitable bags, which are often sturdy plastic bags with a seal. Before leaving for the field, write the sample numbers on small pieces of heavy-duty adhesive tape, and then attach the tape to an acetate sheet if you really want to be organized. On one piece of tape to be attached to the sample and the other to the plastic bag the sample is placed in, each number is written. This procedure has the benefit of ensuring sample numbers are always readable, even in the worst-case scenarios. It also prevents sample numbers from being accidentally missed or duplicated.

5.4.1. Provenance Studies

An analysis of the various clast types present can provide information about the provenance of the material or the source of the sediment (Pettijohn, 1975). The provenance of a clast in a sediment can be determined if its petrology or chemistry can be used to identify it as typical of a specific source area. In certain cases, this allows the paleogeographical location of a source area to be determined and offers details on the rate and mechanisms of erosion in areas that have been uplifted.

In coarser clastic sediments, provenance studies are usually not too difficult to conduct because it is easy to identify a pebble or cobble as having been eroded from a specific bedrock lithology. Numerous types of rocks can be confidently identified thanks to their distinctive compositions and textures. Where all the clasts are sand-sized, it is more challenging to identify the provenance because many of the grains might be individual minerals that could have originated from different places. Though very common, quartz is frequently of little use in determining provenance because the grains can have originated from granite bedrock, a variety of different metamorphic rocks, or reworked from older sandstone lithologies. Certain heavy minerals have been found to be highly reliable markers of the sand's origin (Figure 5.10). As a result, provenance investigations in sandstones are frequently conducted by isolating and characterizing the heavy minerals from the majority of the grains (Mange & Maurer, 1992). The heavy mineral analysis process is a useful technique for identifying the sediment's source (Morton et al., 1991).

Since some clay minerals are only found in specific bedrock types—basaltic rocks, for instance, weather to produce the smectite group of clay minerals—clay mineral analysis is also occasionally used in provenance investigations. The average chemical composition of vast continental regions can also be ascertained by analyzing mud and mudrocks. Big rivers may carry and deposit material eroded from the same area since they drain a significant amount of a continental landmass. Therefore, rather than attempting to gather representative and proportionate rock samples from the same area, a sample of mud from a river mouth serves as a convenient stand-in for sampling the continental landmass. This is a good tool to compare various continents and can be applied to old mudrocks to compare possible detritus sources. For this purpose, geochemical fingerprinting with Rare Earth Elements and isotope dating with the neodymium-samarium system are particularly useful (Wüst & Bustin, 2003; Collinson & Thompson, 2006).

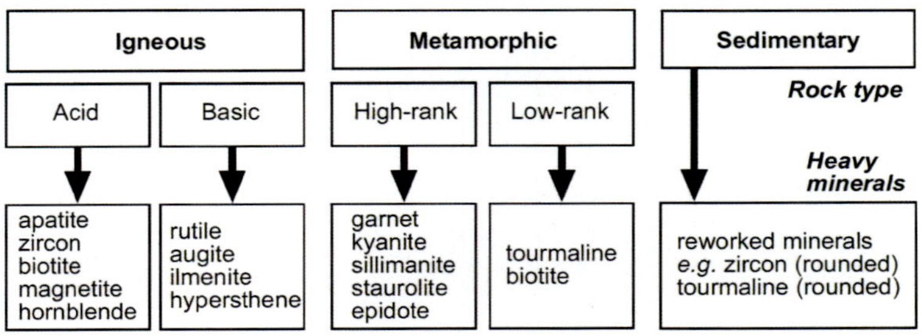

Figure 5.10. Some of the heavy minerals that can be used as provenance indicators.

Source: Gary Nichols; Sedimentology and Stratigraphy; ISBN 978-1-4051-9379-5.

5.5. DESCRIPTION OF CORE

Sedimentary rocks are where most of the world's mineral and fossil fuel resources are found underground. Although there are methods for "remotely" ascertaining the composition of subsurface strata, boring holes is the only way to obtain concrete proof of the makeup of strata that are tens, hundreds, or thousands of meters below the surface.

Companies prospecting for coal and mineral resources, the oil and gas sector, companies looking for water resources, and universities conducting pure research all drill. Drilling a hole does not guarantee that the core will be cut. When analyzing the strata, oil companies typically use geophysical techniques; they only cut the core when specific horizon details are needed. On the other hand, because coal exploration programs need to know the exact location of coal beds and sample them for quality, they usually entail drilling core through the whole hole. The core is kept in boxes in lengths of approximately one meter after it has been cut. The oil companies usually cut a core with a diameter of 100 to 200 mm, which is split vertically to create a flat face (Figure 5.11), but the coal core is typically left whole and has a diameter of only 60 mm. The obvious disadvantage of any core, when compared to an outcrop, is its extreme narrowness, which results in a one-dimensional sample of the strata. Only when looking at the core can one imagine features that are discernible by looking at two-dimensional exposure across a quarry or cliff face, such as river channels, reefs, or even some cross-bedding. This restriction usually makes it more difficult to interpret the strata. Nonetheless, the core has a clear benefit in that it typically offers a vertical section continuity over tens or hundreds of meters. Due to weathering, vegetation, or scree covering the beds, even some of the best natural exposures of strata do not offer this complete coverage.

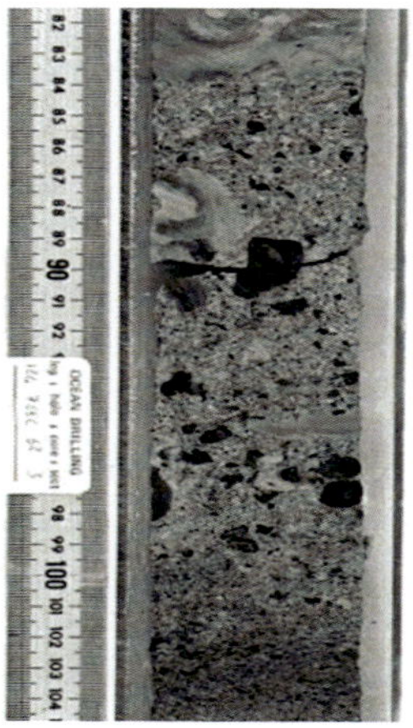

Figure 5.11. When drilling through strata it is possible to recover cylinders of rock that are cut vertically to reveal the details of the beds.

Source: Gary Nichols; Sedimentology and Stratigraphy; ISBN: 978-1-4051-9379-5.

5.6. INTERPRETING PAST DEPOSITIONAL ENVIRONMENTS

Sediments amass in a variety of locations that are geomorphologically defined, including rivers, lakes, coasts, shallow seas, and so forth. Studies of physical geography and ecology have provided extensive knowledge of the physical, chemical, and biological processes that form and define those environments. The nature of the sediment deposited in these environments is determined by the same processes. Determining the environment in which sedimentary rocks were deposited and interpreting them in terms of the transport and depositional processes are essential components of sedimentology. By doing this, a sedimentologist seeks to determine the conditions that existed on Earth's surface at various points in time and space, thereby constructing a picture of the planet's surface history (Wüst & Bustin, 2003; Collinson & Thompson, 2006).

5.6.1. The Concept of 'Facies'

In geology, the word "facies" is frequently used, especially in the study of sedimentology, where the term "sedimentary facies" refers to the totality of a sedimentary unit's attributes. The dimensions, grain sizes and types, color, and biogenic content of the sedimentary rock are some of these features. One example of a rock with cross-bedding as the main sedimentary structure is "crossbedded medium sandstone," which is primarily composed of medium-grade sand grains. Not every feature of the rock is mentioned in the facies name, and in some cases, it might be crucial to highlight certain traits. In other cases, if the color and grain types were thought to be more significant than the grain size and sedimentary structures, the facies name for a very similar rock might be "red, micaceous sandstone." Any study of sedimentary rocks would include a facies description, which would list every property a rock possesses.

This is known as the lithofacies if the description is limited to a rock's physical and chemical properties. A study that focuses on the trace fossils in the rock would be an ichnofacies description, whereas observations that center on the fauna and flora present are referred to as biofacies descriptions. A single rock unit, for instance, might be characterized as having a "Cruziana" ichnofacies, a biofacies of echinoid and crinoids, and a grey bioclastic packstone. These traits combined with others would make up the sedimentary facies.

5.6.2. Facies Analysis

The facies concept serves as the foundation for facies analysis, a rigorous, scientific method of strata interpretation, in addition to being a useful way to characterize and classify sedimentary rocks that are observed in the field. The physical and chemical processes of sediment transport and deposition define the lithofacies' characteristics, while ichnofacies and biofacies offer details on the paleoecology both during and after deposition. It is feasible to reconstruct paleoenvironments by analyzing the sediment in terms of the physical, chemical, and ecological conditions present at the time of deposition, i.e., historical surroundings.

Facies analysis can be used to reconstruct past sedimentary environments in a relatively straightforward manner in some cases, but in other cases, a thorough analysis of numerous variables may be necessary before a rough conclusion can be drawn. When the rock possesses properties specific to a given environment, the process is simple. Hermatypic corals have only ever been known to grow in shallow, clear, and relatively warm seawater; as a result, the presence of these fossil corals in their living state within a sedimentary rock may serve as evidence that the sediments were deposited in such conditions. If the sediments are the end result of processes that can take place in various environments, the analysis becomes more difficult. A "cross-bedded sandstone" lithofacies, for instance, would not provide us with an indication of a particular environment because crossbedded sandstone can form during deposition in deserts, rivers, deltas, lakes, beaches, and shallow seas.

Facies interpretation ought to be impartial and limited to acknowledging the processes that shaped the beds. Therefore, it can be inferred that the bed was formed beneath shallow water, with wind over the water's surface creating waves that stirred the sand to form symmetrical wave ripples, based on the presence of symmetrical ripple structures in a fine sandstone. Although wave ripples do not form in deep water, the interpretation of "shallow water" is made because the ripples do not tell us whether the water was in a lake, lagoon, or shallow-marine shelf environment. Hence, the facies should be described as "symmetrically rippled sandstone" or possibly "wave rippled sandstone," rather than "lacustrine sandstone," as that interpretation cannot be made without more data (Wüst & Bustin, 2003; Collinson & Thompson, 2006).

5.6.3. Facies Associations

The assortment of processes that take place in an environment determines its characteristics. For instance, a lagoon is an area of low energy, shallow water that receives sand from the sea on a regular basis. It is a particular ecological niche where only particular organisms can survive because of changes in salinity. These processes will result in the formation of mud from standing water, sands with wind-generated wave ripples over shallow water, and a biofacies with limited fauna. A facies association made up of these various facies represents the depositional environment.

When a succession of beds are analyzed in this way, it is usually evident that there are patterns in the distribution of facies. For example, in Figure 5.12, do beds of the 'bioturbated mudstone' occur more commonly with (above or below) the 'laminated siltstone' or the 'wave rippled medium sandstone'? Which of these three occurs with the 'coal' facies? When attempting to establish associations of facies it is useful to bear in mind the processes of formation of each. Of the four examples of facies just mentioned the 'bioturbated mudstone' and the 'wave rippled medium sandstone' both probably represent deposition in a subaqueous, possibly marine, environment whereas 'medium sandstone with rootlets' and 'coal' would both have formed in a subaerial setting. Two facies associations may therefore be established if, as would be expected, the pair of subaqueously deposited facies tend to occur together, as do the pair of subaerially formed facies.

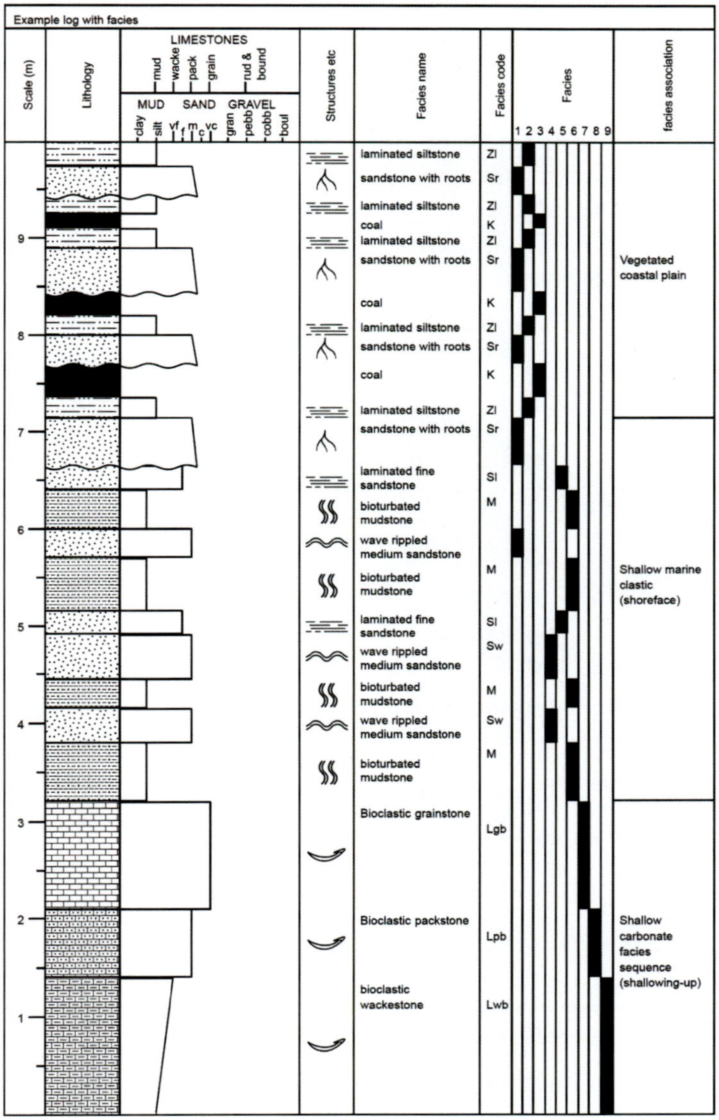

Figure 5.12. A graphic sedimentary log with facies information added. The names for facies are usually descriptive. Facies codes are most useful where they are an abbreviation of the facies description. The use of columns for each facies allows for trends and patterns in facies and associations to be readily recognized.

Source: Gary Nichols; Sedimentology and Stratigraphy; ISBN: 978-1-4051-9379-5.

The procedure of facies analysis therefore can be thought of as a two-stage process. First, there is the recognition of facies that can be interpreted in terms of processes. Second, the facies are grouped into facies associations that reflect combinations of processes and therefore environments of deposition (Figure 5.12). The temporal and spatial relationships between depositional facies as observed in the present day and recorded in sedimentary rocks were recognized by Walther (1894). Walther's Law can be simply summarised as stating that if one facies is found superimposed on another without a break in a stratigraphic succession those two facies would have been deposited adjacent to each other at any one time. This means that sandstone beds formed in a desert by aeolian dunes might be expected to be found over or under layers of evaporates deposited in an ephemeral desert lake because these deposits may be found adjacent to each other in a desert environment (Figure 5.13). However, it would be surprising to find sandstones formed in a desert setting overlain by mudstones deposited in deep seas: if such is found, it would indicate that there was a break in the stratigraphic succession, i.e., an unconformity representing a period of time when erosion occurred and/or sea level changed.

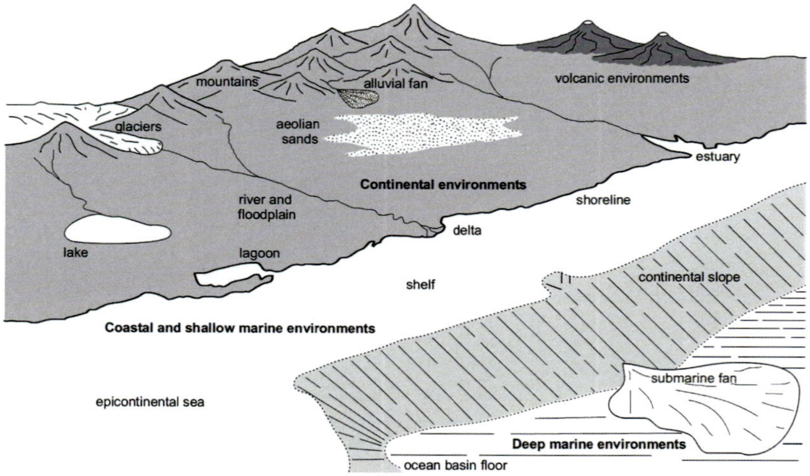

Figure 5.13. A summary of the principal sedimentary environments.

Source: Gary Nichols; Sedimentology and Stratigraphy; ISBN: 978-1-4051-9379-5.

5.6.4. Facies Sequences/Successions

A facies association in which the facies occur in a specific order is called a facies succession or sequence. They happen when a sequence of actions is repeated in reaction to frequent alterations in the environment. For instance, in the event that a bioclastic wackestone facies consistently rests on top of a bioclastic packstone facies, which is subsequently atop a bioclastic grainstone (Figure 5.12), it is possible to classify these three facies as a sequence of facies. Recurrent shallowing-up brought on by deposition on shoals of bioclastic muds and sands in a shallow marine environment could be the cause of this pattern. Facies patterns can be identified by visual inspection of graphic sedimentary logs or by applying statistical methods, like Markov analysis, to ascertain the sequence in which facies occur in a succession. For this technique to work, all of the facies must be arranged in a transition grid along the table's vertical and horizontal axes. At each time a facies transitions to another (e.g., from bioclastic packstone facies to bioclastic wackestone facies). Sequences or successions of facies are observed as transitions from one facies to another that are higher than average.

5.6.5. Facies Names and Facies Codes

Facies are given names after they have been defined. Although there are no set guidelines for naming facies, it makes sense to use terms like "bioturbated mudstone," "trough crossbedded sandstone," or "foraminiferal wackestone" that are generally descriptive. This is better than "Facies A," "Facies B," "Facies C," and so forth, since the latter letters don't reveal anything about the type of facies. It is necessary to strike a balance between choosing a name that accurately describes the facies and one that is not overly complicated. Generally speaking, you should use just enough adjectives to set the faces apart from one another. For instance, if only one mudrock facies is identified in the succession, the term "mudstone facies" is more than sufficient. However, when analyzing successions of shallow marine sandstone, it may be useful to distinguish between "trough crossbedded coarse sandstone facies" and "planar crossbedded medium sandstone facies." As a result, faces schemes can take on different meanings and names based on what the rocks under study require.

While it is generally acceptable to refer to facies associations in terms of the interpreted environment of deposition, the names for facies should normally be strictly descriptive. Based on facies characteristics and possibly biofacies information suggesting freshwater fauna, an association of facies such as "symmetrically rippled fine sandstone," "black laminated mudstone," and "grey graded siltstone" may have been interpreted as having been deposited in a lake. Thus, this facies association can be called a "lacustrine facies association" to differentiate it from other continental facies associations that are deposited as overbank deposits (called a "floodplain facies association") and in river channels (called a "fluvial channel facies association").

Shorter versions of the facies names can be useful in some situations, such as when annotating sedimentary logs (Figure 5.12). A system of letter codes for fluvial sediments was proposed by Miall (1978) and can be modified to correspond with any kind of deposit. The first letter in this scheme, "S" for sand and "G" for gravel, for example, indicates the grain size. One or more suffix letters are used to represent other features, like sedimentary structures; for example, "cross-laminated sandstone" is represented by the letter Sxl. There are numerous variations on this theme, with code letters not having any set rules (some workers use the letter "Z" for silts, for instance), and comparable schemes for carbonate rocks based on the Dunham classification. Generally speaking, it is preferable to create a system that is easily understood and is consistent, such as having all sandstone facies begin with the letter "S."

A different graphical scheme is available for showing facies on sedimentary logs (Figure 5.12): Each facies' vertical extent is indicated by columns that are used alongside the log. This type of presentation has the benefit of allowing trends over time to be seen on the logs if the column order is properly selected, for example, placing deeper marine environments on the right and shallower marine environments on the left for shelf environments (Wüst & Bustin, 2003; Collinson & Thompson, 2006).

CASE STUDY

SEDIMENTOLOGICAL ANALYSIS OF MALLEWADI AREA: A CASE STUDY OF LOWER REACHES OF MAN RIVER BASIN

- **Introduction:** Sedimentology and sequence stratigraphy analysis have for many years been disciplines rather separate from diagenesis which is concerned with processes occurring after deposition. Prediction of rock properties as a function of burial depth in sedimentary basins requires that these disciplines become more integrated. Compaction of sedimentary rocks is driven towards increased density (lower porosity) and higher rock velocity as functions of burial depth (effective stress) and temperature. Both the mechanical and chemical compaction of sedimentary rocks is functions of the primary textural and mineralogical composition of the sediments at the time of deposition and after shallow burial diagenesis. This is controlled by the provenance, transport and depositional environment. Many published sedimentological studies, however, contain little information about the mineralogical and textural composition of the sedimentary sequences. Near the surface, sediments are in an open geochemical system due to groundwater flow, diffusion, and evaporation.

The approach consists of two main phases, the recognition of the grains and the assessment of some of its characteristics. The method is validated on the dimensional features by comparison with a sieving procedure through the analysis of eight coarse-sized sediments of different locations.

- **Study Area:** The study area is a part of Man river, (a right side tributary of Bhima river) which Originates in Mahadeva hill range and flowing NW-SE direction. This region is the southernmost part of the basaltic lava flows of the Deccan Traps which is the major continental tholeiitic province of India. A thick, extensive, quaternary deposit is found at meander along the right bank of Sina, at Mallewadi settlement.
- **Methodology:** The observation of the same was carried out and as per the standard procedure of stratigraphy demarcation of the different beds with stratigraphical succession of different assemblages of flora and fauna with vivid thicknesses of different sizes and shaped sediment were studied in detail in the field. The observations were noted in the field diary and the sediment samples from different strata in a succession according to the standard procedure of collection of samples were collected in the polythene bags. The same samples were treated in the Sedimentological lab to understand the depositional environment of different sediments in a stratigraphical sequence by using the standard method of Sedimentological analysis for the stratigraphical sediments.

The sediment samples were collected from the different terraces along the bank of the Bhima river. Location co-ordinates were established using a Garmin e-Trex 30 GPS receiver. These samples were collected by scooping individual surfacial sediments from different scars at various selected locations and then secured in zip-lock bags. For the silt and clay size particles, pipette analysis was carried out. Finally, the probability curves and Visher's curves were plotted and accordingly, conclusions were drawn. Cumulative grain size distribution curves reflect the total grain size distribution (GSD) of each sediment sample.

- **Geology:** The area under investigation is part of the Bhima basin and channel of the Man river, which is one of the major NW-SE trending right-side tributary of Bhima originates in the Mahadeva hill range. This region is the southernmost part of the basaltic lava flows of the Deccan traps which is the major continental theolitic province of India. The present archaeological monument that is Mahadeva temple is bounded and founded on the thick quaternary deposit of Man along its right bank. As usual, this quaternary deposit is spread during the flood of the Man river occupies a large area on both banks. During the construction activity of the same; the previous age of the temple is affected and filled with these sediments along all sites.

During the field surface exploration, a result of detailed investigations indicated that a huge cutting section along the right side (towards the temple) was found. The observation of the same was carried out and as per the standard procedure of stratigraphy demarcation of the different beds with a stratigraphical succession of different assemblages of flora and fauna with a vivid thickness of different sizes and shaped sediment were studied in detail in the field.

- **Geomorphology:** Physiographical the area belongs to the Deccan volcanic plateau with high peaks ranging in height from... m to... m AMSL. The topography of the area is nearly plane with the deep to moderate dissection of Man River, with high mountains in the source area of the Man River. Therefore, Physiographical point of view the basin can be divided into three units, (1) the high relief area in the catchment or source of Man river, (2) Plane to slightly undulating and (3) few valley fills.
- **Sediment Sample Collection:** These samples were collected by scooping individual surficial sediments from different scars at various selected locations and then secured in zip lock bags. Samples were pre-treated in the lab by using HCl and H_2O_2 to remove carbonate as well as organic matter present within the consolidated material. Then these were oven-dried, initially weighed, then wet sieved through 240 mesh (63 microns) and oven dried again. Then they are dry sieved, with the portion of the pan fraction obtained during the same also assigned to the fine fraction. For the silt and clay-sized particles, pipette analysis was carried out.
- **Data Analysis and Result:** At this Mallewadi site; three samples were collected for the sedimentological Analysis. For these three samples, the minimum mean grain size is 3.783 and the maximum mean grain size is 4.183. The average

mean grain size of those samples is 3.983. The minimum median (Md_0) is 3.6 whereas the maximum median (Md_0) is 3.85. The average median (Md_0) is 3.725. The minimum standard deviation, i.e., sorting is 1.66 whereas the maximum standard deviation, i.e., sorting is 1.95. The average standard deviation is 1.805. The minimum skewness is 0.289 and the maximum skewness is 0.114. The average skewness is 0.202. The minimum Kurtosis is 1.040 and the maximum Kurtosis is 1.524. The average Kurtosis is 1.282 (Figures 1 and 2; Table 1).

Table 1. Dry Sieving Data for Different Terraces at Mallewadi (T1 = Terrace 1; T2 = Terrace 2; and T3 = Terrace 3)

Sieve	(T1)	Proceedings of ASTRA 2015	
4 mm	–	–	0.600
3.55 mm	–	–	0.140
2.8 mm	0.080	0.040	0.420
2.38 mm	–	–	0.100
2 mm	–	–0.98	0.070
1.7 mm	0.050	–	0.240
1.4 mm	0.080	0.14	0.150
1.18 mm	0.040	0.10	0.180
1 mm	–	0.08	–
850 mic	0.170	0.50	0.320
710 mic	–	0.39	–
600 mic	0.120	0	0.130
500 mic	0.530	0.130	0.300
425 mic	0.660	0.180	0.310
355 mic	0.550	0.140	0.180
300 mic	2.700	0.880	1.280
250 mic	–	3.11	–
212 mic	3.180	1.130	1.570
180 mic	0.440	6.76	0.220
150 mic	5.250	4.18	3.140
125 mic	3.880	4.18	2.490
106 mic	7.960	3.070	5.550
90 mic	3.430	6.32	2.470
75 mic	6.030	3.180	5.310
63 mic	4.960	7.1	5.110
53 mic	4.150	2.820	3.190
45 mic	0.530	0.510	0.150
37 mic	2.570	0.830	0.210
25 mic	–	0.890	0.170

Source: https://www.researchgate.net/publication/358877473_Sedimentological_analysis_of_Mallewadi_area_A_Case_study_of_lower_reaches_of_Man_River_basin.

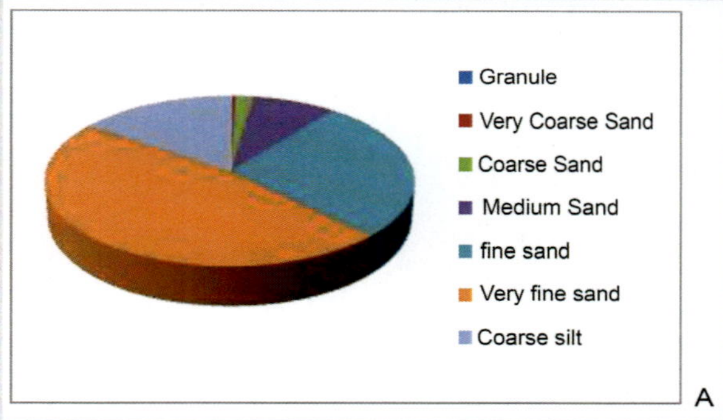

It is inferred from the above pie chart that; the percentage of very fine sand is more in number followed by fine sand, coarse silt and medium sand respectively.

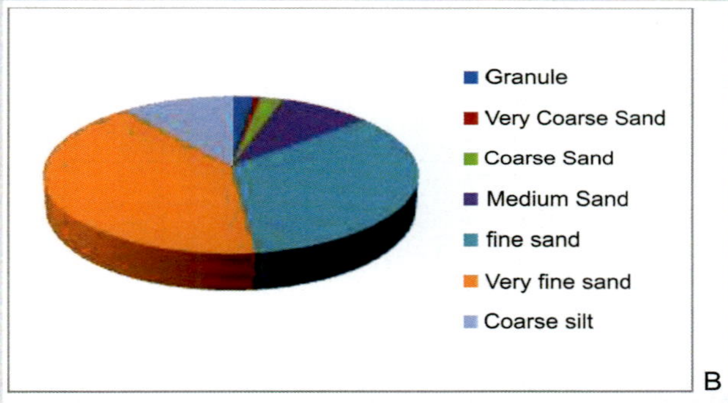

It is inferred from the above pie chart that; the percentage of very fine sand is more in number followed by fine sand and coarse silt, medium sand and granule respectively.

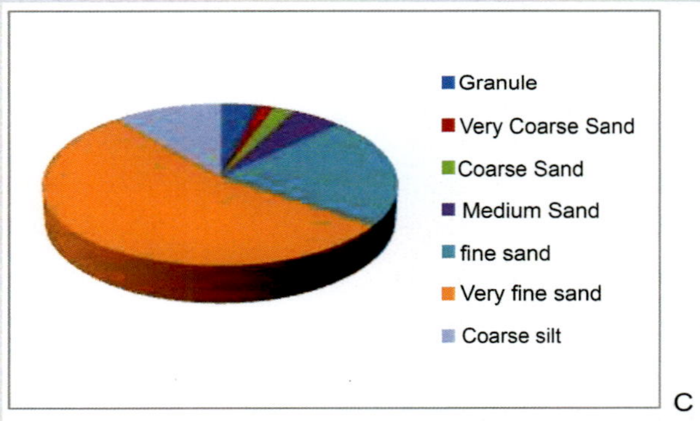

It is inferred from the above pie chart that; the percentage of very fine sand is more in number followed by fine sand, coarse silt and medium sand and granule respectively.

Figure 1. *Pie chart showing grain size distribution for different terraces ((A) Terrace-1; (B) Terrace-2; and (C) Terrace-3).*

Source: https://www.researchgate.net/publication/358877473_Sedimentological_analysis_of_Mallewadi_area_A_Case_study_of_lower_reaches_of_Man_River_basin.

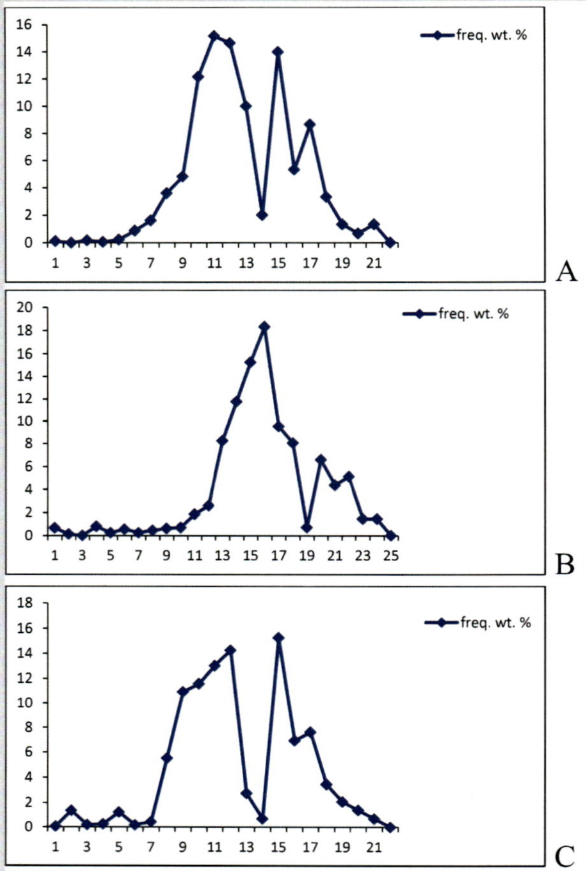

Figure 2. *Frequency plots of different terraces with 'Size in phi' on X axis and 'Cummulative frequency %' on Y axis ((A) Terrace-1; (B) Terrace-2; and (C) Terrace-3).*

Source: https://www.researchgate.net/publication/358877473_Sedimentological_analysis_of_Mallewadi_area_A_Case_study_of_lower_reaches_of_Man_River_basin.

- **Conclusion:** On the basis of several sedimentological observations based on field and lab studies of Mallewadi area, following conclusion were made.
  i. All the sediment in different stratigraphical layer shows variation in fluvial flood energy during their deposition.
  ii. On the basis of their surface texture; they are smooth, polished and at places the shows roughness, grooves, etc.
  iii. These sediments are ranging average the size 212–53 micron, 250–53 micron and 150–53 micron for terrace 1, 2, and 3.
  iv. On the basis of subrounded to rounded nature of the grains and sub angular to angular shape of the sediment; clearly indicate moderate rate of transportation from their source area.

CLASS ACTIVITY

SEDIMENTARY ROCK IDENTIFICATION AND ANALYSIS:

Examine any fossils or trace fossils present in the rock. What can they reveal about the paleoenvironment in which the sediment was deposited? How might the presence of certain organisms indicate specific environmental conditions?

SUMMARY

- Field sedimentologists analyze various sedimentary structures and features such as bedding, ripple marks, and cross-bedding, which provide insights into past environmental conditions and sedimentary processes. The study often requires extensive fieldwork, where geologists collect samples and make detailed observations to understand the spatial distribution and characteristics of sedimentary deposits.
- The data for these paleoenvironmental or stratigraphic studies need to be collected systematically and efficiently, and for this purpose the graphic sedimentary log is the main method of recording data. A sedimentologist may spend a lot of time recording and drawing these logs, in conditions that vary from sunny beaches to wind-swept mountainsides, but the methodology is essentially the same in every instance.
- The vertical scale used is determined by the amount of detail required. If information on beds a centimeter thick is needed then a scale of 1:10 is appropriate. A log drawn through tens or hundreds of meters may be drawn at 1:100 if beds less than 10 cm thick need not be recorded individually.
- Computer-aided graphic log presentation has become widespread in recent years, including both dedicated log drawing packages and standard drawing packages. These can provide clear images for presentation purposes, and are used in publications, but must be used with some care to ensure that logs do not become over-simplified with the loss of detailed information.
- The measurement of palaocurrents is a vital part of the study of sedimentary rocks since they provide information on the palaogeography, palaoslope, current and wind directions and they are useful in facies interpretation. The measurement of paleocurrents in the field should become a routine procedure; a paleocurrent direction is an important attribute of a lithofacies and necessary for its complete description.

- Flute casts provide a current azimuth; some tool marks will also provide a current azimuth, if not at least a line of movement; groove casts provide a line of movement. Flute casts are generally all oriented in the same direction and so several measurements from one bed together with measurements from other beds at the exposure will be sufficient.
- Paleocurrent measurements are grouped into classes of 10°, 15°, 20°, or 30° intervals (depending on number of readings and variability) and then plotted on a rose diagram, choosing a suitable scale along the radius for the number of readings.
- Field studies only provide a portion of the information that may be gleaned from sedimentary rocks, so it is routine to collect samples for further analysis. Material may be required for paleontological studies, to determine the biostratigraphic age of the strata, or for mineralogical and geochemical analyses.

REVIEW QUESTIONS

1. Describe the importance of sedimentary structures such as bedding, ripple marks, and cross-bedding in field sedimentology.
2. What is a graphic sedimentary log, and what information does it typically include?
3. How are graphic sedimentary logs used in the interpretation of sedimentary sequences?
4. What is a paleocurrent, and why is it significant in sedimentological studies?
5. What methods are used to determine the direction of paleocurrents in sedimentary rocks?
6. Explain how paleocurrent data can be used to infer the depositional environment of a sedimentary rock
7. What are the best practices for collecting rock samples in the field to ensure they are representative of the study area?
8. Describe the steps involved in documenting and preserving rock samples collected during fieldwork.

REFERENCES

1. Collinson, J., Mountney, N., & Thompson, D. (2006). Sedimentary Structures. Terra Publishing.
2. Gong, Q., Yan, T., Wu, X., Li, R., Wang, X., Liu, N., Li, X., Wu, Y., & Li, J. (2022). Geochemical gene: A promising concept in discrimination and traceability. Applied Geochemistry, 136, 105133. https://doi.org/10.1016/j.apgeochem.2022.105133.
3. Hakim, K., Bower, D. J., Tian, M., Deitrick, R., Auclair-Desrotour, P., Kitzmann, D.,

Dorn, C., Mezger, K., & Heng, K. (2021). Lithologic controls on silicate weathering regimes of temperate planets. The Planetary Science Journal, 2, 1–29. https://doi.org/10.3847/PSJ/abcbcb.

4. Hodson, A., Porter, P., Lowe, A., & Mumford, P. (2002). Chemical denudation and silicate weathering in Himalayan glacier basins: Batura Glacier, Pakistan. Journal of Hydrology, 262, 193–208. https://doi.org/10.1016/S0022-1694(02)00036-7.
5. Morton, A. C., Todd, S. P., & Haughton, P. D. W. (1991). Developments in Sedimentary Provenance Studies (Special Publication 57). Geological Society Publishing House.
6. Pettijohn, F. J. (1975). Sedimentary Rocks (3rd ed.). Harper and Row.
7. Reading, H. G. (1996). Sedimentary Environments: Processes, Facies and Stratigraphy. Blackwell Scientific Publications.
8. Sun, S. S., & McDonough, W. F. (1989). Chemical and isotopic systematics of oceanic basalts: Implications for mantle composition and processes. In Geological Society of London Special Publications (Vol. 42, pp. 313–345). Geological Society.
9. Torres, M. A., West, A. J., Clark, K. E., Paris, G., Bouchez, J., Ponton, C., Feakins, S. J., Galy, V., & Adkins, J. F. (2016). The acid and alkalinity budgets of weathering in the Andes–Amazon system: Insights into the erosional control of global biogeochemical cycles. Earth and Planetary Science Letters, 450, 381–391. https://doi.org/10.1016/j.epsl.2016.06.003.
10. Wüst, R. A. J., & Bustin, R. M. (2003). Kerogen. In G. V. Middleton (Ed.), Encyclopedia of Sediments and Sedimentary Rocks (pp. 400–403). Kluwer Academic Publishers.

CHAPTER

6

Continental Environments and Sources of Sediment

LEARNING OBJECTIVES

After studying this chapter, you will be able to:

- Define continental environments;
- Explain continental landform;
- Discuss from the source of sediment to the formation of strata;
- Describe mountain-building processes;
- Explain about global climate;
- Define weathering processes;
- Focus on erosion and transport;
- Illustrate denudation and landscape evolution;
- Describe tectonics and denudation;
- Discuss measuring rates of denudation.

KEY TERMS FROM THIS CHAPTER

Bedrock	Clastic detritus
Clastic sediment	Continental land surface
Continental landform	Denudational system
Erosion	Glacial meltwater
Lithologies	Lycopods
Orogenic belts	Orogenic system
Plate-tectonic subduction system	Sediment

INTRODUCTION

Continental environments are diverse terrestrial landscapes that play a crucial role in the generation and distribution of sediments. These environments include mountains, rivers, deserts, glaciers, and lakes, each contributing uniquely to sediment formation and transport. Understanding these environments is essential for comprehending the geological processes that shape the Earth's surface and influence sedimentary deposits.

Mountains, as dynamic continental environments, are significant sources of sediment. The processes of weathering and erosion break down rocks into smaller particles, which are then transported by gravity, water, or wind. Mountain streams and rivers carry these sediments downstream, where they are deposited in various forms such as alluvial fans, riverbanks, and floodplains. The constant uplift and erosion of mountain ranges ensure a continuous supply of sediments to adjacent lowland areas.

Rivers are vital conduits for sediment transport across continental landscapes. They erode materials from their banks and beds, carrying sediments over long distances before depositing them in deltas, estuaries, and coastal plains. The sediment load of a river can vary depending on the geology, climate, and vegetation of the drainage basin. Flood events often lead to significant sediment redistribution, shaping the riverine landscapes and creating fertile floodplains.

Deserts, characterized by arid conditions and sparse vegetation, are another important sediment source. Wind action, or aeolian processes, dominates sediment transport in deserts. Sand dunes and loess deposits are common features formed by the accumulation of wind-blown sediments. These environments are highly dynamic, with shifting sands continually reshaping the landscape.

Glaciers, found in polar and high-altitude regions, are powerful agents of erosion and sediment transport. As glaciers move, they grind underlying rocks into fine particles known as glacial till. When glaciers melt, they release these sediments, which are transported by meltwater streams and deposited in outwash plains and glacial lakes.

Lakes, acting as sediment traps, accumulate materials from surrounding catchments. Sediments in lakes can originate from rivers, glacial meltwater, or direct runoff from the land. The stratified deposits in lake beds provide valuable records of past climatic and environmental conditions.

Continental environments such as mountains, rivers, deserts, glaciers, and lakes are integral to the sediment cycle. They generate, transport, and deposit sediments through various geological processes, shaping the Earth's surface and influencing sedimentary records. Understanding these environments helps geologists reconstruct past environments and predict future changes in landscape dynamics.

6.1. CONTINENTAL ENVIRONMENTS

The collection of physical, chemical, and biological characteristics and processes related to sediment deposition that result in a unique suite of sedimentary rocks is known as a depositional environment. Geologists interpret sedimentary environments using cues found within, such as rock types, sedimentary structures, fossils, and trace fossils. Then, we can reconstruct ancient environments by comparing these hints to contemporary settings. The many depositional environments on Earth can be divided into three categories: common environments found on land (continental environments), common environments found in the ocean (marine environments), and common environments found at the interface between the two (transitional environments).

The environments found on the continents are diverse, but once more, we are only able to explore those where sediment deposition predominates over erosion. Although continents are topographically elevated overall in comparison to oceans, erosion occurs in high-altitude regions, and distinct depositional properties can be found in a number of different areas within a continent. Clastic sedimentary rocks predominate in continental depositional environments, primarily due to their close proximity to the sediment source.

The weathering and erosion caused by glaciers and glacial meltwater primarily regulate the environments found in glacial depositional settings. High latitudes and/or high elevations are the most common locations for glaciers. Although glaciers typically only move a few centimeters per day and cover small distances from their source, they can significantly accelerate mechanical weathering. Glaciers pulverize and crush rock, resulting in moraines—piles of sediment that have been improperly sorted. A system of braided rivers is created by the varying amounts of melted water that glaciers produce and the flow through the moraines. These rivers transport the fine sediments farther from the glacier terminus into an area known as the outwash plain, where the sediment is ill-sorted.

The term "fluvial depositional environments" refers to sediments that are deposited by rivers. A river's discharge and gradient have a significant impact on how the river flows, deposits sediment, and takes on its overall shape. Rivers sort the sediment because they can only carry so much particle size before changing it chemically and physically. There are various types of sediments found in meandering rivers, ranging from stones and pebbles in the channel to sandbars at the outer edge where the water slows. Additionally, we observe a variety of sedimentary structures connected to both flood events and the river's flow.

Additionally, there are sediments found in lake sedimentation environments or lacustrine depositional environments. Because lakes do not have swiftly moving water like rivers do, there is a lot less movement of sediment. Lake sediment can build up from a variety of sources, such as rivers that flow into the lake, rainwater that carries sediment into the lake from the shores, and wind-borne sediment. We observe thin layers of fine sediment with different concentrations of trace fossils because the sediment is undisturbed once it reaches the lake.

Last but not least is the eolian depositional environment, where wind currents—rather than water—dominate. Only smaller particles can be carried because air is less dense than water. Furthermore, unlike water, wind is not constrained by specific channels, so eolian deposits have more characteristics than fluvial deposits. There are some places where the predominant wind patterns are fairly consistent in terms of both direction and strength, and this can lead to the formation of notable sedimentary structures. When water is present in these settings, it frequently takes the form of seasonal lakes that evaporatively lose a significant amount of their water content, occasionally leaving behind chemical sedimentary rocks known as evaporite deposits.

6.1.1. Transitional Environments

Winds, waves, tides, rivers, and ocean currents can all have an impact on the complex regions that make up the interface between continents and oceans. Furthermore, a combination of materials from the ocean (chemical and biochemical) and the continents (clastic and organic) make up the sediments that are found in these areas. Lastly, there are a lot of sedimentary structures and trace fossils in these areas due to the abundance of life, abundant air and water currents, and sediments influenced by the elements. But we can identify transitional zones where various forces like rivers, tides, and ocean currents predominate.

Coastal areas with strong tidal currents are known as tidal mudflat depositional environments. These currents are caused by the gravitational forces of the moon and the rotation of the earth. In areas with high tidal currents and shallow seafloors, large portions of the shoreline may be underwater at high tide and exposed at low tide. These areas typically have finer sediment than other shorelines because the tidal currents bring in marine sediments. The constant bidirectional currents, daily exposure and drying out, harsh environmental conditions, and rich biodiversity serve as distinctive traits of these unique environments.

Beach depositional environments are shorelines where ocean currents predominate. Because of the differences in how the land and water heat and cool throughout the day, shorelines have constant wind both on and offshore. These winds create the famous waves that you see at the beach, but when the waves come ashore, they curve to match the shore's shape, creating a current that runs parallel to the shore. Sand is carried and deposited along the beach by this current. Furthermore, dunes created by the wind support a complex and varied ecosystem. Deltas are the places in beach depositional environments where rivers empties into the sea. The water of a river that is carrying material slows down and deposits sediment into the ocean. The majority of the sediment is deposited at the river's mouth, with some also seeping into the surrounding areas to form a recognizable fan of sediment. The delta is primarily a thick sequence of clastic sediment with signs of the river's strong flow because the sediment is coming from the river.

6.1.2. Marine Environments

Although there are many differences between marine depositional environments, the production of rocks is influenced by several factors, including the depth of the water, the organism community in the area, and the supply and proximity of continental sediment. An environment's concentration of chemical and biological sedimentary rocks that are formed in the ocean will increase with distance from the shore, while clastic sediment will be less prevalent. Furthermore, certain organisms can generate enormous amounts of skeletal material given the correct environmental circumstances.

Areas near the coast that are constantly submerged are known as shallow marine depositional environments. These regions contain a large amount of mature clastic sediment, marine algae (seagrass, for example), and animal skeletons from mollusks (clams and snails), echinoderms (sea urchins, sand dollars), and coral. The energy level of these regions can vary greatly, from very shallow places affected by waves to deeper areas only affected by strong storms. The community of organisms and different types of trace fossils, which can be highly sensitive to depth, as well as the sedimentary structures, can frequently be used to determine a better understanding of the relative depth. Reefs are often found in warm tropical shallow waters, forming through the growth of coral colonies that build large three-dimensional structures from calcite skeletons. Corals can thrive in various marine environments, but they can only create reefs when the symbiotic algae within their tentacles can effectively photosynthesize, providing more energy for faster growth. These reefs act as barriers between shallow-water environments and ocean currents, creating lagoons – shallow, low-energy areas. Lagoons contain thin layers of fine sediment typical of quiet water, as well as chemical sediment deposits resulting from evaporation.

Environments of deep-marine depositional systems make up the majority of the ocean. Apart from the dust carried by the wind, most clastic sediments are not present in these areas. As a result, the ocean is producing the sediment both chemically and biologically. Skeletal remains from some of the tiniest marine creatures are the main source of sediment in these deep environments. Both calcite and silica are used to make the shells of many kinds of unicellular organisms. The majority of these shells are made in surface waters that receive sunlight, which enables photosynthesis. These shells fall into deeper water after the organism dies, and the slow accumulation of sediment creates fine layers of biochemical sedimentary rocks. Sometimes, these shells dissolve or change before they fall to the bottom—which may be miles away—and then precipitate as sedimentary rocks with chemical properties. Given the gradient of the ocean floor, it is evident that there is no distinct line separating shallow and deep-water environments. Deep marine depositional environments are typically located thousands of feet below the surface, out of the reach of even the strongest storms (Ollier, 1984; Anderson & Anderson, 2010).

6.2. CONTINENTAL LANDFORM

A continental landform is any notable topographic feature found on the Earth's largest landmasses. Examples that are commonly encountered are valleys, plateaus, and mountains, including volcanic cones. (Also, similar features found on the ocean floor, such as seamounts, mid-oceanic ridges, and submarine canyons, can also be referred to as landforms. These structures are made distinctive by the tectonic processes that produce them as well as the weather-regulated deformational processes that alter them over time. The resulting topographic features typically exhibit characteristics of both the deudational and tectonic processes.

Mountainous terrain, which is either formed somewhat farther inland by rifting and faulting or along continental margins by collisions between the slablike plates that make up Earth's lithosphere, is the most dramatic expression of tectonism. The large continental regions with little relief and elevation that are impacted by mild uplift, subsidence, tilting, and warping are examples of far more subtle tectonic expressions. The tectonic "stage set" is subject to denudational processes, which have the power to alter its characteristics to the extent that reflects the forces that have predominated over time. As a syn-tectonic phenomenon, volcanism has the power to alter any landscape through vent eruptions that result in individual volcanoes or fissure-erupted flood basalts that can create regional lava plateaus. The climate, whose fluctuations in heat and moisture produce vegetated, desert, or glacial expressions, controls the characteristics of the denudational processes, which include rock weathering and the erosion and deposition of rock debris. Rather than experiencing a single, long-lasting state, the majority of regions have experienced numerous climate changes. Climates can shift considerably more quickly due to variations in elements like solar radiation and much more slowly due to continental drift.

A given landscape is typically the result of a confluence of the aforementioned factors. The complete suite of landforms may occasionally be under the control of volcanism, tectonism, or a unique combination of denudational processes. The type of denudation determines the high-elevation topography where tectonism manifests as orogenic uplift. The rocks are carved into mountainous relief akin to that of the southern Andes or the Alps in humid or glacial environments where geomorphic agencies can take advantage of lithologic variations. The processes of aggradation and planation in arid orogenic settings frequently produce alluviated intermontane basins that merge with high plateaus broken or surrounded by mountains, like the central Andes or Tibet and Colorado in the western United States.

Denudational processes act on rocks that are only moderately elevated and, if sedimentary, slightly deformed in continental regions where mountainous uplifts are absent. Wide basins, ramps, swells, and plains are the results of this. In environments with rain and rivers, these are most thoroughly examined (occasionally achieving localized relief in the form of mountains on uplifts). In other places, they might be extensively scoured and aggraded where glacial, or broadly alluviated and pedimented where primarily arid.

On top of the major features already mentioned, there are smaller denudational landforms. They include pediments, pans, dune complexes, dry washes, alluvial veneers, bajadas, and fans in areas where aridity has predominated. In areas where humidity has predominated, ridge-ravine topography, integrated drainage networks, and related thick soils are found. Anywhere that arid and humid conditions have alternated, combinations of these features are common, and either category may laterally merge with the intricate suite of erosional and depositional landforms created by continental glaciers at higher latitudes.

6.2.1. General Observations

6.2.1.1. Basic Concepts and Considerations

The term "landform evolution" refers to the gradual alteration of topography from a designated initial morphology to a modified form. Only energy that is available to perform work within the relevant geomorphic system can cause changes, and it follows that evolution will come to an end when that energy is depleted or can no longer be effectively used to create new changes. After that, there won't be any topographic changes in the latter steady state, or dynamic equilibrium, condition until the dominant conditions end or are disturbed, allowing for the start of a new evolutionary sequence.

- The hills are shadows, and they flow
- From form to form, and nothing stands;
- They melt like mist, the solid lands,
- Like clouds, they shape themselves and go.

Tennyson's poetry effectively conveys the geomorphic requirements of time and shifting landforms. It has long been known that, given enough time and in the absence of opposing forces, gravity would eventually pull the roughness of the Earth's surface down to form a featureless subaqueous spheroid. Even the ancients were well aware of the ongoing effects of gravity. Such an evolution would be incredibly simple, and it might even predict what will happen to terrestrial landforms in the future—that is, when internal processes that produce relief stop working billions of years from now in response to the system's increasing entropy.

Even today, the lands have been reduced by denudation to low, frequently nearly featureless plains in areas where the mechanisms responsible for elevating and providing relief have not functioned for several hundreds of millions of years. However, it is evident that the possibility of a periodic regeneration of continental elevations, especially of large-scale relief features, must be included in any contemporary theory of landform evolution. Because of the current rates of erosional destruction, there wouldn't be any continents or mountains today if there wasn't any regeneration.

Landscape evolution theory has a history of evolving ideas in response to fresh data showing ever-increasing complexity.

The way scientists and thinkers have approached the Earth's processes that defy gravity and alter land elevation and roughness makes this situation abundantly clear. Xenophanes of Colophon's writings implied the existence of these processes (c. 570–c. Around 478 BCE, Herodotus (c. Between 484 and 420 BCE), and 1452–1519 for Leonardo da Vinci. The ideas of relief genesis and continental renewal culminated in the theories of isostatic theory developed by George Biddel Airy and John Henry Pratt in England in the middle of the 1800s, and in the theories of plate tectonics proposed by Harry H. Hess and Robert S. Dietz from the US in the early 60s. The Earth's surface roughness periodically rising is a phenomenon that geologists continuously predict, accept, and gain more insight into.

Numerous other theories have arisen over time that have complicated geomorphic theory. Proposed by Georges-Louis Leclerc, comte de Buffon, circa 1750, were noteworthy theories regarding continental submergence by seas. These theories carried implications for changes in relative sea level and sedimentary leveling of submerged areas.

During the 19th century, there were theories that suggested glacial "drift" was caused by icebergs rafting gravel during Noah's Flood, which further complicated theoretical matters. The Noachian Deluge's geomorphic allure has largely faded since then. However, it is generally accepted that sedimentary deposits from ancient inland seas make up the majority of continental bedrock. In places like Australia, these deposits underlie and form enormous structural plains.

The understanding of the impact of volcanism on landforms dates back to the 1700s, but it was not fully incorporated into modern tectonic theories until 1961. In contrast, the study of climate's influence on landscapes began in the early 1800s with the introduction of glacial theory. However, the prevailing theory of landform evolution from the 19th century, as proposed by William Morris Davis, largely overlooked the role of non-glacial climates. It was not until around 1950 that research began to demonstrate the significant impact of a variety of climatic conditions on landscapes, showing that climate change is a frequent and powerful force in shaping the Earth's surface. This research has disproven the idea that climatic effects are accidental and has highlighted the importance of considering climate in geomorphic studies (Ollier, 1984; Anderson & Anderson, 2010).

6.2.2. Constraints on Modern Landform Theory

Instead of just recounting the haphazard evolution of geomorphic concepts since their inception approximately 200 years ago, it would be more appropriate to list here the rigorously established conditions that form the basis of any contemporary theory of landform evolution. After a brief summary of the earlier theorists' postulates, it will be clear what they achieved, disregarded, neglected to take into account, or were unaware of. In sum, a modern theory of landform evolution must contend with the following well-established factors:

- Continents consist of a craton of crystalline rocks 1 billion to 3 billion or more years old, have been periodically submerged by epicontinental seas, and are in most cases locally covered with veneers of nearly flat-lying sedimentary rocks.
- Where orogenic events were involved less than 500 million years ago, mountainous

elevations and relief containing deformed rocks exist on continents.

- Lowering of the land by denudational processes is accompanied by essentially continuous isostatic adjustment by load-compensating uplift.
- Mountainous relief of the continent-to-continent collision type (e.g., the Appalachian Mountains of eastern North America) can eventually be eliminated by erosion, whereas trench-type mountains (e.g., the Andes of western South America) probably cannot as long as the associated trench subduction system endures.
- Climates on lands vary through time in response to lateral continental drift of 0–12 centimeters (0–5 inches) per year. North America, for example, is moving northwest at a rate of about three centimeters per year. On the other hand, Antarctica is hardly moving and has been in a polar position undergoing glaciation for about 30 million years.
- Over most lands, climates also vary with atmospheric, oceanic, and solar factors in cycles lasting thousands of years (the Milankovitch solar radiation cycle, for example, has a duration of ±25,800 years).
- In select hydrographically favored sites on time scales not influenced by continental drift (e.g., Antarctica), climates on continents or portions thereof can remain essentially constant for periods of millions of years.
- Since geomorphic processes under arid, humid, glacial, and possibly other climate conditions can induce particular landforms, areas subject to periodic climate change often show polygenetic landform associations.
- Landforms exposed for millions of years to a constant environment may display a climax (steady-state) landform association that is essentially timeless and in which landform evolution through denudation is reduced to mere negative allotropic growth.
- Since volcanism is seemingly localized in accordance with mobile heat-dispersal patterns within the Earth, eruptive effects may be imposed on any surficial geomorphic system at any stage of development.
- Similarly, mobile tectonic patterns involving rock deformation may be brought to bear on any surficial geomorphic system, with resulting relief, elevation, and topographic changes.
- Impacts on planetary surfaces by falling meteoroids, asteroids, and cometary bodies are periodic but are capable of generating landforms of mountainous proportions.
- Surficial geomorphic agents

of denudation responsible for many, if not most, landforms include mass wasting, running water, glacial ice, and wind. They are not all of equal significance in every climatic setting, however.

- The geomorphic agents respond to various climates, changing in character and effect. They also respond in some degree to altered conditions of elevation and relief.
- The behavior of denudational agencies and related geomorphic processes is neither constant nor linear in nature. Rates vary from long-term, imperceptible, and gradual to brief, rapid, and catastrophic.
- Changes produced by geomorphic agents vary in magnitude but not directly with time—i.e., the same change involving the same energy expenditure may be either slow or fast. (Studies of river systems, for example, suggest that greater changes in channel morphology occur during brief infrequent floods than during protracted low-flow periods.)
- Perturbations in geomorphic processes or environments cause accelerated changes in most landform configurations, soils, and deposits, which eventually slow down as new equilibrium forms develop.
- A given landform or deposit is only stable in association with its formative process and environment, and in any subsequent alternative setting, it begins to change toward a new equilibrium morphology.
- In a denudational setting, slope as an influence over process rate may be subordinated to such factors as runoff volume, soil-moisture content, bedrock coherence, ground-cover type, channel roughness, channel cross-section, weathering type, sediment caliber, and sediment quantity.
- When the climate in a region changes, the elimination of relict landforms and deposits causes a disequilibrium phase, which is followed by a dynamic equilibrium phase as new geomorphic equilibria are established. The disequilibrium phase may range from a few score or 100 years for certain organic responses to many thousands of years for soil, hillslope, or drainage adjustments.
- Some landforms or deposits, once formed, strongly resist subsequent changes regardless of climatic history—e.g., entrenched meanders such as those that exist in parts of the Appalachians, chert felsenmeers (accumulations of rock blocks) like those in the southern Ozark region of the United States, and duricrusts of the type commonly found in Australia.
- No such thing as an "average" terrestrial climate seems to

exist, and certainly a climatic "norm" for one continental configuration would differ from that for another—e.g., the supercontinent Pangaea of pre-Cretaceous times (more than 146 million years ago) differed climatically from its subsequent fragments for both the Cretaceous (approximately 146 million to 65.5 million years ago) and the present.

- Sea level has been found wanting as a stable limiting datum for erosional processes or as an influence on stream behavior. Glacioeustatic fluctuations on the order of 130–150 meters appear to have been commonplace during the continental glacial sequences of the Carboniferous (approximately 359 million to 299 million years ago) and Pleistocene (approximately 2.6 million to 11,700 years ago) and at other times, and periodic desiccation of restricted ocean basins has occasionally permitted major rivers to deepen their courses thousands of metres below mean sea level.
- The Earth and the solar system as a whole are at least 4.5 billion years old. This is long enough for some geomorphic phenomena to occur several times but probably not long enough for others to happen even once. Certainly it is doubtful if more than nine collision-type mountain systems can have been eroded away in one spot, even if it were possible for them to form there.

6.2.3. Landform Theories of the 18th and 19th Centuries

6.2.3.1. Catastrophism

The German mineralogist Abraham Gottlob Werner was the main proponent of this theory in the late 18th and early 19th centuries. Werner claimed that all of the rocks on Earth were created by quick chemical precipitation from a "world ocean," which he then catastrophically disposed of. His catastrophic theory of Earth's changes had two main geomorphically significant effects, despite not being coherently focused on the origin of landforms. Firstly, it somewhat precipitated the Scottish scientist James Hutton's formulation of a less radical alternative viewpoint in 1785. Secondly, it was partially true: natural disasters do happen on Earth and they do alter its topography. Landforms can be created or destroyed by asteroid impacts, volcanic eruptions similar to Krakatoa, hurricanes, floods, and tectonic erosion of mountain systems, among other natural disasters. Not all change, though, is disastrous.

6.2.3.2. Uniformitarianism

The concept proposed by Hutton, that the Earth has evolved into its current state through ongoing processes, is now known as the doctrine of uniformitarianism. This doctrine focuses on geology rather than just geomorphology and is more concerned with surficial changes related to landforms compared to Werner's ideas. Hutton's concept laid the foundation for what is now known as process geomorphology, which emphasizes studying the effects of contemporary geologic forces like running water. Over time, the influence of moving ice, gravity, and wind in shaping valleys and hillslopes was also recognized. Uniformitarianism became a guiding principle for geologic historians like William Smith and Sir Charles Lyell in the 19th century, as they argued for incremental and gradual geological change. Lyell specifically needed a longer time scale to support his theory, which geologic historians were able to provide through their research.

6.2.3.3. Gradualism

It was firmly established by Lyell's gradualism concept and related process observation on an extended time scale that small forces operating continuously for extended periods of time could accomplish a great deal. Even modern thinking agrees with that conclusion. Still, Lyell took an extreme stance when he rejected virtually all abrupt and catastrophic-sounding geologic processes. Studies have indicated that changes can happen quickly as well as gradually.

Within the framework of Hutton's uniformitarianism and Lyell's gradualism, 19th-century geomorphologists achieved a great deal of noteworthy successes. The most famous of these was the research done by Grove K. on regional denudation in the American West and by Johann von Charpentier and Louis Agassiz on glacial phenomena in Europe. Gilbert along with Clarence E. Dutton, which highlighted the function of water in motion. The majority of the conclusions about glaciers remain valid, and Gilbert's hydraulic research served as a foundation for more recent theories. However, neither he nor Dutton offered thorough theoretical explanations of terrestrial morphogenesis on a scale comparable to that of the previously mentioned W. M. Davis (Ollier, 1984; Anderson & Anderson, 2010).

6.2.4. Davis's Erosion Cycle Theory

6.2.4.1. The Geographic Cycle

Beginning in 1899, Davis proposed that denudation of the land occurs in what he called "the geographical cycle." According to Davis, this cycle is initiated by an uplift of an area above sea level, followed by a wearing down of the surface through the action of running water and gravity until either the region is worn away (base leveled) or the events are interrupted by renewed uplift. It was further explained that such a cycle of erosion occurs under conditions of a rain-and-rivers environment (what present-day investigators would call a humid climate), which were assumed to reflect the normal climate for the Earth. The fact that Davis dismissed glacial phenomena as accidents of climate and viewed climatic areas as geographically fixed afforded his theory more latitude. Furthermore, Davis proposed the idea of a separate arid geographical cycle in 1905. In all cases, erosive power was presumed to be controlled primarily by slope; hence, the cyclic system was slowed down as the land was leveled and relief and elevation were diminished. The end point of a low-inclination landform was termed a peneplain, and it was said to be locally surmounted by erosionally resistant highs called monadnocks. The peneplain as a whole was presumed to be graded to regional base level (in all likelihood mean sea level) by denudational agencies (e.g., running water), which were supposedly controlled by this datum.

At least half of the 25 limitations on theories of landform evolution listed above conflict with the ideas in Davis' erosion cycle. There are three main factors that undermine the Davisian erosion cycle theory:

(1) the currently accepted requirement for constant isostatic uplift during erosion; (2) the variability of most lands' climate; and (3) the river hydraulic behavior that Gilbert observed, which limits base-level influences over interior slopes and prevents valley alluviation in normal humid conditions.

Within established bounds, the idea of an erosion cycle started by uplift remains viable. But unlike what Davis believed, such a cycle can only occur under very specific climatic conditions and in very restricted geographic and hydrographic settings. Additionally, Davis chose morphological groups of landforms as the chronological "mile posts" for his cycle of landform change (i.e., stages of development) have been discovered to comprise unique, typically polygenetic arrays of landscape features that have little to no sequential temporal significance and represent the interaction of multiple environments.

6.2.4.2. Davisian Dynamic Equilibrium

The concepts of process slowing in a late stage of process evolution as energy is consumed and process interruption as a way to accelerate change (rejuvenation) are also ideas that Davis contributed to the theory of landform evolution. The latter notion parallels contemporary theories of entropy relationships and is quite similar to the description of dynamic equilibrium, or the achievement of a steady-state (climax) environment.

Davis introduced his concept of landscape-development phases shortly after Charles Darwin's theory of organic evolution. His labels of "youth," "maturity," and "old age" for landscapes are clearly anthropomorphic. While these terms were popular at the time, their lack of chronological relevance may be why their continued use today is puzzling.

6.2.4.3. The Geomorphic Concepts of Penck and King

Davis's theoretical foundation for geomorphic evolution was expanded upon in a rather unique way by German scientist Walther Penck in 1924, and it was later supported with modifications by Lester C. in 1953. South Africa's King. Both Penck and King kept some Davisian devices, such as the graded stream, base-level control of erosion surfaces, and peneplain. Each theory used the idea of parallel retreat of fluvial-structural escarpments to create plains, and it punctuated the erosion cycle by starting fresh stream incision. The name "inselbergs" (separate hills rising above plains; the word comes from the German word for "island mountains"), not monadnocks, topped King's "pediplains," which he named after the planation process. The term Tectonic Geomorphic School has been applied to its proponents because the resulting stair-stepped landscapes (Treppen, the German word for "steps") of scarps and flats were thought to reflect tectonics and to be correlatable.

The theory of plate tectonics has greatly harmed geomorphologists' belief that deformational landforms reflecting tectonic pulses were globally synchronized enough to be correlatable. More evidence has been found to support the distinct theory that hillslopes, once developed, retreat laterally to create a low-inclination surface deserving of a unique name (pediment–pediplain).

Although Penck's Treppen concept is less clever overall, it appears to suffer from much of the same theoretical flaws as Davis's geographical cycle in retrospect. Penck and King, like Davis, did not utilize climate influences dynamically; in fact, King even went so far as to say that climate had no bearing on anything. Furthermore, King and Penck did not recognize the isostatic consequences of erosion that had been established almost a century earlier, similar to Davis. King proposed that sheetfloods "mold" the surfaces of pediments and labeled semiarid areas as places where this might be feasible based on their sparse vegetation. Recent research indicates that the "flat" terrain that sheet floods occur on may be a result of the floods rather than their cause. As will be discussed below, the so-called molding would seem to be the result of desert stream-flood processes operating to local base levels in the absence of significant plant cover.

Penck and King proposed an implied landform "chronology" for a geomorphic system associated with intermittent uplift, but dating such events is challenging. Moreover, King based his planation technique on a regional sea-level erosional datum, which is called into question by the previously mentioned constraints. The Penck-King theoretical ensemble's main contribution may lie in the idea of lateral escarpment retreat rather than Davis' preferred method of wearing down lands. It is a well-known fact that certain landforms serve as pediments. They have a planar shape, truncate many different kinds of bedrock, and are best understood in terms of scarp retreat in the absence of vegetation. The ideal amount of moisture to promote this process is still up for debate. However, the overall character of the mechanism appears to have been determined (primarily through comprehensive research in the field of process geomorphology), and the hydraulic limitations set forth by Gilbert and associates appear to be met.

Essentially, it has been discovered that runoff deposits sediment in deserts where volume loss from evaporation and infiltration absorbs the excess transport energy. The long-term locus of the deposit establishes the depositional base level, below which runoff upslope from that level cannot erode and form a deposit. The energy expended by such overland flow against non-vegetated hillslopes causes their backwearing.

The sedimentation phenomenon is undoubtedly one of the more perceptive geomorphic discoveries, even though the specific hydraulic and sedimentologic parameters were not determined until much later. The three most well-established morphogenetic systems on Earth today are this type of land planation in conjunction with alluvial aggradation in deserts, stream incision that creates regional drainage networks and enhances relief under humid conditions as described by Davis, and glacial scour and deposition as clarified by Charpentier, Agassiz, and others (Ollier, 1984; Anderson & Anderson, 2010).

6.2.5. Climatic Morphogenesis

6.2.5.1. Morphogenetic Area

Although there was evidence in the first 10 years of the 20th century that climate was a major dynamic factor in the evolution of landforms, formalized theory did not emerge until the mid-1900s. Julius Büdel, a German geographer, and a number of French geomorphologists, including Jean Tricart, André Cailleux, Jean C. Peltier introduced the idea of a morphogenetic area, which is characterized as an area where a specific set of landforms is developing under a specific climate. It was only gradually, primarily through research conducted in tropical regions, that the extent of the regional climate shifts from humid to arid on the various continents became apparent. Davis recognized the differences between the geomorphic mechanisms of arid and humid lands a long time ago. However, the recognition of the morphogenetic, or geomorphic, system was compelled by recent evidence of widespread geographic mobility for such environments. A system of this kind is described as a collection of organizations and procedures working together in a specific setting to create a landscape. Morphogenetic regions and their systems have the ability to push one another aside, which means that remnant soils, deposits, organisms, and other features would be left behind.

For geomorphologists, the finding of widespread climatic dynamism and the correlating recognition of plate-tectonic phenomena opened up entirely new theoretical possibilities. There was not a single theory of regional landform development in existence in 1950 that could have reasonably explained the limitations imposed by the new tectonic and climatic discoveries.

6.2.5.2. Interactions between Geomorphic Systems

Climates shift over time, forcing one of the previously mentioned geomorphic systems onto the remnants of another. Furthermore, each climatic type's areas export matter to nearby morphogenetic areas, changing the resulting landforms. For instance, deserts use eolian processes to export dust, and the ensuing deposits change the soil profiles in areas downwind, like the eastern United States, or actually form loess depositional landforms, like in China's Shansi Province, which is located in the lee of the Ordos Desert. Where their flow will not sustain their sediment transport to some distant ocean, river systems that originate in humid regions may encroach on downslope deserts to create alluvial riverine plains. On the other hand, when rivers' sediment loads and flows are adequate and the débouché (point of emergence) is protected, they create deltas in response to climate change. Glaciers undergo transformations on areas covered in ice before exporting their outwash deposits into the environment below them (Ollier, 1984; Anderson & Anderson, 2010).

6.2.6. Tectonic Geomorphology

Orogenic tectonism, which includes volcanism, adds its obvious dimensions of elevation and slope to any surficial environment it encounters, in addition to the usual climatic imprints. It is now evident that early orogenic realms provide gravitational opportunities for Earth sculpture that are rare in other places. In orogenic belts, the typical processes for simultaneously gradualistic denudation by ice, wind, and flowing water are set aside by relatively quick uplifts of material that range from nearly unconsolidated sediment to semicoherent but severely deformed masses of igneous and metamorphic rocks. In such circumstances, thousands of cubic kilometers of rock are torn loose by gravity and then fall and/or slide, frequently traveling hundreds of kilometers in a single "geologic instant" to a lower resting place that is occasionally lubricated by subaqueous avenues. This kind of event seems best described by the word catastrophic.

Thus, the shaping of the Earth is understood to involve more than just the slow removal of weathered debris by mechanisms governed by climatic regimes. Over 100 active volcanoes are reported to exist on the Kamchatka Peninsula, located in the far eastern region of Siberia. It should come as no surprise that volcanic landforms predominate in the area. In an almost entirely tectonic landscape, the Afar Triangle at the base of the Red Sea is shaped by recently formed faults that cut through unweathered basaltic lava flows on a recently emerging seafloor. Ridge-ravine topography can be found in the Appalachians, south of the glaciated knobs. The ancient mountain system is covered in thick saprolitic soils on its upper slopes, and it may have been in a humid climatic nucleus for 100 million years. However, after an arid-to-humid climate shift at the end of the Jurassic Period (roughly 145–55 million years ago), the same region still has water gaps and entrenched meanders that echo drainage patterns established long ago, most likely on alluvial cover masses of Early Mesozoic age (roughly 225 million years old). In the same region, weathering conditions that were established 150 million years ago, when southeast North America was still in the tropics before recent northwesterly continental drift, are reflected in the tropical soils and ridge-top lateritic deposits of Georgia and Alabama.

6.2.6.1. Consideration of Unique Landforms and Other Distinctive Topographic Features

Of course, there are situations in which particular bedrock types react differently with specific weathering and erosion processes to create distinctive landscapes and landforms. The solutional effects represented as karst topography are arguably the most well-known. This is most noticeable in limestone regions, like those in northeast China, parts of northeastern China, and on islands like Puerto Rico and Jamaica. Other examples of limestone terrain include the Karst plateau in Yugoslavia and Kentucky in the United States. Similar landforms may form on other types of sedimentary rock or on igneous or metamorphic types in tropical regions where silica is more soluble. One example of such a landform is quartzite, which is found in the isolated plateau remnants of the Venezuelan Guiana Shield. In such tropical regions, it is easy to observe the humid climatic conditions that encourage solution production and dripstone formation.

As the Australian geomorphologist C. pointed out, granitic terrain also gives rise to a distinctive array of landforms, including domed erosion residuals, which frequently occur in patterns closely tied to joint spacing in bedrock. R. Twidale. Massive boulder piles of excavated core stones can be found in areas where human activity or alternating humid and dry climates have caused erosional stripping of weathered zones. These characteristics are particularly noteworthy in the St. Helena Islands, southern Brazil, parts of India and Australia, and the island of Hong Kong. The American state of Missouri contains the Francois Mountain region.

The complexities of terrestrial surface change demand a theoretical overview that is both flexible and multifaceted. Oversimplified, sweeping landscape generalizations that apply to the whole Earth such as the postulates of Davis and King can hardly be employed when dealing with a planet where virtually every geomorphic element constitutes a potential interruption or complication to every other system. Nevertheless, there do seem to be certain kinds of activity that are repeated sporadically in both tectonic and climatic realms. These repetitions encourage the re-creation of particular suites of landforms and could be taken to imply a certain rationality to events. However, they probably are no more rational than eddies in a river that develop only where possible.

When determining what is truly geomorphically repeatable and possible, issues of geographic and chronological scale also come into play. Lighter materials that make up the lithosphere and the Earth's core, once formed, must remain firmly fixed due to the interaction between density variations in matter and gravity. A discussion of landform evolution must start at this level since the concentration of the least dense solids in the continents is part of a complex process that is currently linked to plate tectonics.

One of the fundamental effects of the plate tectonics theory is that continents are landforms of a particular type, even though this was not specifically mentioned by the theory's creators. First-order landforms, continents appear to have only one cycle of continental denudation in Earth's history. It started with the first surface concentration of continental lithosphere, and as previously mentioned, it is likely to conclude when the final endogenic forces (i.e., those inside the Earth) run out and gravity and entropy take control as the planet's internal systems break down. In the context of this cycle, which lasts between 8 and 10 billion years, the specifics are essentially meaningless because the outcome is predetermined—unless, that is, the Sun goes nova and causes havoc.

The majority of second-order features on continents are mountains and the comparatively low-elevation regions that are created as the mountains rise. Mountains and the geomorphic systems that act upon them are distinct among continental landforms in that the uplift generates a far greater excess of potential energy than the remaining land area. The evolution of landforms in mountains is inherently distorted by this unique form of excess energy. Although Davis appeared to recognize this in his theories, he was unaware of the range of climatic and tectonic forces at play as well as the limitations of slope as a deformational influence.

6.2.7. Orogenic and Epeirogenic Morphogenesis

6.2.7.1. Orogenic Geomorphic Systems

Different types of mountain-building systems develop under specific circumstances related to their type, location, and characteristics. The main categories recognized are (1) mountains resulting from the collision of continents, formed through interactions of lithospheric plates along continental edges, (2) mountains formed by collisions near oceanic trenches, typically along a single continental margin with nearby plate-tectonic subduction, and (3) rift-type mountains that extend into continental interiors, where transcurrent faults impact cratons and deform sediment layers, or where spreading zones create fault-block mountainous landscapes. Over geological time, multiple mountain-building events of each type have occurred, with unique rules governing the geomorphic development of each.

Mountains of the continent-continent collision type have special attributes that direct their geomorphic evolution. These distinctive characteristics are the following:

- The collision creating the mountains incorporates a finite volume of rock that is not augmented following the collision.
- The orogenic rock mass is subject to isostatic uplift during denudation; in general, sedimentary rock types are exposed first, followed by crystalline varieties.
- The collision that initiates such orogenesis ultimately adds rock to the adjacent craton, and in thickening the adjacent crust often initiates nearby cratonic tilting and/or uplift.
- Because such mountains develop between continents and are thus elevated in

the midst of a consequent megacontinent (Pangaea in the case of the Appalachians), they are far from oceanic evaporation sources and therefore often undergo initial denudation under arid geomorphic systems in the manner of the present mountains of central Asia.

- As the climatic setting of such mountains is largely established tectonically, it may endure in the same climate for scores of millions of years and, as noted in 1901 by the American geomorphologist Douglas W. Johnson, a desert mountain range tends to bury itself in its own waste.
- Re-exposure of such mountains to nearby precipitation sources by plate adjustments may result in dramatic climate changes from arid to humid so that perennial fluvial erosion is widely initiated on a relict arid, alluvial cover mass with resulting transverse drainage by superimposition. In illustration, one can compare the Appalachian Mountains of North America and the Zagros Mountains of Iran, as described by the American geomorphologist Theodore M. Oberlander in 1965.
- Because of their finite initial rock volume, mountains of the continent-continent collision type can be lowered by erosion, somewhat in the manner visualized by Davis. No such structures more than 500 million years old show mountainous relief.
- Volcanic landforms are rarely a part of the topography during orogenesis of this mountain type.

Mountains of the collision type associated with oceanic trenches have their own distinct attributes that control evolution. These are as follows:

- The merging of a pair of lithospheric plates along a deep-sea trench initiates orogenesis tied to the subduction process (i.e., the sinking of one plate beneath another at convergent plate boundaries).
- Rock mass is added to the orogenic belt via subduction as long as the trench remains "operational."
- Denudation accompanies uplift and may reduce rock mass in the orogenic system in the long run, but whether the total mass is growing, shrinking, or static depends on the budget established by additions from subduction versus losses from erosion.
- Mountainous elevations tend to increase through much of the life of the orogenic system, since rock lost through erosion is generally removed locally and linearly by rivers and glaciers (the Andes exemplify the type bordering a continent, and they appear to be higher now than at any time since they began to form 150 million years ago).

- Because mountains of the trench-associated subduction type develop and endure adjacent to an ocean on at least one side, they are subject to climatic variability tied to such factors as latitudinal position, orientation with respect to prevailing wind patterns, ocean surface temperatures, and progressively increasing elevations.
- Examples such as the Andes that border a continent can show alternating segments that are highly volcanic.
- Andean types also may display highly contrasting denudational systems under a variety of climatic conditions on opposite sides as well as along the length of the range.
- Although an erosion cycle resulting in the overall lowering of a trench-associated mountain system does not appear viable as long as the trench endures, a complex steady-state mass situation would seem to be one potential development during this time.
- Occasionally orogenesis related to trench-continent interaction may extend far inland; the parts of the Andes exhibiting this trait display mechanical rock deformation but little volcanism, and a similar genetic mechanism has been suggested for the Rocky Mountains of North America.
- During their early years, the Rocky Mountains displayed volcanic phases accompanied by upthrusting but now seem tectonically quiescent and are apparently experiencing denudational lowering.

Rift-type mountains are primarily of the block-fault variety. They have the following set of special attributes:

- Block-fault mountains appear to originate where a spreading ridge of the plate-tectonic type develops.
- On continents, the spreading is expressed in high-angle faulting and may be accompanied by volcanism of tholeiitic basalt type.
- Rifting may be limited to linear zones, as in the Rift Valley system of East Africa, or maybe more broadly expressed, as in the Basin and Range Province of the western United States.
- The extent of rifting may be limited to mere surficial fracturing of the continental crust, or it may extend to actual rupturing of a lithospheric plate and renewal of seafloor spreading, as occurred along the Atlantic seaboard of North America at the end of the Jurassic.
- Because block-fault mountains are of endogenic origin, they may occur in and experience a variety of denudational environments. The examples from Africa and North America cited above are in settings ranging from arid to humid. The highest such mountains show glacial effects.

6.2.7.2. Climatically-Dominated Epeirogenic Realms

The areas of a continent that are epeirogenic (i.e., those that have avoided orogenesis during the last 500 million years) undergo denudation in a scenario where the slope factor, if it is tectonic in origin, is localized and so mild as to have minimal impact other than directing the flow of ice or water. These are the same regions that reveal, in shield areas, the roots of worn-down mountain systems or exhibit varying degrees of sedimentary rock veneers that have mostly accumulated in epicontinental seas over the past 500 million years. It is not surprising to find that climate dominates morphogenesis on stable cratons in the absence of significant tectonism. Large stretches of cratons located away from mountain belts are either scorched by wind and heat from the desert or are home to tropical and temperate forests and grasslands. Currently, only Antarctica is home to a continental ice sheet, though evidence suggests that North America and Eurasia may have once supported one as well. Climate change most noticeably interrupts morphogenesis in these epeirogenic regions. Nearly all of the landforms are polygenetic. The early stages of soil development that started under humid conditions just a few 1,000 years ago are hardly visible in many of the most recent glacial deposits. Furthermore, patches of cacti and alluvium that were left behind when the areas were deserts can still be seen in widely forested, humid regions. The prominent slopes there are all dendrogenic in origin; the steeper ones were typically created by stream incision, while the gentler ones were typically created by pedimentation and/or alluviation (Ollier, 1984; Anderson & Anderson, 2010).

6.2.8. A Unified Landform Theory

When considered comprehensively, the various theories related to landform development that have been discussed thus far—such as orogenic and epeirogenic morphogenesis, uniformitarianism, gradualism, erosion cycle, dynamic equilibrium, disequilibrium, geomorphic system, morphogenetic area, and tectonic geomorphology—have up until now been treated by theorists as distinct conceptual constructs rather than as geomorphic components of a single, all-encompassing theory. This circumstance bears a striking resemblance to the story of the several blind men who identified an elephant by touching only specific parts of the creature. While there is some merit to each of their geomorphic theories, the earlier theories were developed using incredibly scant data. The theoretical components start to make more intuitive sense when viewed in the context of the entire solar system, which is home to a collection of planetary geomorphic entities. The planetary location and early differentiation of cosmic material impose a degree of variability, but the directional factors imposed by radiation, gravity, and increasing entropy leave the solar system with an incomplete degree of randomness. There are two possible geomorphic factors for any given planet: (1) endogenic phenomena, which are related to internal heating and manifest as tectonism and volcanism, as on Earth, or radiation phenomena primarily tied to the Sun and resulting in climatic influences and biological activity; or (2) exogenic impact phenomena from solar debris that may have been modified by tidal disruption caused by nearby planetoids. Morphogenesis takes place as a result of interactions between planetary subsystems connected to the aforementioned elements.

6.2.8.1. Behavior of Geomorphic Systems

Geomorphic systems driven by gravity have the potential to be cyclical in the sense that they eliminate excess elevation and relief. They show two phases of activity graphing: an initial disequilibrium where relief and free energy are maximal (and often catastrophic), and a subsequent dynamic equilibrium where relief and elevation are almost completely eliminated and the amount of free energy available to perform work is so low that change is almost imperceptible. It is evident that the latter behavior is gradualistic. For the cycle to be broken or restarted in such systems, external disturbances are necessary.

The solar system's accretionary, gravity-driven impact morphogenesis cycle, which produces high relief and cratered surfaces, is clearly coming to an end. It appears that this activity peaked in the first billion years following the formation of the planetary system and is unlikely to resume. The surface of bodies like the Moon and the planet Mercury, where endogenic tectonic forces are almost nonexistent and impact effects are mostly preserved, is a prime example of its expression. Internal heating drives orogenesis on Earth and a few other planets (or satellites), reviving gravity-driven geomorphic cycles on a regular basis. There will only be one cycle of continent formation in Earth's history, as was previously mentioned.

Radiation-driven geomorphic systems are dependent on the fluctuations in the Sun's nuclear fusion processes. Solar effects are primarily manifested on Earth as morphogenetic areas, which are distinguished by a specific climate and related processes, due to atmosphere and organisms. When the morphogenic areas shift, remnant features exposed to the system are destroyed, and new deposits and landforms that are in morphological balance with the new system are created. These geomorphic changes in these areas are primarily cyclical. Landform, deposit, and process changes also show a two-phase graph following the start of a system or following a disturbance in an existing one. These landform alterations are initially indicative of time, and the partially altered relict features may allow reconstruction of landform evolution events, provided that morphogenesis has not reached a dynamic equilibrium phase.

As can be seen from the above, in the dynamic equilibrium phase of radiationally driven geomorphic systems, form and process are closely related. Form (which is heavily influenced by relict features) may show little to no consistency with the process in morphogenetic areas in states of disequilibrium; the process may have just begun. Hybrid features are remnants of past processes of change, like a desert or a glacial alluvial deposit in a valley being reshaped by a perennial stream. In contrast, the stream valley of Davis's mature stream, which was previously discussed, has a level floor, whereas the cross profile of a late-phase humid valley is V-shaped. In addition, the alluvium is altered after being partially reworked by the "hybrid" stream, and it does not act as it would if there were no alluvium.

A form that is alien to the related geomorphic system and processes can occasionally be preserved by a series of geomorphic events working together. Rather than expressing the current flow patterns, the winding routes of entrenched meanders carved into bedrock in areas like the Appalachians represent the granular surface and sediment-water volume relations that were in place when the flow pattern was first initiated in the Mesozoic.

6.2.8.2. The Concept of Periodic Random Dominance

On Earth, the interactions between gravity- and radiation-driven geomorphic systems can occur independently, leading to a mix of activities dominated by one or the other at random intervals. This can result in peak energy expenditures from each system happening in different geographic locations. The most significant changes in landforms occur where active mountain-building intersects with changing climates, while minimal changes occur in areas where stable land surfaces are affected by ongoing morphogenic processes. This balance of geomorphic systems allows for both sudden catastrophic events and gradual changes over time. Different forms of erosion cycles and dynamic equilibrium play important roles in shaping Earth's surface in various ways, whether as a result of long-term climate influences or the gradual reduction of relief and elevation due to erosion following mountain-building events. The idea of polygenetic landforms and landscapes where geomorphic system dominance fails to develop is implied by the concept of periodic random dominance as a feature of landform evolution. In fact, because dominance depends on a specific combination of tectonic and/or climatic factors over an extended period of time in a particular area, it becomes the special case. According to one estimate, polygenetic landforms cover about 80% of the land area on Earth. Less than 10% of the region experiences any kind of geomorphic system dominance; this number drops to 20% when Antarctica is taken out of the equation.

6.2.8.3. Process Geomorphology and Systems Equilibria

Process behavior and terrain response are key factors in determining the specifics of landform evolution within a given geomorphic system. When it comes to the dominance of geomorphic systems over systemic alternation, there are two general scenarios: (1) agencies that are working with remnants that they are altering, frequently very quickly; and (2) agencies that are working with equilibrium features that they have produced but are either unable or not able to alter further. In any given geomorphic system, the four main surficial geomorphic agencies on Earth—wind, running water, glacial ice, and gravity—induce processes that tend to evolve toward a situation of least work. Polygenetic terrain is typically a mix of hillslopes and "flats," and depending on whether the system tends to produce relief or reduce it, either topographic type may predominate in the later stages of a geomorphic cycle.

Since natural geomorphic systems on Earth's surface rely on outside energy sources for their operation, they are categorized as open. Since the rates of exogenetic and endogenetic energy input differ, there are changes in the coordinate agencies that are comparable to power surges in an electrical system. Rivers that receive too much runoff consequently occasionally flood. Storms are the result of the transfer of excess heat that the atmosphere accumulates locally. Glaciers, which are typically the definition of slowness, can experience an excess of mass and energy and subsequently rise. The energy required for sediment deposition, transportation, and erosion varies substantially over time in all cases. Additionally, turbulence, eddy formation, shearing, and vortex activity, and periodic local stagnation are produced by the interaction of solids, fluids, and gases.

Process associations within individual geomorphic systems respond to the aforementioned scenarios by displaying typical systems phenomena, such as "feedback," "threshold reactions," and evolution toward dynamic equilibrium (least-work) modes. Processes can switch back and forth between disequilibrium and steady-state conditions quite frequently in systems that experience periodic perturbations.

It's possible that an individual agency's actions and apparent process direction don't correspond with how the geomorphic system as a whole has evolved. For instance, a 10,000-year-long episode that resulted in the formation of an alluvial fan may be observed to have involved multiple fan-head trenching incidents, each of which was destructive in and of itself, but was subordinated to the dominant depositional events of the trend. A river like the Mississippi may also be observed depositing gravel on point bars inside bends as it reworks an old alluvial deposit in a valley. However, unless there is a systemic disruption, the long-term effect of the river's activity will be the removal of all the alluvial deposit in its path, including the point bars. Naturally, through engineering changes, humanity has "short-circuited" the Mississippi's and many other rivers' natural evolutionary processes).

It is apparent from the above that understanding the nature and function of remnant deposits and landforms is essential to understanding the direction of landform evolution through the study of geomorphic processes. This is a glaring flaw in the uniformitarianism theory put forth by Hutton.

Under specific circumstances, the idea of periodic geomorphic system dominance offers the logical potential conclusion for the evolution of landforms. It should produce either altered or unaltered tectonic landscapes. In turn, these could be epeirogenic or orogenic. When altered, they might exhibit glacial, arid, or humid morphogenesis in addition to marine influences. On the other hand, in regions where polygenetic morphogenesis is more prevalent, a combination of tectonic, marine, or climatic factors are imposed upon the environment, leading to the formation of a hybrid suite of landforms.

6.3. FROM SOURCE OF SEDIMENT TO FORMATION OF STRATA

The bedrock that is exposed on the continents is the ultimate source of sediment in the formation of sediments and sedimentary rocks. The uplift of pre-existing bedrock with igneous, metamorphic, or sedimentary origins serves as the foundation. After being raised, the bedrock experiences weathering at the surface, resulting in the formation of clastic detritus and the release of ions into the surface and nearby waters. Following weathering, erosion is the process of removing the material from the bedrock surface, enabling material to be transported as particulate matter or dissolved by a number of mechanisms. Physical, chemical, and biological processes in a sedimentary environment on land or in the sea will eventually deposit the sediment. The last phase is the lithification of the sediment to create sedimentary rocks, which tectonic processes may eventually cause to surface. These activities are a part of the series of occurrences known as the "rock cycle."

Initial steps in the sequence of events depicted in Figure 6.1 are examined, beginning with the uplift of the continental crust and moving on to the weathering and erosion processes that cause the landscape to become denuded. Next, the relationships among weathering, erosion, climate, and lithospheric behavior are examined in terms of the Earth Systems that provide the sedimentary material.

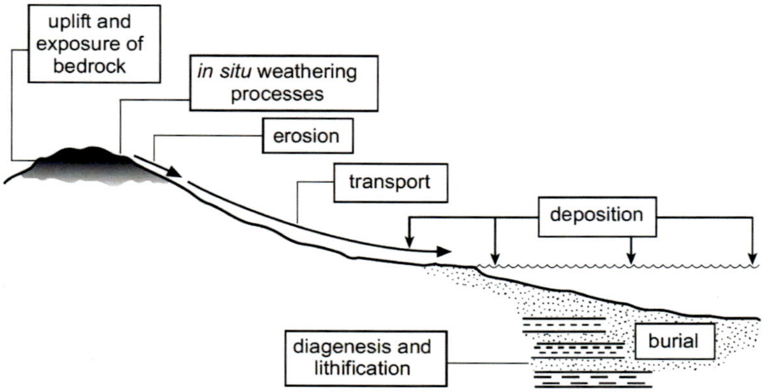

Figure 6.1. The pathway of processes involved in the formation of a succession of clastic sedimentary rocks, part of the rock cycle.

Source: https://raregeologybooks.wordpress.com/wp-content/uploads/2014/09/sedimentology-and-stratigraphy-by-gary-nichols.pdf.

6.4. MOUNTAIN-BUILDING PROCESSES

Plate tectonic theory offers an explanation for the formation of all major crustal morphological features over most of Earth's history, as well as a framework for comprehending the processes that result in the formation of mountains (Kearey & Vine, 1996; Fowler, 2005). The topographic features of the Earth's surface are shaped by plate movements and related igneous activity, which are subsequently altered by erosion and deposition. Areas of elevated terrain on Earth's surface today can be linked to plate boundaries. For instance, the Andes contain an igneous rock core that originated from the subduction of oceanic crust from the East Pacific beneath South America, and the Himalayas are an orogenic belt, a mountain range formed by the collision of the continental plates of Asia and India. On the sides of significant rifts where the crust is breaking apart, like the East African Rift Valley, high ground can also be found. The Appalachians in the northeastern USA and the Highlands of Scotland are remnants of plate collisions caused by the closing of previous oceans, according to an interpretation of the stratigraphic record that suggests the same processes that created the mountains have happened in the past. In the same way, historical subduction zones and associated magmatic belts are identifiable in western North America's Western Cordillera. Therefore, the main processes that cause the crust to rise and create high-ground areas that serve as a source of clastic sedimentary material are plate tectonic processes.

However, not every vertical movement of the crust can be connected to a horizontal plate movement. There are regions of the crust where the mantle beneath is relatively hot, and there are other regions where the temperature of the mantle beneath is cooler. The mantle has an uneven temperature distribution throughout. The hot spots are referred to as "plumes," which are upwelling masses of buoyant mantle that can occasionally be very large in size. These are called "superplumes," and they most likely originate from the core–mantle boundary. Large-scale continental crust uplift occurs above the hot, buoyant mass of a superplume, creating high plateau regions like those found in southern Africa today. Even though these plateaus are far from any plate boundaries, they are nonetheless significant sites for erosion and the production of debris that feeds sedimentary basins (Ollier, 1984; Anderson & Anderson, 2010).

6.5. GLOBAL CLIMATE

The main factor influencing the global climate belts is latitude. There is a temperature gradient from each pole to the equator because the polar regions receive less solar energy per unit area than the equatorial zones. The atmospheric pressure belts are determined by these temperature variations: low pressure is found at the Equator, where the air is heated, expands, and rises, and high pressure is found at the poles, where cold air sinks. Winds are created by these pressure differentials, which transport air masses from high-pressure regions in the subtropical and polar zones to low-pressure areas that lie between them. These air movements are influenced by the Coriolis force, which the Earth's rotation imparts, and this results in a fundamental wind pattern. Any moving object, such as an air mass, water in the ocean, or an airplane, will be deflected to the left in the southern hemisphere and to the right in the northern, due to the Coriolis force, which is the result of any body moving towards or away from the poles over the surface of a rotating sphere (Figure 6.2).

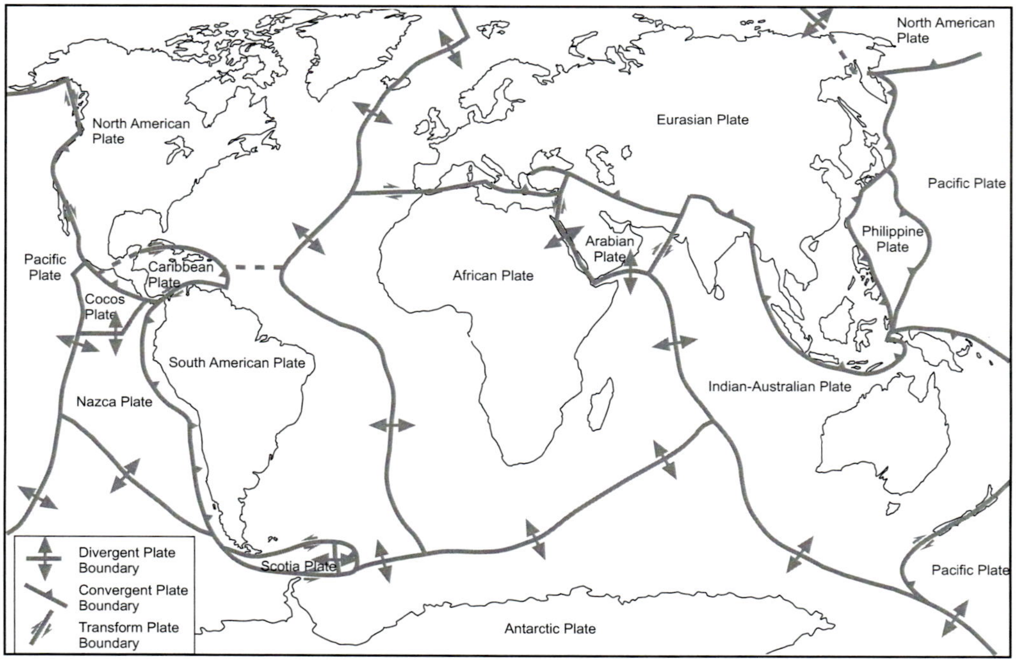

Figure 6.2. The boundaries of the present-day principal tectonic plates.

Source: https://raregeologybooks.wordpress.com/wp-content/uploads/2014/09/sedimentology-and-stratigraphy-by-gary-nichols.pdf.

Four primary climate zones are created by the interaction of wind belts and temperature distribution. Mostly located north and south of the Arctic and Antarctic circles are the polar regions. These are areas of low pressure and high temperatures where it rarely, if ever, gets above freezing. The humid, temperate mid-latitude climate belts, with strongly seasonal temperatures and moderate precipitation, are located between roughly 60° and 30° on either side of the Equator. The arrangement of land masses in the latitudes of the tropics of Cancer and Capricorn determines the width of the dry subtropical belts. These arid regions, which cover sizable continents, are characterized by high pressure, high temperatures, and little precipitation. The wet equatorial zone, with lots of rainfall and warm temperatures, is located in the center.

The extent of continental land masses and the elevation of the land determine the local climatic characteristics of these climate zones, which vary in width globally. Because plate movements cause variations in the height and positions of continents over geological time, it is only possible to relate paleoclimate belts to modern belts in a relatively simplistic manner unless extensive climate modeling is done.

6.6. WEATHERING PROCESSES

Numerous weathering processes can alter the physical and chemical makeup of rock that is near the surface of the earth. When the rock gets close to the surface, these processes typically begin with water seeping down into joints created by stress release. These processes are strongest at the surface and in the soil profile. To produce a regolith—a layer of loose material—that can be transported away from the site, weathering is the mechanical and chemical breakdown and alteration of bedrock (Figure 6.3).

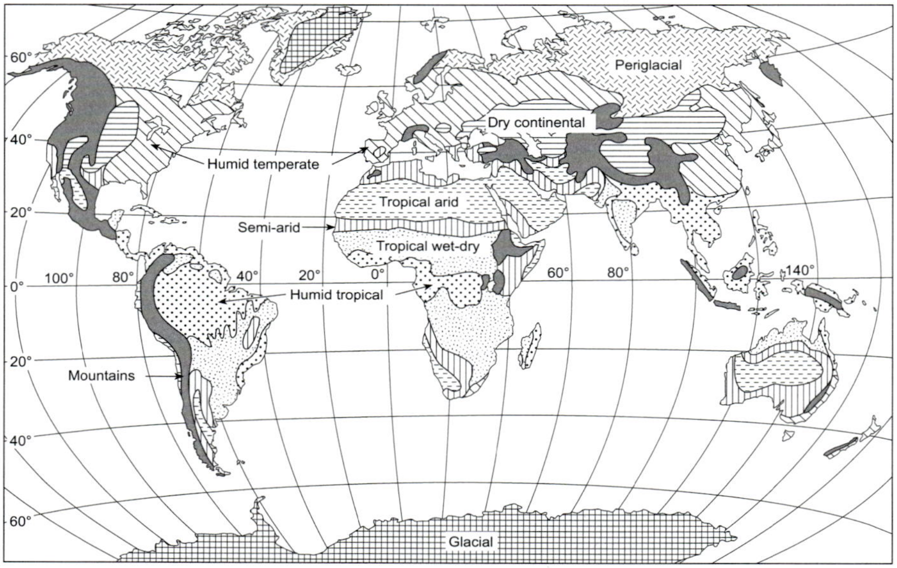

Figure 6.3. The present-day world climate belts.

Source: https://raregeologybooks.wordpress.com/wp-content/uploads/2014/09/sedimentology-and-stratigraphy-by-gary-nichols.pdf.

6.6.1. Physical Weathering

These procedures, which may also separate the various minerals without requiring any chemical reactions, fracture the solid rock into fragments. The following are the key players in this process.

6.6.1.1. Freeze–Thaw Action

This process, also referred to as frost shattering, is very effective in places where temperatures regularly fluctuate around 0°C, such as high mountains in temperate climates and in polar regions. Water entering cracks in rock expands upon freezing, forcing the cracks to widen.

6.6.1.2. Salt Growth

Water with dissolved salts, like seawater, can seep into cracks, particularly in coastal regions. As the water evaporates, salt crystals are left behind and their expansion creates concentrated forces that can widen the cracks in the rock.

6.6.1.3. Temperature Changes

Temperature fluctuations likely contribute to the deterioration of rock by causing it to split. In desert regions, where temperatures can vary significantly between day and night, different minerals within the rock may expand and contract at different rates, creating internal forces that lead to the rock breaking apart. This phenomenon, known as exfoliation, results in thin layers of the rock's surface peeling off (Figures 6.4 and 6.5).

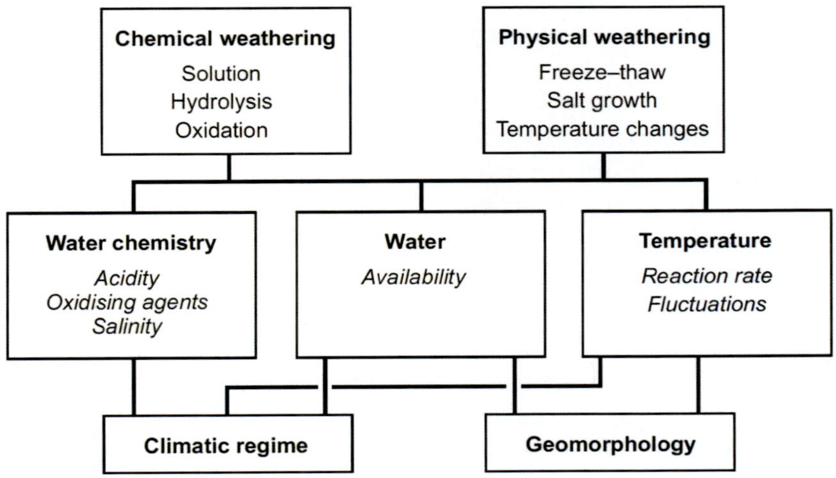

Figure 6.4. The principal weathering processes and their controls.

Source: https://raregeologybooks.wordpress.com/wp-content/uploads/2014/09/sedimentology-and-stratigraphy-by-gary-nichols.pdf.

Figure 6.5. Frost shattering of a boulder (50 cm across) in a polar climate setting.

Source: https://raregeologybooks.wordpress.com/wp-content/uploads/2014/09/sedimentology-and-stratigraphy-by-gary-nichols.pdf.

6.6.2. Chemical Weathering

These processes involve changes to the minerals that make up a rock. The reactions that can take place are as follows.

6.6.2.1. Solution

At Earth's surface temperatures, the majority of silicate minerals that form rocks have very low solubility in pure water, making most rock types resistant to rapid solution. Only very alkaline waters can cause silica to become somewhat soluble in them. Carbonate minerals have a moderate solubility, particularly in the presence of acidic groundwater—that is, water that percolates through bedrock near the surface. Evaporite minerals, like gypsum and halite (sodium chloride), are the most soluble and can contribute significantly to sedimentary bedrock in certain areas.

6.6.2.2. Hydrolysis

Hydrolysis reactions rely on the separation of H_2O into H^+ and OH^- ions, which occurs in the presence of an acidifying agent. Essential natural acids that promote hydrolysis include carbonic acid (created by dissolving carbon dioxide in water) and humic acids, various acids produced by bacteria breaking down organic material in soil. Several silicates also undergo hydrolysis reactions, such as the transformation of orthoclase (a feldspar) into kaolinite (a clay mineral) through a reaction with water.

6.6.2.3. Oxidation

The most common indication of oxidation is the production of iron hydroxides and oxides from iron-containing minerals. Even though there may not be much iron present, many rocks exposed at the surface exhibit the characteristic reddish-orange rust color of ferric iron oxides.

6.6.3. The Products of Weathering

Terrigenous material, derived from continental land masses, is created through weathering and erosion processes. This material can include minerals removed from bedrock, lithic fragments, and new minerals formed through weathering. It is a key component of the regolith found on top of bedrock in many locations. One way to classify minerals that form rocks is based on how stable they are in the surface atmosphere. Physical weathering merely divides the quartz crystals from other minerals in the rock and from each other. Stable minerals like quartz are largely unaffected by chemical weathering processes. Micas and orthoclase feldspars are comparatively resistant to these processes, while olivines, amphiboles, pyroxenes, and plagioclase feldspars all react quickly at the surface and are hardly ever carried away from the weathering site in their original state. Clay minerals are the main byproducts of silicates' chemical weathering. The breakdown of various bedrock minerals under various chemical conditions results in a wide range of clay minerals, the most common of which are kaolinite, illite, chlorite, and montmorillonite. Iron oxides (primarily haematite) and aluminum oxides (bauxite) can also form in environments with severe chemical weathering (Figure 6.6).

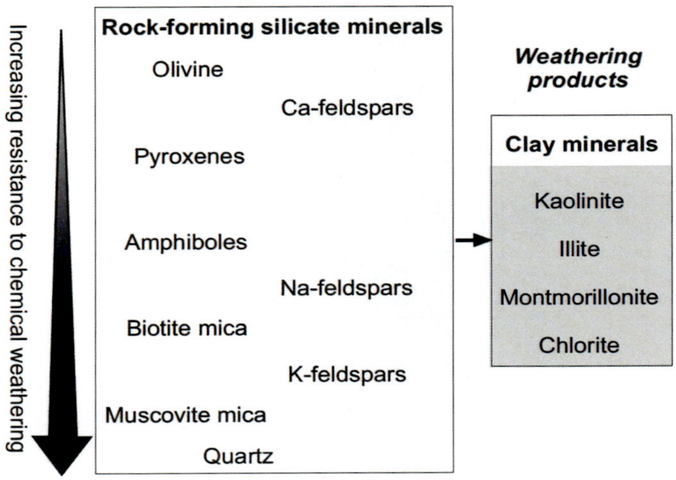

Figure 6.6. The relative stability of common silicate minerals under chemical weathering.

Source: https://raregeologybooks.wordpress.com/wp-content/uploads/2014/09/sedimentology-and-stratigraphy-by-gary-nichols.pdf.

Where chemical weathering is less pronounced, lithic fragments could make up a significant portion of the debris produced by physical processes. These fragments' characteristics, which can include any lithology found at the Earth's surface, will directly reflect the type of bedrock. Certain lithologies are easily dissolved into rocks composed of evaporite minerals, and other lithologies are easily broken apart due to their extreme fragility. Fragments of basaltic lithic detritus can accumulate around volcanoes, and fragmented limestone can constitute a significant clastic element in certain shallow marine habitats.

6.6.4. Soil Development

Soil formation is a critical process in the transformation of bedrock and regolith into detritus that can be transported and deposited. Physical and chemical weathering of bedrock in place creates soil that can be further altered by biogenic processes. Plant roots breaking into bedrock can help break it apart, while vegetation accumulation changes the composition of surface water by forming humic acids. Over time, soil profiles thicken as bedrock breaks down and organic matter builds up, but erosion can also occur. Gravity and water can cause soil profiles to be removed completely or partially. Erosion is more prevalent on slopes and flat ground, leading to soil disaggregation and the contribution of detritus to rivers. In temperate and humid tropical regions, most sediment in rivers likely came from a soil profile at some point (Figure 6.7).

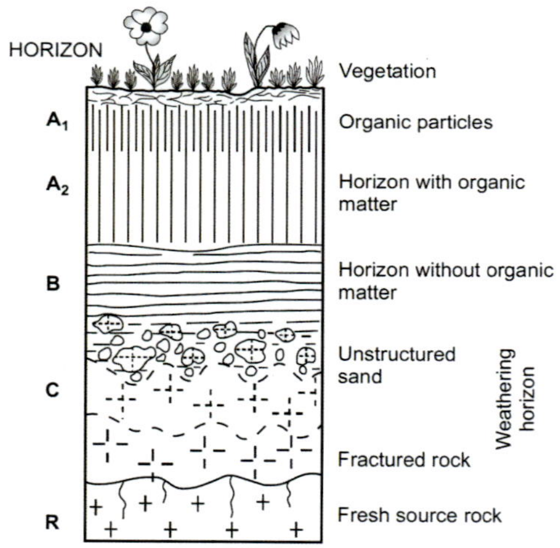

Figure 6.7. An in situ soil profile with a division into different horizons according to presence of organic matter and degree of breakdown of the regolith.

Source: https://raregeologybooks.wordpress.com/wp-content/uploads/2014/09/sedimentology-and-stratigraphy-by-gary-nichols.pdf.

The floodplains of rivers in continental depositional environments are important for soil formation. Over time, these soils may become covered by layers of sediment and are preserved in the stratigraphic record as fossil soils (Ollier, 1984; Anderson & Anderson, 2010).

6.7. EROSION AND TRANSPORT

Weathering is the process of breaking down bedrock in its original location, while erosion involves the removal of loose material. Material on the surface of the land can be moved downhill by gravity, water, wind, ice, or a combination of these forces. Falls, slides, and slumps can move large amounts of material in mountainous areas, but typically only transport it to the valley floor. Water is the main agent for transporting material over long distances, although ice and wind also play significant roles in certain environments.

6.7.1. Erosion and Transport Under Gravity

Movements of weathered material downslope under gravity on steep slopes in mountainous areas and along cliffs are typically the initial stages of erosion and transport.

6.7.1.1. Downslope Movement

There are a range of ways in which materials can move downhill. A landslide involves a mass of bedrock moving without significant disintegration, maintaining its structure and layers as it travels downhill. On the other hand, if the rock breaks apart during movement, it is called a rock fall, resulting in a jumbled pile of material at the bottom of the slope. Gravity alone can trigger these movements, whether from an earthquake, erosion at the base of the slope, or saturation of the slope by heavy rainfall (Figure 6.8).

In addition, soil creep and lubrication of the regolith by water can cause movement downslope. This process is much slower than that of falls and slides, and it might not be noticeable until a hillside is observed over an extended period of time. Slumping is a process that can be thought of as an intermediary between slides and creep movement. Similar to slides, slumps happen instantly, but because the material is saturated with water, it is plastic and deforms as it moves downslope. A slump could fragment into a debris flow if there is enough water present.

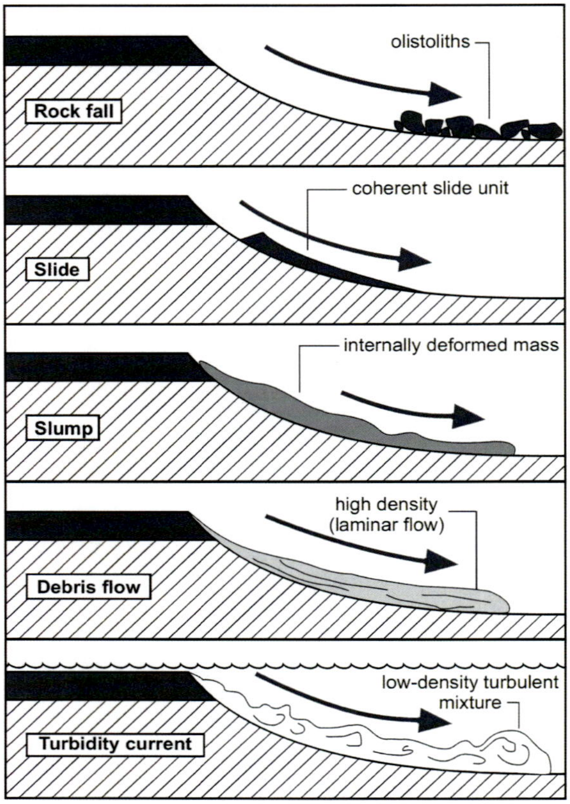

Figure 6.8. Mechanisms of gravity-driven transport on slopes. Rock falls and slides do not necessarily include water, whereas slumps, debris flows, and turbidity currents all include water to increasing degrees.

Source: https://raregeologybooks.wordpress.com/wp-content/uploads/2014/09/sedimentology-and-stratigraphy-by-gary-nichols.pdf.

6.7.1.2. Scree and Talus Cones

In mountainous regions, weathered detritus such as grains, pebbles, and boulders fall down mountainsides and accumulate at the base of slopes. This accumulation, known as scree, can be modified by water, ice, and wind, or remain preserved as talus cones – concentrations of debris at the bottom of gullies. Due to the short transport distances of only a few 100 meters, the clasts in scree deposits are angular, with minimal abrasion. Some sorting and stratification may occur as water moves smaller particles downward, but overall scree deposits are poorly sorted and crudely stratified. Talus deposits do not display distinct bedding patterns, but the layers are usually close to a 30° angle. Unlike alluvial fans, talus deposits are not influenced by water in their transportation and deposition processes.

6.7.2. Erosion and Transport by Water

Water erosion on hillsides begins as a sheet wash, i.e., uncontrolled surface runoff after rain that descends a slope. The regolith may be eroded by this overground flow, which could also pick up loose debris from the surface. The amount of water involved and its carrying capacity are determined by the surface's characteristics as well as the amount of rainfall. For example, water moves more quickly down steep slopes, vegetation tends to slow flow and collect debris, and surface water infiltrates porous substrates. Because of this, surface run-off is best suited to transport debris during flash floods on steep, impermeable slopes in arid regions with sparse vegetation. In temperate and tropical climates, vegetation cover and more porous, thick soils tend to lower surface runoff's transport capacity (Figure 6.9).

Figure 6.9. A scree slope or talus cone in a mountain area with strong physical weathering.

Source: https://raregeologybooks.wordpress.com/wp-content/uploads/2014/09/sedimentology-and-stratigraphy-by-gary-nichols.pdf.

Sheet wash turns into concentrated rills and gullies, which eventually combine to form channels and create the beginnings of streams and rivers. These waterways gradually erode the ground beneath them, wearing away at both regolith and bedrock. The constant movement of the water scours the channel, causing pieces of material to break off and fall into the stream. In areas where the bedrock is soluble, such as limestone, the water dissolves the material over time. The eroded material is then carried downstream in the form of bedload, suspension, or in solution. As streams merge together, they form larger rivers that can feed different environments like alluvial fans, areas of deposition, lakes, or seas.

6.7.3. Erosion and Transport by Wind

The temperature of water masses that move with ocean currents, heat absorbed by land masses, and cold air over high glaciated mountain regions are some of the local variations in atmospheric pressure that cause winds. Global temperature distributions also play a role in these variations. Winds are produced across the Earth's surface by a complex and ever-changing pattern of high pressure (anticyclones) and low pressure (depressions) regions. Today's winds can reach speeds of up to 100 km/h or more in storm force, with hurricanes reaching twice that speed.

Sand-sized debris, silt, and loose clay can all be picked up by winds from the ground. Since wind erosion is most effective in environments without plant boundaries, it is common in dry deserts, high alpine regions, and frigid climates with little vegetation, such as those near the poles. In any climate, exposed sand banks in rivers, sandy beaches, and dry floodplains of rivers can all be subject to wind erosion. The wind is capable of carrying eroded fine material hundreds or thousands of kilometers away, up to sand grade (Schutz, 1980; Pye, 1987). The strength (velocity) of the air current affects the amount of material carried.

6.7.4. Erosion and Transport by Ice

In areas of temperate mountains, glaciers play a major role in the erosion and movement of regolith and bedrock. In comparison to similar unglaciated regions, the rate of erosion is 2 to 10 times higher in mountainous areas covered by glaciers (Einsele, 2000). On the other hand, because the ice in these polar "cold-based" glaciers is frozen to the bedrock, they tend to prevent material erosion. The main mechanism of ice movement in these glaciers is shearing within the ice body. There are two ways that ice erodes bedrock in temperate (warm-based) glaciers: abrasion and plucking.

The frictional action of material blocks (also referred to as "tools") embedded in the ice on the bedrock causes glacial abrasion. These instruments create a few millimeter-deep, elongating glacial striae in the bedrock that run parallel to the direction of the ice flow. Because of this, striae can be used to map out the paths of ice flow even after the ice has melted. Rock flour, clay, and silt-sized debris are produced during the scouring process and are mixed into the ice.

Plucking by glaciers is most frequently seen when they encounter an obstruction in their path. As the glacier moves over the obstacle, abrasion occurs on the uphill side, while on the downhill side, blocks ranging in size from centimeters to meters are dislodged by the ice. These blocks, once picked up by the glacier, are usually loosened by subglacial freeze-thaw processes. The resulting landforms from this combination of abrasion and plucking are known as roche moutone´e, named for their resemblance to sheep when viewed from a far distance (Ollier, 1984; Anderson & Anderson, 2010).

6.8. DENUDATION AND LANDSCAPE EVOLUTION

Denudation is the term used to describe the lowering of the land surface caused by erosion and weathering together. The processes of weathering and erosion are somewhat interdependent; however, the combination of these processes—that is, the rates and magnitudes of denudation and the effects on the supply of material to sedimentary environments—is most relevant to sedimentary geology. A combination of topographic and climatic factors influence rates of denudation. These factors also affect vegetation and soil development, which in turn affect weathering and erosion. Additionally, the responses of various bedrock lithologies to these mixes of chemical, biological, and physical processes vary.

6.8.1. Topography and Relief

It is important to distinguish between a terrain's altitude and relief, or the variation in the area's ground height. Even though a plateau region is located thousands of meters above sea level, its denudation rates may not differ significantly from those of a lowland area with a similar climate if the plateau is level. The mechanical denudation rate rises with increasing relief because erosion processes become more effective. Stream flow and overland water flow are faster across steeper slopes and therefore have more erosive power; rock falls and landslides are obviously more common on steep slopes than in areas of subdued topography. The most steep slopes for bedrock and regolith erosion are found in a deeply incised topography made up of steeply sided valleys divided by narrow ridges.

Areas experiencing uplift from tectonic activity and thermal doming typically have the highest relief. This process is most pronounced around plate boundaries, especially in regions with convergent margins like orogenic belts. Conversely, regions that are tectonically stable have lower relief due to weathering and erosion, resulting in gently sloping topography. Cratonic centers in continental plates usually have low relief and slow rates of denudation.

6.8.2. Climate Controls on Denudation Processes

Factors governing the rate and pathway of the reactions have an impact on chemical weathering processes. First of all, since water is necessary for every chemical weathering process, these reactions are inhibited in areas with low water availability in arid regions. Temperature is also significant because hot climates encourage chemical weathering because most chemical reactions are more vigorous at higher temperatures. Lastly, the chemistry of the water influences the reactions: acids promote hydrolysis, while dissolved oxidizing agents ease oxidation reactions (Einsele, 2000). Different types of bedrock have varied reaction rates and efficiencies.

Erosion rates are influenced by climate, as the presence of water plays a critical role in the breakdown of soil and rock by sheetwash, as well as the erosion caused by rivers and streams. Additionally, temperature is a key factor, with ice playing a significant role in mountainous regions where glaciers are more effective at moving debris than rivers. Denudation rates vary based on the climate in a particular region, with identifiable patterns in different global climate zones.

6.8.2.1. Wet Tropical Regions

Chemical weathering is more intense in hot, humid, tropical regions due to the combination of high temperatures and abundant water. As a result, bedrock in these areas tends to be heavily weathered and transformed on the surface. Even rocks like granite can be broken down into quartz and clay by the weathering of feldspars and other silicate minerals. This process generates fine particles and partially dissolves the bedrock. Despite high rainfall leading to increased stream flow, the dense vegetation in these regions helps prevent soil erosion on steep slopes.

6.8.2.2. Arid Subtropical Regions

Chemical weathering processes are suppressed in arid regions due to the scarcity of water. Because the bedrock is often devoid of soil or vegetation, rainfall has a short residence time on the surface of the earth, which reduces the amount of time it has to undergo chemical alteration. Significant mechanical breakdown can occur, particularly in arid regions where warm days and cold nights encourage freeze-thaw action, utilizing whatever water is available. Temperature variations also lead to exfoliation. However, because there is no soil or vegetation, intense but rare rainstorms can be very effective at clearing surface debris; flash floods, in particular, carry more debris than equivalent volumes of water that occur gradually over a longer period of time. Wind ablation is a significant process that removes fine-grained debris from the regolith, especially in areas of barren desert.

6.8.2.3. Polar and Cold Mountain Regions

Chemical weathering is not as significant in cold, dry regions where chemical reactions occur at a slower pace. Physical weathering processes are more effective in these areas, but they still depend on the presence of water. In cold mountain regions, weathering typically results in broken pieces of bedrock with little to no change in mineral composition. For example, granite breaks down into gravel clasts, quartz, feldspar, and other minerals. Physical weathering mostly produces coarse material without much clay or rock dissolution. Mountain glaciers are strong erosive agents as they move downhill over rocks, but in polar regions, the ice remains frozen to the bedrock and causes minimal erosion. Periglacial regions, which border glaciers, experience seasonal snow cover that melts in the summer. However, the ground may stay frozen at depths of a few meters throughout the year (permafrost), causing water near the surface to saturate the regolith and trigger slope slumping. Freeze-thaw cycles can also lead to downslope creep. Wind erosion is significant due to limited vegetation in subarctic areas.

6.8.2.4. Temperate Regions

In temperate climates, physical and chemical weathering processes are typically less intense. Erosion is more active in wetter climates, but dense vegetation in humid areas helps stabilize the surface and reduce erosion. The rate of erosion in limestone terrains is greatly influenced by climate, with humid temperate and tropical regions experiencing 10 times higher rates of denudation compared to arid subtropical and subarctic regions.

6.8.3. Bedrock Lithology and Denudation

One important factor influencing the rates and patterns of denudation is the type of bedrock. The rate at which weathering processes erode the rock and release materials for erosion is the primary determinant. Because different lithologies erode and decompose at wildly disparate rates in humid climates where chemical weathering processes predominate, there is the greatest variability in these conditions. The primary factor is the proportions of the silicate minerals that form the rock; mafic rocks, like basalts, weather, and erode more quickly than quartz-rich rocks, which are less prone to breakdown. The denudation of terrains like volcanic arcs, which are primarily made of basaltic to andesitic rocks, produces significant amounts of clay minerals. Deep lateritic soils, or laterites, are created when silicate rocks undergo severe chemical weathering. Laterites are red soils primarily made of iron and aluminum oxides.

Since dissolution is the main process that weathers limestone bedrock, the formation of karst scenery dominates the pattern of denudation. Cave systems beneath the surface and deep, steep-sided canyons above are the result of solutions related to rock fractures and joints. Denudation of limestone terrains produces little clastic detritus; near the erosion site, conglomerates of limestone clasts may form, but the majority of the material is in solution and sandsized detritus is mainly absent.

When the bedrock is poorly lithified, it creates a unique landscape known as badlands, which are formed by the deep erosion of soft sandstones and mudstones as significant amounts of sediment are washed away.

6.8.4. Soils and Denudation

The development of the soil is crucial to the weathering processes. First, because soils retain water, the thickness of the soil profile affects the amount of water that is available. A thin soil profile does not allow for adequate water retention for chemical weathering reactions in the bedrock, while a thick soil profile allows for water to be stored and lost through plant evapotranspiration, which lowers the amount of water available for weathering reactions (Figures 6.10 and 6.11).

Figure 6.10. Erosion by solution in beds of limestone results in a karst landscape.

Source: https://raregeologybooks.wordpress.com/wp-content/uploads/2014/09/sedimentology-and-stratigraphy-by-gary-nichols.pdf.

Figure 6.11. Badlands scenery formed by the erosion of weak Mudrock beds.

Source: https://raregeologybooks.wordpress.com/wp-content/uploads/2014/09/sedimentology-and-stratigraphy-by-gary-nichols.pdf.

Second, humic acids—a group of acids produced by biochemical reactions in soils—increase the rates at which carbonate bedrock dissolves. Third, because soils are home to both plants and animals—particularly those with deep-rooted roots that can enlarge fissures—they also contribute to the erosion of bedrock. While a lot of soil processes can help with weathering, soil development can stop erosion by supporting a layer of vegetation that shields the bedrock.

6.8.5. Vegetation and Denudation

The climatic regime, which is influenced by latitude and altitude, determines the kinds of vegetation and their coverage over the land surface. The bedrock and the regolith that covers it can be effectively shielded from erosion by raindrop impact and water flowing overland by a dense cover of vegetation. Plants can stabilize even the steepest mountain slopes. In tropical regions, anthropogenic activities (man-made effects) like logging or the destruction of vegetation cover by natural events like fires can have a catastrophic effect on erosion because the bedrock beneath plant roots may be deeply weathered and the regolith is more prone to being washed away by rainfall or floods. Where a vegetation cover binds fine detritus into the soil, the effects of wind action on the regolith are also mitigated. In cold or arid areas, there is little vegetation cover, leaving the regolith vulnerable to wind and water erosion. Overland flow after storms may be extremely rare in desert regions, but because there is little vegetation, a large amount of loose debris may be washed away in a single flash flood.

The nature of the vegetation colonizing the land surface has changed considerably through geological time. Four stages in the development of land plants are significant in terms of sedimentological processes (Schumm, 1968).

1. Pre-Silurian: There was no land vegetation at this time so it can be assumed that the denudation rates of continental areas were generally higher than they are today.
2. Silurian to Mid-Cretaceous: The main plant groups were ferns, conifers, and lycopods with relatively simple roots systems with a limited binding effect on the regolith.
3. Mid-Cretaceous to Mid-Cenozoic: Angiosperms (flowering plants) became important and had more complex root systems that were more effective at binding the soil.
4. Mid-Cenozoic to Present: The evolution of grasses meant that there was now a widespread plant type which covered large areas of land surface with a dense fibrous root system that very effectively binds soil (Figure 6.12).

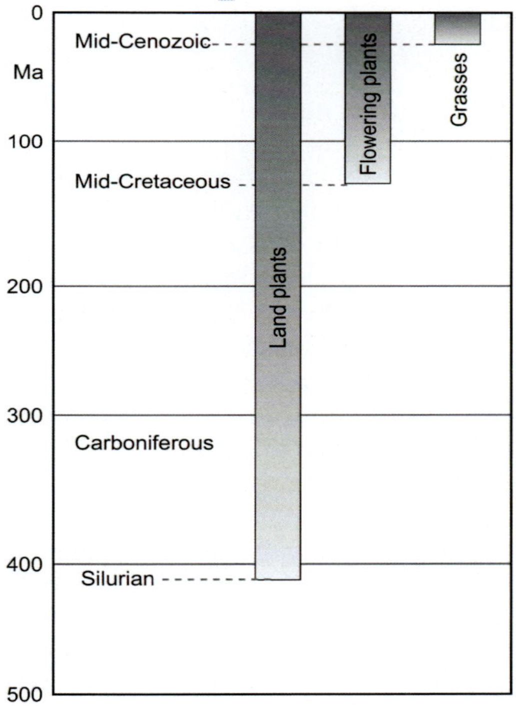

Figure 6.12. The development of land plants through time: Grasses, which are very effective at binding soil and stabilizing the land surface, did not become widespread until the mid-Cenozoic.

Source: https://raregeologybooks.wordpress.com/wp-content/uploads/2014/09/sedimentology-and-stratigraphy-by-gary-nichols.pdf.

6.9. TECTONICS AND DENUDATION

Although plate tectonics and mantle behavior play a major role in the formation of the topography of the continental land surface processes—particularly erosion—have a significant impact on how the landscape is altered. There is a climatic control on the relationship between erosion and tectonics because climate has a significant impact on weathering and erosion processes (Burbank & Pinter, 1999).

Denudation causes material to be removed from the bedrock that has been raised, which lowers the mass of material in these areas. Isostatic uplift is the result of this mass removal. Comparable to a block of ice floating on water, this process in relatively rigid crust overlying mobile mantle occurs when 10% of the ice is above the water line. However, if some of the exposed ice is removed from the top, the entire block will move up in the water, leaving 10% of the mass above the water line. An underlying mass of thickened crust forms a bulge or root down into the mantle in mountain belts; as material from the top is eroded, an isostatic readjustment occurs, causing the entire crustal mass to move upward. Steep relief areas typically experience higher rates of denudation, and as a mountain belt expands, the tectonic uplift that created the steep topography is more susceptible to erosion. The topography is reduced by surface processes after the tectonic uplift stops. Denudation and isostatic readjustment would gradually remove mass from the top of the root until its base was level with the surrounding crust. Denudation lowers the area to normal crustal thickness, thereby obliterating the mountains within an orogenic belt. Denudation, however, does not happen uniformly, and it is conceivable to imagine an apparently paradoxical scenario in which uplift is brought about by denudation. If a big plateau is the initial topography formed, erosion will begin with rivers cutting into the plateau and removing material from the valleys, but the areas in between won't be significantly eroded. Since there will be less mass in the region, the entire plateau—including the parts between the rivers that haven't been depleted—will experience isostatic uplift. Uplift caused by denudation will persist until the valleys widen and the interfluve regions begin to erode at a rate equal to that of the valleys.

Denudation rates may be influenced by the climate, but topography can also affect an area's climate. When moisture-laden air is driven upward and produces rainfall on the range's upwind side, a mountain belt may produce a rain shadow effect. The resulting dry winds over the mountains cause the climate on the range's downwind side to become more arid. Because of this orographic effect, a mountain range has a distinct climatic division, which affects how much erosion occurs on each side. It is clear from the above that tectonic forces and surface processes both contribute to the current and historical form of the Earth's surface, and that these two systems interact with one another through a number of feedback mechanisms. As an illustration, it has been proposed that the formation of the Himalayas altered Asian weather patterns, enhancing the Asian monsoon, which is characterized by heavy seasonal rainfall throughout southern Asia (Raymo & Ruddiman, 1992). As a result, the Himalayas experienced increased erosion, which led to isostatic uplift. The Tibetan plateau, which is located to the north and is considerably drier and less eroded due to its location in the rain shadow of this weather system, has been raised but, unlike the Himalayas, has not seen significant river erosion (Figures 6.13 and 6.14).

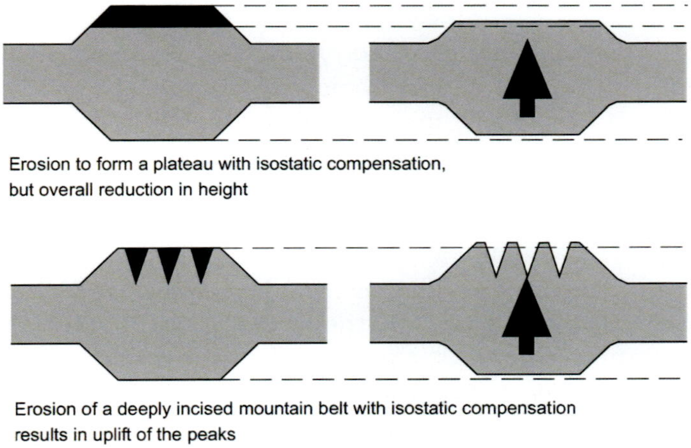

Figure 6.13. Uplift due to thickening of the crust followed by erosion results in isostatic compensation as the load of the rock mass eroded is removed. If the erosion is uneven then locally the removal of mass from valleys can result in uplift of the mountain peaks between.

Source: https://raregeologybooks.wordpress.com/wp-content/uploads/2014/09/sedimentology-and-stratigraphy-by-gary-nichols.pdf.

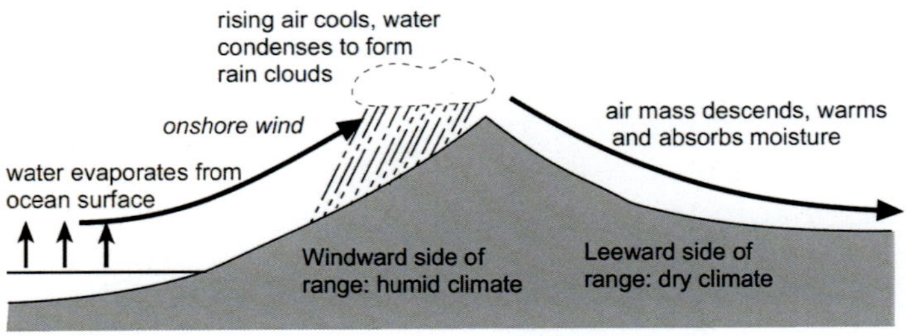

Figure 6.14. The rain shadow effect in mountain belts: Moisture in air blown from the sea falls as rain as the air mass cools over a mountain range.

Source: https://raregeologybooks.wordpress.com/wp-content/uploads/2014/09/sedimentology-and-stratigraphy-by-gary-nichols.pdf.

6.10. MEASURING RATES OF DENUDATION

The long-term rates of erosion in the past have been estimated thanks to the development of techniques that enable thermochronology, or the temperature history of rocks, to be performed. The main method, called fission-track dating, is applied to minerals like zircon and apatite, which are found in sandstones as heavy minerals and as accessory minerals in many igneous and metamorphic rocks. The fundamental idea behind fission-track dating is that alpha particles are released during the radioactive decay of uranium isotopes in mineral grains. These particles travel through the crystal's lattice and leave behind "fission tracks," or traces of their passage. The lines of the tracks become obfuscated as the heat anneals the lattice, reaching over 110°C for apatite and 300°C for zircon. New tracks begin to form as the crystal cools, and the longer the crystal has cooled below the annealing point, the more fission tracks it contains. Zircon Fission Track, which is less frequently used, and Apatite Fission Track Analysis (AFTA).

Thermochronological analysis is a technique used to determine the date at which a crystal was at a specific temperature in the past.

Understanding the geothermal gradient—that is, the variation in temperature with crustal depth—is necessary to translate thermochronology data into rates of erosion. The temperature rises by roughly 25° for every 1,000 meters of depth in many parts of the world, or a geothermal gradient of 25° per kilometer; however, in areas with volcanic activity, the temperature rises even more. Assuming a surface temperature of 20°C and an increase of 25°C per kilometer, a rock at a depth of 4 km will be approximately 120°C, annealing all of the fission tracks in apatite crystals. The rock body may be raised by tectonic forces, and as the rock above the sample is eroded away by erosion, the rock will cool and become closer to the surface. Once all of the rock above has been eroded away and the sample is at the surface, ready for collection and analysis, fission tracks may then begin to form in the apatite crystals in the sample. As a result, we can determine the rate of erosion by measuring the fission tracks, which can also tell us when the sample was at a particular depth and how long it took to erode the rocks above. In addition to Ar-Ar dating, other relatively recent thermochronological techniques include U/Th-He. It is feasible to ascertain the dates on which the rocks were at various temperatures and, consequently, at various depths by combining different techniques. A geothermal history of rock samples and, consequently, an erosion history in a region can be produced through statistical modeling of fission track data (Ollier, 1984; Anderson & Anderson, 2010).

CASE STUDY

EXPERIMENTAL PILOT PLATFORM FOR THE TREATMENT OF DREDGED SEDIMENTS:

SOLINDUS ERDF had the aims to find a solution to reduce the volume of Walloon dredged (river) sediments to be landfilled. The final goal was to prove that size classification can isolate pollutants inside the fine fraction. The other fractions are sufficiently clean to be valorised. In many cases, wet techniques are more suitable to treat river sediments, even if a more advanced separation is required. This has been demonstrated through a pilot station, running about 1 ton per hour.

The separation included few steps: 1) anthropogenic and coarse materials removal by using scrubbing trommel and vibrating screens; 2) after buffering storages, coarse sand (2 mm to 250 µm) through a curved grid; 3) fine sand (250 to 63 µm) through hydrocyclones followed by a screw classifier 4) Silt (63 to 15 µm) from clay and fine silt (under 15 µm) through 2 parallel finer hydrocyclones. The last silt separation is innovative, as it cannot be found in industrial plants or demonstrators. After separations, clay and silt are recovered from water suspensions. Clay fraction is recovered by decanting with flocculants and filter pressing. Silt fraction is recovered using filtrating big bags. These separations concentrate the heavy metal pollution in the finest fraction (approximately 40%). Sand fractions (10–20%) can be valorized as sand. About 40% of the sediment (mainly between 63 and 15 µm) can be used as non-polluted sediment (mainly as embankment), in terms of pollutant levels (heavy metals and organic pollutants).

The main challenge is constituted by the quality of the sand fractions and non-polluted sediment fraction. These have to be sufficiently clean to valorize them. Sand fractions can be washed using attrition and spiral techniques. Froth flotation techniques can be applied to remove some heavy metals present as pollutants in the silt fraction (between 63 and 15 µm). The water used in the platform circulates in a closed loop, allowing saving water resources. The experimental platform is now available for industrial tests or for projects. An economic study for a 1,00,000 tons/year (dry matter) plant was made in 2010, with the assumptions of building it in Wallonia and with this distribution, typical of river sediments: 18% as sand, 40% as non-polluted sediment and 42% as sediment to be landfilled. For an investment of 7.3 M€, operational costs were 2.9 M€/year, giving a cost of 29 €/ton.

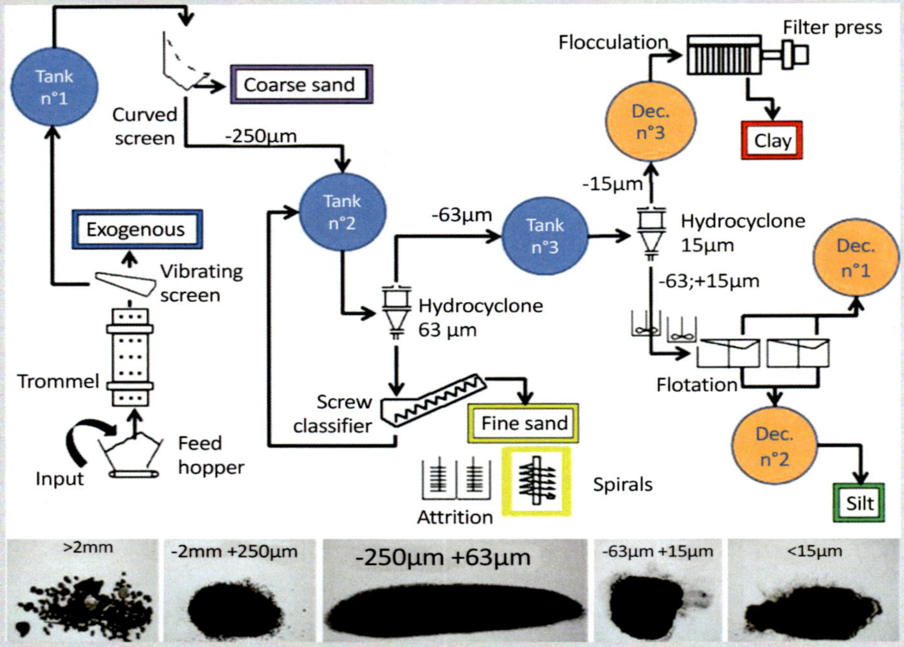

CLASS ACTIVITY

To explore and understand the different sources of sediment in continental environments and the processes that transport and deposit these sediments. Students will spend 30 minutes researching their assigned environment. They should cover:

- The main processes of sediment generation in their environment.
- Examples of sediments typically found in this environment.
- How sediments are transported and deposited.

SUMMARY

- Continental environments are diverse terrestrial landscapes that play a crucial role in the generation and distribution of sediments. These environments include mountains, rivers, deserts, glaciers, and lakes, each contributing uniquely to sediment formation and transport.
- A depositional environment is the accumulation of chemical, biological, and physical properties and processes associated with the deposition of sediments that lead to a distinctive suite of sedimentary rocks.
- Marine depositional environments differ in multiple ways, but the controlling factors in the rocks that are produced is related to the proximity and supply of continental sediment, the water depth, and the community of organisms that live in the area.
- A continental landform, any conspicuous topographic feature on the largest land areas of the Earth. Familiar examples are mountains (including volcanic cones), plateaus, and valleys.
- Plate tectonic theory provides a framework of understanding the processes that lead to the formation of mountains, as well as providing an explanation for how all the main morphological features of the crust have formed throughout most of Earth's history
- Most rock-forming silicate minerals have very low solubility in pure water at the temperatures at the Earth's surface and so most rock types are not susceptible to rapid solution.
- Soil formation is an important stage in the transformation of bedrock and regolith into detritus available for transport and deposition.
- In mountain areas, weathered detritus falls as grains, pebbles, and boulders down mountainsides to accumulate near the bottom of the slope.
- Glaciers in temperate mountain regions make a very significant contribution to the erosion and transport of bedrock and regolith.
- Climate plays an important role in weathering and erosion processes, and hence there is a climatic control on the interaction between erosion and tectonics.

REVIEW QUESTIONS

1. What do you understand by continental environment?
2. Explain Davis's erosion cycle theory.
3. What is physical weathering and its important agents?
4. How climate controls on denudation processes?
5. Write short notes on soils and denudation.

REFERENCES

1. Anderson, R. S., & Anderson, S. P. (2010). Geomorphology: The Mechanics and Chemistry of Landscapes. Cambridge University Press.
2. Burbank, D., & Anderson, R. (2000). Tectonic Geomorphology. Blackwell Science.
3. Einsele, G. (2000). Sedimentary Basins: Evolution, Facies, and Sediment Budget (2nd ed.). Springer-Verlag.
4. Molnar, P., & England, P. (1990). Late Cenozoic uplift of mountain ranges and global climate change: Chicken or egg? Nature, 346, 29–34. https://doi.org/10.1038/346029a0.
5. Ollier, C. D. (1984). Weathering. Longman.
6. Ritter, D. F., Kochel, R. C., & Miller, J. R. (2006). Process Geomorphology (4th ed.). McGraw-Hill.
7. Selby, M. J. (1994). Hillslope sediment transport and deposition. In K. Pye (Ed.), Sediment Transport and Depositional Processes (pp. 61–88). Blackwell Scientific Publications.

CHAPTER 7

Soft Sediment Deformation

LEARNING OBJECTIVES

After studying this chapter, you will be able to:

- Describe the soft-sediment deformation structures;
- Learn about deformation structures related to gravitationally unstable vertical profiles of bulk density;
- Understand deformation structures related to non-uniform confining loads;
- Discuss deformation structures related to a downslope body-force component

KEY TERMS FROM THIS CHAPTER

Ball-and-pillow structures	Clastic dikes
Convolute bedding	Convolute laminations
Dish structures	Flame structures
Load casts	Pillar structures
Pore fluid	Sand volcanos
Sedimentary rocks	Sedimentary structures
Seismites	Slump structures

INTRODUCTION

Soft-sediment deformation is a term that encompasses a variety of processes and resulting structures, the extent of which is only now being identified. As a result, the phrase frequently conveys the nature of the process or structure being reported in an unintentionally misleading manner. Rigid definitions are not possible due to a number of issues, most notably the later changes that obscure the nature of the material at the time of deformation; however, more cautious use is advised. The type of structure needs to be mentioned. It is proposed that a material's cohesion, which can be determined from the structure's form in near-surface sediments, is a good indicator of how soft the material is. Using terms like early or late would help make the timing more clear. An indication of the generating force, which may come from tectonism, gravity, or some local movement—all of which are now known to act on unlithified material—is especially needed.

Soft-sediment deformation structures emerge during the initial phases of the sediment's consolidation, either during or soon after deposition. This is due to the fact that for the deformation to happen, the sediments must be liquid-like or unsolidified. Lowe (1975) has also classified these formations into a category known as water-escape structures. Rivers, deltas, shallow marine areas affected by storms, and deep water basins with turbidity currents are the most frequent locations for soft-sediment deformations to occur. The reason for this is that the high rates of deposition in these environments allow the sediments to pack loosely (Allen, 1982).

7.1. SOFT-SEDIMENT DEFORMATION STRUCTURES

Many years of stratigraphic and sedimentological research have demonstrated that soft sediments frequently undergo non-tectonically driven deformation. The resulting structures come in a variety of shapes and are increasingly referred to as soft-sediment deformations. Field evidence makes it abundantly evident that they were formed either during deposition or not long after burial began, which is consistent with experiment and theory. The majority of soft-sediment deformation, if not all of it, is most likely related to the early phases of sediment consolidation, when the deposit is at its weakest and pore fluid is being ejected from the sediment most quickly.

The factors causing soft-sediment deformation can also abruptly and significantly accelerate the normally gradual process of pore-fluid expulsion, assuming that pore fluid is mobilized. Lowe classified the majority of soft-sediment deformations as water-escape structures as a result, referring to aqueous environments as the primary source of soft-sediment deformations. However, it should be made abundantly evident that pore fluid mobilization is almost always a result of deformation rather than its primary cause.

The only forces that could cause soft-sediment deformation under the described conditions are weak, in terms of geology, at or near the sediment-fluid interface. Therefore, the impacted deposits had to have been either liquid-like or solids with negligible yield strength in comparison to regular rocks at the time. There are two ways to convert naturally occurring sediments into a liquid-like state: liquefaction and fluidization. It is reasonable to assume that liquefaction plays a major role in the creation of numerous types of soft sediment deformation. Here, we'll apply the liquefaction, fluidization, and sedimentation synopsis we previously provided, along with additional data, to comprehend soft-sediment deformation structures. However, because natural sediments are more complex than the relatively simple systems that civil and chemical engineers have studied thus far, the application is not complete. Because very few varieties seem to be specific to a particular depositional setting, soft-sediment deformation structures reveal more about the early consolidation history of sediments than about their depositional environment. However, the structures are most common in water-laid sediments and occur most frequently in areas where high sedimentation rates are present to encourage the loose packing of sands and silts, such as in river floodplains, deltas, deep-water basins impacted by turbidity currents, and shallow-marine environments affected by storms. First, it is important to provide an overview of the characteristics of the materials that are involved in soft-sediment deformation, framed within the progression of potential states during the process (Figure 7.1).

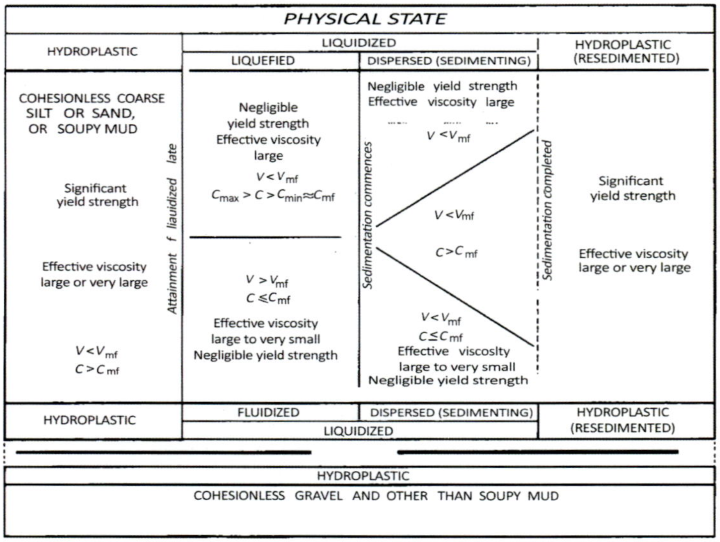

Figure 7.1. Summary of the rheological history of unconsolidated sediments that become liquidized.

Source: John R. L. Allen. Developments in sedimentology. Elsevier. ISBN: 0-444-41945.

With its notable yield strength and effective viscosity, the original material is appropriately described as hydroplastic (Shrock, 1948; Elliott, 1965; Lowe, 1975). Since the pore-fluid superficial velocity's vertical component, V, is smaller than the minimum fluidization velocity, V_{mf}, it falls within the seepage regime. As we have seen, cohesive mud and gravel resist both liquefaction and fluidization, so it is unlikely that they will lose their hydroplastic state during deformation. With loosely packed cohesionless sands or coarse silts, liquidization is practically a given and can happen as a direct result of deformation or as a consequence of another external factor. The liquidized sediment has a high effective viscosity and a negligible yield strength. This is unlikely to be less than 0–1–0 Ns m^{-2} for a sand sample that has been fluidized by water in conditions that are nearly identical to minimum fluidization, or for a liquefied quartz sand that has been wet. However, as superficial velocity increases, the effective viscosity of fluidized sand decreases sharply. A liquefied or fluidized sediment can no longer be said to exist once the liquefying or fluidizing mechanism is removed; instead, what remains is a dispersion that needs to settle out. Liquidized and hydroplastic materials will coexist until this process is finished and a hydroplastic condition is fully restored.

Theoretically, none of the key time points highlighted in Figure 7.1 have to align with the limits in time on deformation. In actuality, though, its beginning and ending during the liquidized state are probably going to align almost precisely with the maximum bounds on the state. Hence, the length of the elements' translation path and the overall strain they experience during soft-sediment deformation are determined by the duration of the liquidized state. The length of the liquefied or fluidized state and the length of the dispersed (sedimenting) condition make up this magnitude.

The liquidized state is reached instantaneously; it's possible that a liquidizing mechanism doesn't start working until dynamic liquefaction. Laboratory work suggests that for the complete liquefaction of loosely packed and moderately packed water-logged sand, shaking times of approximately 10s and 100s are required in the case of cyclic liquefaction, which is the more significant scenario. A deposit must fully fluidize over time, as a "fluidization wave" must progressively move through it. Therefore, a sediment's rheological characteristics may gradually alter as a liquidizing mechanism begins and as full liquidization is achieved. As a result, deformation might start before the liquidized state is completely reached.It is likely that the liquefied state, as depicted in Figure 7.1, can only last for a split second before sedimentation begins. Shaking wet sand in a jar to dissolve it starts the sedimentation process almost immediately and continues until the entire mass solidifies. After liquefaction, the bed contracts more when shaking is continued than when it is stopped, but settlement does not start much later. On the other hand, the availability of fluid supplied externally, from a different bed or set of beds, or from a different area within the same bed, is the only constraint on the length of the fluidized state. As a result, this state could last for a few seconds, minutes, or even hours, during which time deformation might happen continuously. An estimate of the minimum unit discharge of fluid from the external source can be obtained by analyzing the grain size of the sediment that is suspected of having been fluidized; however, only a positive source identification (e.g., a slumped bed) will make it possible to determine the flow's duration.

By extending Kynch's over the dispersed layer's thickness, it is easy to determine the length of the dispersed (sedimenting) state as shown in Figure 7.1. When grains are evenly distributed and have a single size and density, the duration is:

$$T = \frac{h_0 \Delta C}{C_D W} \quad (1)$$

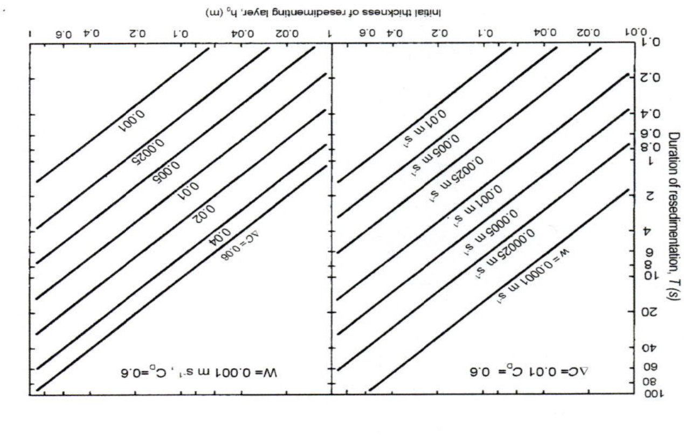

Figure 7.2. Time required for the re-sedimentation of a bed of liquidized sand, as a theoretical function (Eqn. (1)) of initial thickness of the layer, particle falling velocity in the dispersion, and concentration difference between dispersion and deposit. Particle concentration in deposit assumed to be 0.60.

Source: John R. L. Allen. Developments in sedimentology. Elsevier. ISBN: 0-444-41945.

where; 'h' is the thickness of the initial layer; 'AC' is the difference in grain concentration between the deposited and dispersed sediment; 'C' is the concentration of the deposited grains; and 'W' is the dispersed particles' falling speed. If characteristic values of AC, C, and W are available, Eqn. (1) provides order-of-magnitude values for non-uniform, polydisperse systems. Alternatively, to enhance the estimate, it could be applied little by little. With W in place of, and for, layer thickness, concentration difference, and free falling velocity, Figure 7.2 illustrates how T changes. In most real-world scenarios, T is typically on the order of 1–100 s, with thick layers of highly concentrated fine sediment having the largest T.

Deformation of soft sediment happens as a result of four main types of stresses: (1) gravitationally unstable vertical profiles of bulk density of the sediment; (2) non-uniform confining loads; (3) the presence of the sediment on a slope (downslope component of body force); and (4) fluid flow, either internally or at the regular interface between the sediment and fluid. Apart from identifying these, Lowe (1975) also made an effort to separate out three deformation modes, which partially correspond to the states depicted in Figure 7.1. The exceptional preservation of the (deformed) laminae and the absence of evidence for anything other than local liquidization identify hydro-plastic deformation. Deformation in the liquidized state seems to correspond with Lowe's concepts of liquefied and fluidized deformation. It is questionable if different types of liquidized deformation can ever be reliably identified, and significant accuracy is lost as a result of his incorrect distinction between the liquefied, fluidized, and dispersed (sedimenting) states. One of the criteria for Lowe's fluidized deformation is the elutriation of fines; however, fines can also be flushed from high-concentration dispersions, which may result from liquefaction. Lowe also proposed the apparent absence of preserved lamination as a criterion of fluidized deformation. Lamination may be lost by percolation and other types of grain diffusion when a high-concentration dispersion owing to liquefaction experiences a large total strain, even though it is almost guaranteed to disappear during fluidization involving a large bed expansion (Scott & Bridgewater, 1975). Later on, in addition to hydroplastic deformation, reference will typically only be made to liquidized deformation due to the seeming ambiguity of the criteria. Regular folds can be produced inside a sedimentation unit or at the interface between two, as a result of some types of soft-sediment deformation that are primarily dependent on liquefaction. This process is entirely internal to the single or paired layers. According to continuity, the anticlines must rise at a rate inversely proportional to their width, and the synclines must descend at a rate corresponding to their width. Grain dispersion in these folds is considered fluidized if the rate of anticlinal ascent matches or surpasses the settling speed of the grains; this is because the grains either stay stationary in relation to the ground or ascend at a slower rate than the fold crests. We can calculate the minimum rate of folding as a criterion for fluidization if it can be shown that fluidization took place in the anticline cores, where the process is typically favored by the narrower folds. The anticlines of width L_{CD} ascend at a speed of V_{CD}, while the synclines of width L_{AB} descend at a speed of V_{AB} in the basic model illustrated in Figure 7.3.

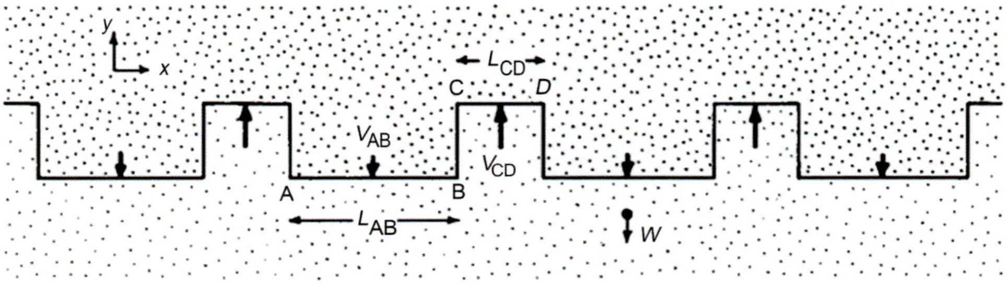

Figure 7.3. Definition diagram for soft-sediment deformation at the interface between two horizontal sediment layers when fluidization is involved.

Source: John R. L. Allen. Developments in sedimentology. Elsevier. ISBN: 0-444-41945.

By using continuity and replacing V_{CD} with the grain settling speed, the following criteria can be used:

$$V_{AB} \geqslant \frac{L_{CD}W}{L_{AB}} \tag{2}$$

is acquired to fluidize. The affected deposit can be used to estimate the lengths involved and the order of magnitude of W (Shrock, 1948; Chakrabarti, 1977; Seilacher, 1984).

7.1.1. Classification of Soft-Sediment Deformation Structures

There are noticeable distinctions between the various soft-sediment deformation structures, but these have not always been taken into consideration in the nomenclature that has emerged for these structures. To date, no exhaustive classification of them has been published; however, Lowe and Ricci Lucchi have made a significant beginning in this regard. However, because Lowe's categories are based on various primary criteria, it can be challenging to classify some structures under more than one category. Elliott and Swarbrick's earlier classifications are too narrow to be used broadly.

For classifications of soft-sediment deformation structures to be effective, a variety of criteria must be used, but only one type of criterion—structural geometry, deformation kinematics, or the source of the deforming stresses—should be prioritized. The classification provided in Table 7.1 groups soft-sediment deformation structures under primary categories that represent causative stresses because we here emphasize the mechanics of origin. Lesser criteria differ between classes based on the circumstances, but since this characteristic is significant theoretically, a lot of attention is paid to the number of sedimentation units or other types of layers involved in deformation. This is especially true for Class I structures, which are associated with bulk density sediment profiles that are gravitationally unstable. In essence, these are acknowledged as manifestations of specific types of Rayleigh-Taylor instability. A continuous or discontinuous density profile is relevant to the theoretical handling of this instability.

Table 7.1. Classification of Soft-Sediment Structures

I.	Gravitationally Unstable Profiles of Bulk Density:		
	1.	Continuous profiles (single sedimentation units) (convolute lamination).	
	2.	Discontinuous profiles:	
		a.	single sedimentation units (dish and pillar structures);
		b.	paired sedimentation units, upper layer not pierced (load casts);
		c.	paired sedimentation units, upper layer pierced (ball-and-pillow structures);
		d.	multilayer complexes of sedimentation units, with more dense layers pierced (sand balls).
II.	Non-Uniform Confining Loads:		
	1.	Single sedimentation units (complexly deformed cross-stratification);	
	2.	Paired sedimentation units (folding and faulting due to laterally spreading sand bodies).	
III.	Downslope Component of Body Force:		
	1.	Slides (internal stratification substantially undisturbed);	
	2.	Slumps (considerable disturbance of internal stratification, usually much over-folding and internal thrusting).	
IV.	Fluid Flow (Usually Sediment-Bearing):		
	1.	Shear stress essentially horizontal (simple deformed cross-stratification);	
	2.	Shear stress steeply inclined to general bedding (eroded and/or bent up laminae).	

Source: John R. L. Allen. Developments in sedimentology. Elsevier. ISBN: 0-444-41945.

7.1.2. Types of Soft-Sediment Deformation Structures

- When beds are folded and crumpled in intricate ways, it results in convoluted bedding. This type of deformation is found in fine or silty sands and is usually confined to one rock layer. In point-bar, delta, flood plain, and intertidal-flat deposits, convolute laminations can be discovered. While they usually measure between 3 and 25 cm, larger formations up to several meters thick have been observed (Boggs, 2006).
- Mud forms into wavy or flame-shaped structures called flame structures. Usually, these flames spread into the layer of sandstone above. Sand is deposited on top

of less dense mud, causing this deformation. The features that form alongside flame structures below are called load casts, which are technically a subset of sole markings. Load casts are the pendulous sand knobs that descend downward into the mud between the flames, whereas flames are the thin fingers of mud injected upward into the surrounding sands.

- Although they can also occur in limestones, sandstones, and evaporites, slump structures are most frequently found in mudstones and sandy shales. They can be found in places with steep slopes and rapid sedimentation rates as a result of the displacement and movement of unconsolidated sediments. These structures frequently have flaws.
- Thin, dish-shaped formations called dish structures are typically found in sandstones and siltstones. Each dish forms as a result of dewatering, and its size typically varies from 1 to 50 cm. Along with dish structures, pillar structures are also formed through dewatering. They are oriented vertically and cut through massive or laminated sands. The diameter of these formations can vary from a few millimeters to more than one meter.
- Sandstones are typically the sedimentary rocks that have sole markings on the underside that cover shale beds. Their directional features allow them to be used for determining the direction of flow of ancient currents. When a bed erodes, a groove is left behind that is eventually filled in by sediment, forming sole markings (Boggs, 2006).
- Seismic waves from earthquakes disturb sedimentary beds, which are known as seismites (Seilacher, 1969). They are frequently employed to interpret the local seismic past. Sand volcanoes, sand blows, and some clastic dikes are examples of soft sediment deformation structures to which the term has also been applied (Bachmann & Aref, 2005).

7.2. DEFORMATION STRUCTURES RELATED TO GRAVITATIONALLY UNSTABLE VERTICAL PROFILES OF BULK DENSITY

7.2.1. Convolute Lamination

This term refers to a laterally extended sequence of roughly regular folds that are either limited to the upper portion of a single sedimentation unit or developed throughout it (Figures 7.4 and 7.5). Generally speaking, there is little to no evidence of dicollement, the folds strengthen upward from the undeformed base or from some unaffected higher level, and they may die away toward the top. The folds have been reported to have horizontal extents of up to 750 km^2, and their degree of development is similar within any given unit. The folds' wavelength increases with the thickness of the bed or deformed zone. The folds, which mostly follow Fleuty, can be soft, open, closed, isoclinal, box-like, or even elastica-like. Seldom are there strong asymmetries and recumbencies; instead, most folds are symmetrical or mildly to moderately asymmetrical, and there are very few instances of a barely overturned limb. Axial surfaces of asymmetrical folds typically lean in the direction of the paleocurrent. Anticlines can have cores of either highly distorted and blurred laminae or even structureless sediment. They are typically sharper than synclines. Numerous laborers characterize the folds in three dimensions as resembling upright cones or domes and basins. In certain cases, these are linked to or beneath cylindrical folds that are oriented either in the direction of the paleocurrent or perpendicular to it.

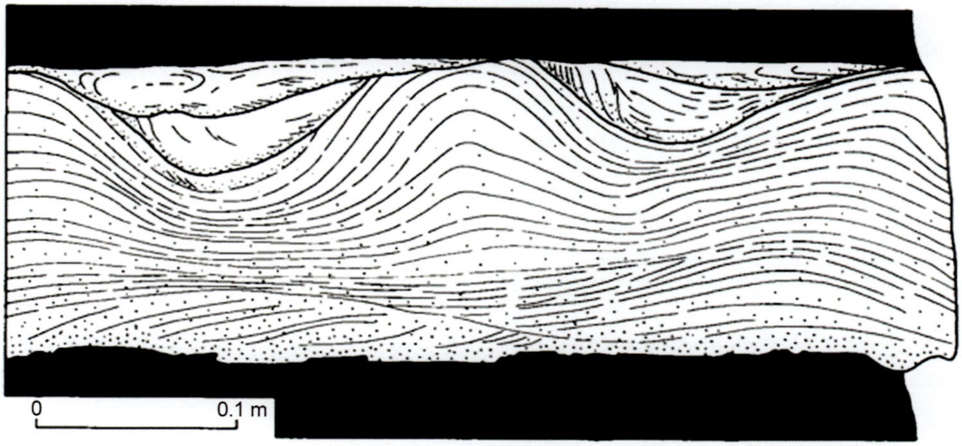

Figure 7.4. Syndepositional convolute lamination (slightly schematic) in turbidite sandstone of Pico formation (Pliocene), Ventura basin, California, U.S.A.

Source: John R. L. Allen. Developments in sedimentology. Elsevier. ISBN: 0-444-41945.

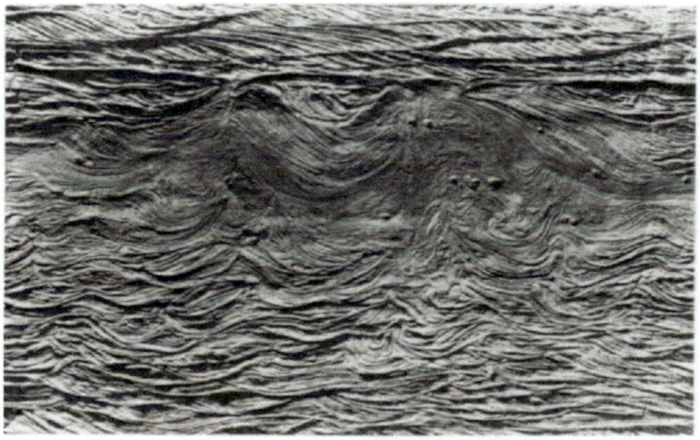

Figure 7.5. Syndepositional convolute lamination (note erosional truncations and evidence of earlier folding within fold limbs), River Ganges alluvium, Bhagalpur, Bihar, India. The deformed layer is normally graded, from sandy up to silty, and sandy sediment, in cross-laminated sets climbing at a low angle, reappears above the erosion surface truncating the folds toward the top of the section. Length of bed shown approximately 0.47 m.

Source: John R. L. Allen. Developments in sedimentology. Elsevier. ISBN: 0-444-41945.

Convolute lamination is generally found in turbidites or their fluvial and deltaic analogs, which are cross-laminated very fine sands or coarse silts that contribute to graded beds that are in the order of 0.1–1.0 m thick. Horizontally laminated fine or very fine sands and interlaminated clean with muddy silts are frequently affected as well. One case of convoluted lamination from a reversely graded bed is reported. Convolute lamination occurs in deposits with no discernible normal grading. There are similarities between convolute lamination and certain soft-sediment deformations found in coarse-grade, frequently cross-bedded sandstones.

Based on the relative timing of deposition and deformation, Kuenen, 10 Haaf, and Nagtegaal's work enables the distinction of three types of convolute lamination (Figure 7.6).

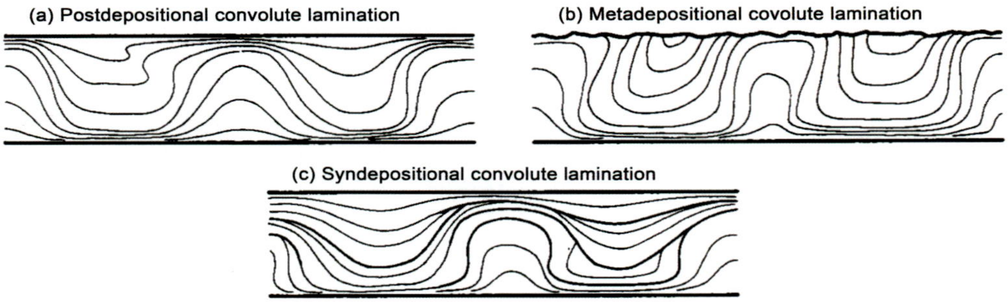

Figure 7.6. Types of convolute lamination.

Source: John R. L. Allen. Developments in sedimentology. Elsevier. ISBN: 0-444-41945.

Post-Depositional Convolute Lamination develops after the burial process has begun. The center of the convoluted zone is where the folds have the largest amplitude, and they die away upward and downward toward the undeformed laminae. They don't exhibit any internal discordances beyond what would be expected (e.g., amid sets of cross-laminated materials).

Either right before or right after deposition stops, metadepositional convolute lamination occurs. The unit top, which may have sand volcanoes above anticlines, or a nearby internal erosion surface truncate the folds. Numerous employees note this structure.

Syndepositional convolute lamination, which forms episodically rather than continuously during bed deposition, is shown in Figures 7.4–7.6(c). The main observation supporting this is the erosional thinning of laminae groups from synclines toward anticlines, which frequently ends with the truncation of anticlinal crests. Similar to the metadepositional variety, the top of the bed or a planar erosion surface immediately beneath it may cut through the folds as a whole. Almost as common as the metadepositional kind appears to be the syndepositional convolute lamination.

Regardless of the cause of convolute lamination, all researchers appear to agree that the deformed sediment was either hydroplastic or liquidized. Based on the extreme thickness variations displayed by individual laminae, the smoothness of folds, and the lack of micro-faults, it is likely that the latter was its state most of the time. We can deduce that the primary deforming forces acted primarily vertically, resulting in an isotropic pattern of shear stress vectors distributed about scattered centers in the horizontal plane, based on the folds' predominant basin-and-dome form. However, a secondary horizontally directed force—possibly fluid drag or a downslope component of the sediment weight—probably also existed in situations where the folds lean toward the vertical.

Migliorini attributes convoluted lamination to the localized expulsion of pore waters from a rapidly accumulating bed, while 10 Haaf and Chakrabarti identify this phenomenon as a contributing factor. Although it is apparent that such expulsion did occur in certain cases, this does not explain the post-depositional structure or offer a motivating factor. Convoluted lamination can result from the expulsion of trapped air in certain situations, such as on beaches. According to Rich and Kuhn-Velten, convolute lamination is caused by an overlying sedimentary layer sliding downslope. The main points of contention are that (1) fold limb movements are rarely horizontally greater than vertically, and (2) folds do not always or consistently lean toward the vertical.

Prentice and Sanders proposed that syndepositional convolute lamination could be explained by the effect of fluid drag on rapidly accumulating sediment. According to Sanders, the deposit becomes locally and sporadically cohesive, with the hardened masses being dragged forward and eroded at the same time to form the anticlines, and grains continuing to accumulate in the constantly replenishing troughs of liquid-like material in between. There was no adequate explanation provided for the cohesive and liquid-like state alternation.

Kuenen provides an additional primarily hydrodynamic explanation for syndepositional convolute lamination, proposing that deformation is driven by high fluid pressure at the crests of current ripples and low pressure in their troughs. Chakrabarti's theory, which Holland elaborates on, is that the structure is related to "some irregular distribution of pressure or suction, drawing up the anticlinal convolute folds and depressing the laminations between." These theories' main drawback is the assumed coupling between folds and pressure fluctuations. Variations are essentially convected in Kuenen's model at the ripple celerity, whereas in Holland's model at the speed of the depositing current. Nonetheless, in every instance, the convection velocity needed to have been significantly higher than the pace at which the anticlines could have developed downstream. Examining syndepositional convolute lamination reveals that multiple ripples crossed each anticline during growth, making it rare for the necessary phase relationship between the folds and bed form to be reached. The fact that (1) fold wavelength correlates with fold-layer or bed thickness rather than ripple wavelength and (2) syndepositional convolute lamination is not limited to cross-laminated deposits are two more criticisms leveled against Kuenen's model. Kuenen's alternative proposal, which suggests that the folds represent differential loading by the ripples, is met with similar objections. Additionally, there is the issue that the folds are not always ripple-like in plan. The theory put forth by Dzulynski and Smith that longitudinal patterns in convoluted folds are connected to secondary flows in the depositing current does not require resolution of these issues.

E. Williams offers a very different explanation, he thinks that the cores of the folds in convolute lamination are caused by the laminar flow of liquidized sediment around unaffected portions of the deposit that cling to the layer's floor and roof. This notion is refuted by the fold geometry (Shrock, 1948; Chakrabarti, 1977; Seilacher, 1984).

Finally, a gravitationally unstable bulk density gradient may be the cause of convolute lamination, an instability (Rayleigh-Taylor) phenomenon that can occur during or after sediment accumulation. The most likely explanation for the unstable gradient was proposed by Anketell et al. (1970), who suggested that vertical variations in grain concentration—possibly imposed during deposition—were the cause of the gradient. Allen developed a related theory unique to normally graded beds, which states that the local grain concentration and bulk density may decrease more quickly in the lower than in the upper part of the bed during sedimentation after liquefaction, reversing an initially stable gradient or steepening an initially unstable one. Although the explanation was created for the post-depositional structure, it still holds true for the other varieties. For instance, in the case of the syndepositional type, each increment of quickly deposited sediment must almost instantly dissolve upon deposition. The vibratory action of pressure fluctuations in the turbulent current could easily accomplish this. These theories are generally supported by field data. Ramberg's calculations provide a constant fold wavelength/layer thickness ratio that is close to three for a viscosity-dominated system (Model V) that is comparable to a graded layer. A ratio of roughly two was measured by Crowell et al. (1966) from the Pico Formation turbidites. The consistency between the folds' three-dimensional form, which essentially has axial symmetry, and the force pattern predicted for a system deforming in the gravity field but not undergoing net horizontal translation is perhaps most telling.

7.2.2. Load Casts

In the field of gravity, a horizontal layer of dense fluid layered over a layer of light fluid is unstable because the system's potential energy does not reach a minimum. When sediment layers are contrasted and arranged in this manner, the Rayleigh-Taylor instability is latent as long as their yield strengths are large enough, but it becomes actual during liquidization. Although load casts were once highly contentious, they are now acknowledged as one of the most prevalent soft-sediment deformations that illustrate the instability of discontinuous layered systems. According to Kuenen, they are "swellings in the shape of slight bulges, deep or rounded sacks, knobby excrescences, or highly irregular protuberances" that are found under gravels or coarse sands that are superimposed on finer sands, or on the undersides of sands that are overlying muds. Fuchs reported load casts for the first time, naming them Fliesswiilste and reproducing them experimentally. Sorby, Kukuk, and Shrock later reported on them as well. Though they can be found in shallow marine, deltaic, and riverine settings, these interfacial patterns are most frequently found in turbidites. Sometimes load casts can be seen in pyroclastic successions, layered igneous bodies, and lake sediments.

In the English Lake District's Borrowdale Volcanic Series, Sorby's specimen of load casts in a vertical section is replicated in Figure 7.7. It displays rounded pockets of coarse ash bulging into volcanic mud, which itself protrudes upward in more pointed, diaper-like shapes. A bolder layering in the mud is generally still conformable to the diaper shapes, with the possible exception of a few areas where slippage may have occurred due to excessive shear. Faint laminae in the coarse ash trace the edges of the pockets. Similar relationships are portrayed by Hills and Ballance. In these examples, the mud at the heads of some diapers is pulled out into long streaks, creating the "anticrests" created by Lamont, the "load waves" created by Sullwold, and the "flame-structures" created by Kelling and Walton. The diapers and pockets in Figure 7.7 are generally symmetrical about the near-vertical axes, but in other instances they lean significantly away from the vertical in a consistent direction, which is typically the direction of the paleocurrent. These asymmetrical load casts are referred to by Prentice as "flow casts" and 10 Haaf as "squamiform." The load cast-related flame structures pierce the upper layers of the involved sediment, but they do not penetrate deeper.

Figure 7.7. Vertical section through load casts beneath volcanogenic sand bed overlying mud, Langdale Slates (Ordovician), English Lake District.

Source: John R. L. Allen. Developments in sedimentology. Elsevier. ISBN: 0-444-41945.

Figure 7.8. Photograph from below of load casts formed experimentally by collapse of a layer of soft sand mixed with Plaster of Paris into a layer of mud, when both became liquidized in response to a sequence of shocks. Specimen 0.3 m across.

Source: John R. L. Allen. Developments in sedimentology. Elsevier. ISBN: 0-444-41945.

Figure 7.8 depicts the appearance of load casts in a plan from an experimental example created by vibrating the layers after sifting sand mixed with plaster of Paris onto mud. Numerous instances of sand protuberances are roughly equant and have a tendency toward polygonal form in the geological record. Structures with an asymmetric vertical profile are typically limited to subparallel elongate forms. Reinhardt and Cleaves describe large features resembling load casts from gravel overlying saprolite. The wavelength of load casts at mud-sand interfaces is typically in the range of 0.01–0.1 m but can reach the order of 10 m. Needham, Francis, and Howells all describe remarkably large load casts from volcanic sequences. It is well known that load casts cannot form unless sediments have little to no strength. When the ash was added, Sorby described the mud in Figure 7.7 as being in a "creamy semi-liquid condition." Shrock characterized the lower sediments as either extremely viscous or hydroplastic. Sims is able to connect past earthquakes, which likely liquefied the sediments, to the creation of load casts and other soft-sediment deformations in contemporary lake deposits.

The forces involved in load casting were not well understood. Shrock states that "hydroplastic sediments, if unequally loaded with sand or gravel, yield to the weight of the superincumbent load by flowing" to explain load casts. However, this explanation makes no mention of the fact that load casts primarily form as a result of gravitational instability in the system. Kelling and Walton presented this concept, but it had a lot of Shrock's ideas mixed in. They likened load casts to salt domes generated by gravitational instability, to which diapiric igneous bodies and delta mud-lumps may be added. However, they asserted that unstable equilibrium was only generated when the upper layer initially projected down into the lower layer.

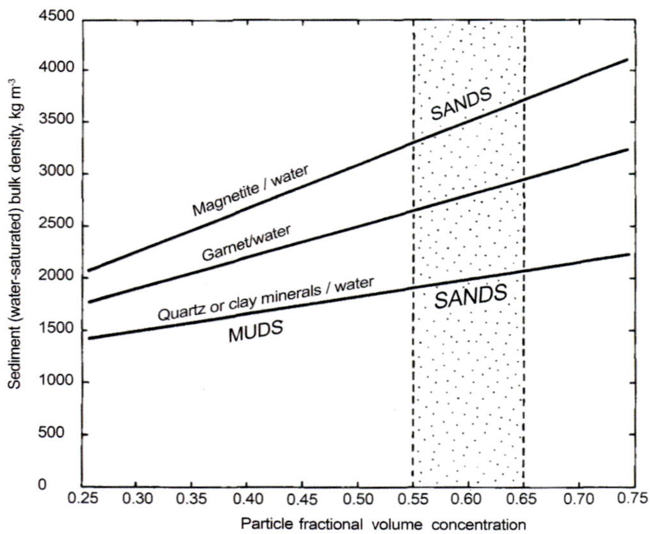

Figure 7.9. Bulk density of water-saturated sediment as a function of particle density (mineralogy) and fractional volume concentration.

Source: John R. L. Allen. Developments in sedimentology. Elsevier. ISBN: 0-444-41945.

While such irregularities cannot usually be the cause of load-casting, they can help localize it, as in the case of load-casted flute and tool marks. Dzulynski and his colleagues have made significant progress in explaining load casts in terms of the post-liquidization gravitational instability of a dense layer overlying a light layer by means of qualitative experiments and theoretical references. The bulk density of water-saturated mineral-density sediments as a function of particle concentration is displayed in Figure 7.9, which indicates that systems with quartz sand covering clay-mineral mud have a significant driving force. Moreover, they could explain the narrow dippers and flame formations by interpreting the lower effective viscosity of the upper layer in relation to the lower layer, as per Artyushkov's theory.

There are currently no experimental or theoretical studies of Rayleigh-Taylor instability in sand-mud systems. For both the viscosity-dominated and the inertia-dominated cases, there are two general theories, though. Which is currently suitable can be determined using Ramberg's standard for disregarding inertia:

$$L \leqslant \frac{2\pi}{k}\left(\frac{\nu^2}{g}\right)^{1/3} \qquad (3)$$

where; 'L' represents the load-cast wavelength; 'g' is the acceleration caused by gravity, k ≈ 1 for sand-mud systems; and 'v' is the kinematic viscosity of the sediment layers (assuming equal viscosities). With a kinematic viscosity of 0.0005 m^2s^{-1}, which translates to a bulk density of 2,000 kg m^{-3} and an effective viscosity of 1 Ns m^{-2}, inertia becomes insignificant at wavelengths less than approximately 0.015 m. Since this value is within the typical range for naturally occurring load casts, a purely inertial theory might not be as satisfying as one that considers both viscosity and inertia.

7.2.3. Pseudo-Nodules and Ball-and-Pillow

Never does load-casting cause such severe deformation that the diapers rising from below pierce the upper sediment layer. However, there are other two-layer systems as well as some multi-layered ones where the denser layers are entirely destroyed, reducing an extensive bed or beds of sand to a collection of isolated masses embedded in a lower-density mud that is more or less continuous and represents one or a combination of several original strata. These masses have a variety of shapes; some even have a cigar or sausage-like appearance. Others are pillow- or bolster-shaped. Two other characteristics set them apart: (1) structures that were preserved on the original sand-mud interface, such as animal tracks, ripples, and tool marks, are wrapped around their outsides; and (2) internal lamination closely resembles the external form, except for areas where the tops of masses are erosively truncated.

There is much debate regarding the proper classification and nomenclature for these intricate structures, but two factors stand out as particularly helpful: (1) the size of the masses, and (2) the way in which they agglomerate. A structure that consists of a single, laterally extensive row of uniformly sized sand masses that resembles a concretion horizon is referred to in this context as a "pseudo-nodule." A sheet of sand masses of different sizes arranged vertically and horizontally in a mud matrix is known as a ball-and-pillow structure. The terms "hassocks," "pillows," "bolsters," or "rolls," which are all appropriate given their size, are frequently used to characterize the larger masses. The term "balls" is typically reserved for the smaller ones. Unlike load casts, pseudo-nodules are more prevalent in shallow-marine and deltaic deposits and have a different environmental distribution. Compared to deposits made in lakes or by rivers, turbidite deposits have fewer examples.

Based on these accounts, Figure 7.10 summarizes the traits and connections of pseudo-nodules. In certain cases, an undisturbed bed can be followed laterally through a zone where loads are cast, into a row of pseudo-nodules, and allegedly even into a ball-and-pillow structure. Pseudo-nodules are similar to the parent bed in that they are 2–4 times wider and up to 8 times longer than thick. Although individual masses several meters wide are known, and many are no more than 0.1 m across, a typical width is approximately 0.5 m. Nodules that are noticeably longer have the same orientation, which is typically parallel to the depositional strike. The axial surfaces of the synclinal folds, which are determined by the internal bedding and external form of pseudonodules, range from vertical to moderately leaning. Within a set, the leaning direction is typically constant. A row of sand masses often has a common top formed by a laterally extensive erosion surface; this relationship suggests that the sand masses are convoluted, as in metadeposition. The laminae internal to a pseudo-nodule are observed to have been severely thinned and bent inward and even downward around the edge when there is no such truncation, as in the lower row of Figure 7.10. Internal discordances can occasionally show an episodic growth that resembles syndepositional convolute lamination.

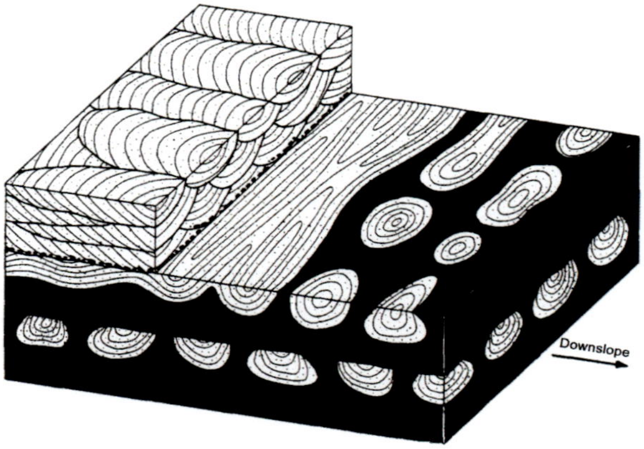

Figure 7.10. Schematic lateral passage of a continuous sand bed overlying mud into pseudonodules.

Source: John R. L. Allen. Developments in sedimentology. Elsevier. ISBN: 0-444-41945.

Part of a ball-and-pillow structure sheet with a lateral extent of at least 1.5 km is depicted in Figure 7.11, which is located in the middle of an upward-coarsening deltaic sequence that is Carboniferous in age. It is one of several so-called "slump sheets" that Kuenen talks about. In this example, a laterally extensive sheet of sandstone truncates the zone containing sand balls and pillows by approximately 2 meters. The sand masses are located in mudstone and have an average diameter of 0.4 meters and an apparent length that varies from less than 0.1 meters to slightly over 1 meter. Though a few are rod-like and some are S-shaped, the majority are shallow, flat-lying, dish-like bodies with overturned edges followed by the internal bedding (Figure 7.12). Ksiazkiewicz discovered comparable forms in disturbed turbidites. This specimen exemplifies a multitude of ball-and-pillow formations, primarily documented from shallow-marine and deltaic environments. There are sand masses in some sheets that are astonishingly large, spanning several meters, while others are thought to cover thousands of square kilometers (Figure 7.13).

Figure 7.11. A zone of ball-and-pillow structures (sandstone in mudstone) overlying shales and mudstones, and overlain (apparently erosively) by sheet sandstone and interbedded sandstone and mudstone. The ball-and-pillow zone is approximately 2 m thick. Coal measures (Carboniferous) east of Amroth Castle, Amroth, Dyfed, Wales.

Source: John R. L. Allen. Developments in sedimentology. Elsevier. ISBN: 0-444-41945.

Figure 7.12. A single saucer-shaped sandstone pillow (note overturned edges) embedded in mudstone. Coin approximately 0.03 m across.

Source: John R. L. Allen. Developments in sedimentology. Elsevier. ISBN: 0-444-41945.

Figure 7.13. Vertically flattened sigmoidally folded sandstone bed in mudstone. Coin approximately 0.03 m across.

Source: John R. L. Allen. Developments in sedimentology. Elsevier. ISBN: 0-444-41945.

Additional instances of ball-and-pillow structures exist, which are primarily made up of sand masses that are small enough to be classified as balls. Many are from areas with shallow water. Certain layered igneous bodies have load casts that are connected to detached, ball-like masses of heavy crystals.

Pseudo-nodules and ball-and-pillow structures are now considered soft-sediment deformation features, rather than concretions. Chadwick referred to them as "storm rollers," suggesting they may be related to intense wave action causing liquefaction. Cooper, Kuenen, Macar, and Antun, and Hubert et al attribute these structures to downslope movement of interbedded sand and mud, with some vertical displacements also noted. Alternatively, Sorauf believes many of these features are due to the gravitational instability of layered sediments with different densities, leading to primarily vertical movements. Kaye and Power, Howard and Lohrengel, and McBride et al propose that differential loading of soft mud by rapidly deposited sands is the main cause, with earthquakes potentially contributing in some cases. Sorauf suggests that isolated sand pillows may result from differential loading by sand filling channels.

It seems unlikely that pseudo-nodules and ball-and-pillow structure are primarily caused by translation on a slope. Researchers Kuenen and Bogacz et al. created axially symmetrical structures in the lab that resembled numerous pillows and pseudo-nodules (e.g., Figure 7.12) by giving unconsolidated multilayers a shock. During deformation, there was no net translation of masses horizontally. A complete sheet of foundered sand masses cannot be translated horizontally using a criterion that could have resulted from the oblique descent of sand rafts, as demonstrated by the shapes in Figure 7.13 and the comparable ones noted by Ksiazkiewicz. The slight leaning of certain pseudo-nodules is the only indication of significant horizontal displacement; other evidence, such as decollement or internal shear planes, is typically absent.

After considering the arguments of Dzulynski, Anketell, and Dzulynski, and Anketell et al., the most likely explanation for pseudo-nodules and ball-and-pillow structure is that they reflect the Rayleigh-Taylor instability of liquidized sand and mud layers. While pseudo-nodules unmistakably depict the two-layer scenario, the ball-and-pillow structure might not always capture the disruption of multilayers properly. A few sheets of sand balls and pillows may have originated from the repeated detachment—or sedimentary stopping—of large amounts of liquidized sand from the underside of a single, thick sand bed, according to findings by Hayashi and Iwai, Dott, and Lindsey. These findings indicate that sand balls appeared to have detached from load casts. A sheet of pillows and balls that makes up Goldring's Grand Slump Member may contain vertical, angular sand blocks, supporting Anketell et al.'s suggestion that "brittle" sands could founder in a similar way.

7.2.4. Heavy Mineral Sags and Drops

It was demonstrated by AD Stewart, Selley, and Lowe that when all of the heavy mineral laminae and bands in quartz sands are liquidized, they deform in response to density instabilities. When sufficiently close vertically, individual heavy mineral layers, which are frequently only a few grains thick and infrequently exceed 0.01 m., deform as a group. They are interspersed with bands of quartz or quartz-feldspar sand. Irving claims that the heavy minerals in Stewart's and Selley's examples, which come from the Torridon Group of Scotland, are almost exclusively specularite, while Selley claims that the main heavy minerals are magnetite, haematite, and ilmenite. Their density is nearly twice that of feldspar and quartz, at 5,000 kg m^{-3}.

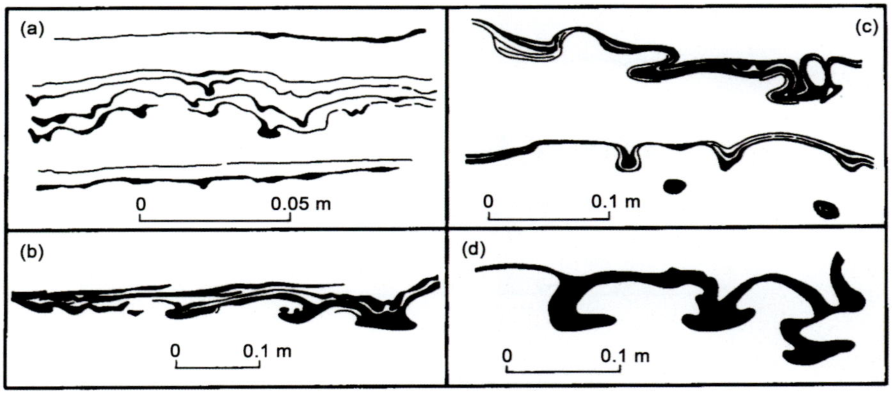

Figure 7.14. Drop and sag structures in heavy mineral bands (black) within sandstones, Torridon Group (Precambrian), northwest Scotland.

Source: John R. L. Allen. Developments in sedimentology. Elsevier. ISBN: 0-444-41945.

AD Stewart divides drop production into four phases. First, the heavy mineral layers sag gently and locally, with a horizontal spacing that is 5 to 10 times the thickness of the layer. During the second stage, the arched-up portions between grow attenuate while the heavy grains flow laterally into the sags, which thicken as a result. At this point, inward overfolds may form at the edges of sags. Occasionally, an arch will slide into a nearby sag and gather into folds that resemble recumbent concertinas. These phases are depicted in Figure 7.14(a) and (b). A sag transforms into a nearly autonomous drop in the third stage, sinking through the quartz sand and staying connected to the parent lamina by a weak wake. These structures are shown in Figure 7.14(c), albeit the section's plane shows no wakes. According to Stewart, in certain cases, the last stage (Figure 7.14(d)) depicts drops spreading sideways to form a vortex ring that gives rise to a secondary drop.

AD Stewart provided compelling evidence for the origin of heavy mineral sags and drops in liquidized sands through density instability, drawing comparisons between two phenomena: (1) Kuenen's experimentally produced drop-like sand masses, which were created by liquidizing a sand layer overlying mud; and (2) Thomson and Newall's blobs and vortex rings, which were created by dropping a liquid into another miscible one. It can be observed that Stewart's fourth stage alone is absent in Figure 7.15, which depicts the stages in the formation of spherical drops—Thomson and Newall's blobs—by the instability of a narrow horizontal cylinder of colored glycerol added to plain glycerol. Its validity in the case of heavy mineral drops is somewhat questionable because Thomson and Newall were only able to create vortex rings in liquids with kinematic viscosities comparable to water (1×10^{-6} m^2 s^{-1}). It is 10^2–10^3 times larger in liquidized sand at a large enough grain concentration to allow lamination retention, when only spherical drops, as those calculated by Lowe, may be produced. If drops hit a rising surface of redeposited grains, they may have flattened instead of exhibiting vortex rings in Figure 7.14(d). However, there is a noticeable difference in scale and density between Thomson and Newall's drops and heavy mineral sags; it is unclear how these differences could impact the formation of vortex rings.

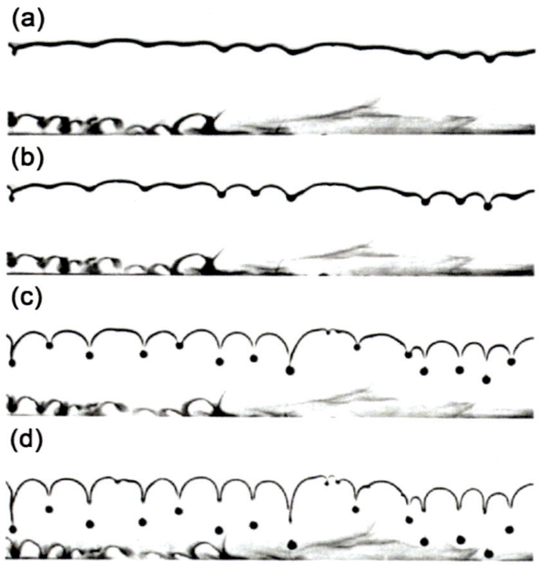

Figure 7.15. Successive stages in the gravitational instability of a narrow cylinder (approximately 0.21 m long, 0.002 m diameter) of dyed glycerol mixed with very little water and introduced by means of a capillary tube into plain glycerol at room temperature.

Source: John R. L. Allen. Developments in sedimentology. Elsevier. ISBN: 0-444-41945.

When assigning a bulk density of 3,200 kg m^{-3} to the liquidized heavy mineral layer and 1,900 kg m^{-3} and an effective viscosity of 1 Ns m^{-2} to the quartz sand, Figure 7.14(c) depicts drops with a diameter of 0.01 m that appear to have fallen approximately 0.05 m. Based on Stokes' law, the drops, once independent, could have fallen this distance in only 0.7s. Even though it took several seconds for the drops to form, the entire amount of time appears to be well within the duration needed for the sedimentation of liquidized sand beds with typical thickness (Shrock, 1948; Chakrabarti, 1977; Seilacher, 1984).

7.2.5. Deformation Involving an Ephemeral Layer: Passively Deformed Beds

Only sedimentary layers—which persist after disturbance in some palpable form—have been included in the deformation structures that have been reported thus far. We now move on to situations in which deformation appears to have been dependent on the existence of a pore fluid layer between sedimentary layers; this layer was eventually expressed from the sequence and is therefore invisible to current inspection.

H is a potential example of deformation associated with an ephemeral fluid layer. D. Johnson's bedding was distorted passively. He reports repetitions of the vertical sequence from the late Precambrian shallow-marine rocks of north Norway: (1) strongly deformed cross-bedded sandstones (passively deformed bed); (2) weakly deformed cross-bedded sandstones; and (3) undeformed cross-bedded sandstones. The severely distorted rocks, known as convolute-bedded, typically range in thickness from 1 to 4 meters, but they can "form unbroken sequences up to 8 meters thick." The passively deformed beds that follow the sharp above curve into elongated shallow basins divided by sharper anticlines, but they do not exhibit any signs of internal liquidization. The upward sand injections that, in certain cases, appear to have penetrated the bed, are localized by the anticlinal hinges. Passively deformed layer's uneven top is buried by undulating sandstones.

The tone and scene of these scenes lend credence to HD Johnson's theory that liquidization—likely caused by a storm—took place in a layer of sand that started just below the sea floor. The unaffected, but hydroplastic, sand above the liquidized layer became the passively deformed bed that covered the liquidized layer. "The quasi-liquid nature of this bed causes the overlying bed to founder, particularly where there is an upward escape of pore water" and sand through the overlying bed is the only explanation provided for the forces causing deformation. Only in the presence of an appropriate prior driving force is a liquid-like state sufficient to cause deformation. Either (1) the passively deformed bed had a bulk density greater than the sand below, or (2) a fluid layer developed between the liquidized sand and the passively deformed bed because the lid was subsiding more slowly than the re-sedimenting dispersion below, could have been a driving force. This easy experiment illustrates why it is difficult to rule out the second possibility. Rapidly sifting fine sand into water creates a layer that is roughly 0.10 m thick. A much thinner layer of extremely fine sand is then deposited on top of it. The sands melt when the container is jostled. Puffs of grain, originating from both seemingly linear and circular vents, appear above the sediment-water interface almost instantly. Small craters are later found on the surface, some of which have modest cones. If the sands are treated independently or if the succession is inverted, none of these effects manifest. The bed that is passively deformed is comparable to an upper layer of extremely fine sand. It exhibits unstable behavior because, as a layer, it sinks more slowly than the fine sand that is accumulating below and has a bulk density greater than that of the plain water layer that is building up above the sand.

7.2.6. Deformation Involving an Ephemeral Layer: Dish Structures

Dish structures are a relatively new finding. It appears that Crook initially referred to them as "discontinuous curved lamination," which contradicts Pedersen and Surlyk's assertion that Natland and Kuenen were misrepresented. Stauffer and Wentworth subsequently published near-simultaneous short accounts under the term "dish structure," but it wasn't until Lowe, LoPiccolo, and Lowe's work that thorough studies became available. I have also observed dish structures in the thicker mid-Ordovician turbidites at Cardigan in Wales. Dish structures are common in the coarser-grained deposits of turbidity currents and related high-concentration flows. Seldom are the structures observed in fluviatile and shallow-marine sediments. In two of the coarser-grained 179 ash layers that were deposited by air falls among Eocene marine diatomites, Pedersen and Surlyk discovered dish structures. These reports result in the description that follows.

Figure 7.16. Vertical section through fine-grained sandstone with small dish structures. Specimen 0.035 m across. Note sharply defined bases and muddy lower zones to dishes, and also the relatively clean tops. Jackfork Group (Carboniferous), southeast Oklahoma.

Source: John R. L. Allen. Developments in sedimentology. Elsevier. ISBN: 0-444-41945.

Dish structures can be found in all sand grades and in laterally extensive sheets of coarse silt, with the majority of examples occurring in normally graded beds. Every structure (Figure 7.16) consists of two components: the sediment enclosed in the dish and surrounded by the dish or dishes above it, as well as the dish itself. Dishes are essentially flat to strongly concave up laminae or bands in sections typical of bedding that are richer in clay, silt, and occasionally organic matter than the surrounding sediment and are typically darker in color. They're lying echelon. Dishes are polygonal, circular, oval, or elliptical in plan, and typically lack a noticeable elongation. Their widths range from approximately 0.02 m to over 0.5 m, and their vertical spacings range from less than 0.01 m to as much as 0.08 m. Dishes are always made with sharp bases. The tops are typically gradational (Figure 7.16), although there are known instances of sharp upper boundaries. Two types of changes were noted by Lowe & LoPiccolo (1974) in each dish and the infill that covered it: (1) size-grading, which was indicated by an increase in the mean sand-grain size, and (2) content-grading, which was indicated by an increase in the amount of interstitial clay, silt, and organic matter (e.g., Figure 7.16). Often, content grading culminates upward in a band of clean-looking, pale-colored sand immediately beneath dishes that are overlaid (Figure 7.16; see also Pedersen & Surlyk, 1977). Seldom does the raised edge of one dish come into contact with the base of the dish above it.

The thickness and coarseness of the bed appear to be reflected in the vertical patterns and relationships to previously formed sedimentary structures displayed by dish structures in deep-water sands. As observed by Lowe and LoPiccolo, increasingly concave dishes lie on top of dish structures with relatively flat bases and narrow, sharply upturned margins in the thinner and finer-grained beds (Sequence A in Figure 7.18(a)). Dish edges frequently break through parallel lamination in the lower portion of the bed, while cross-lamination or convolute lamination in the upper portion is frequently broken (Figure 7.17). Lowe and LoPiccolo observed that the previous laminae are frequently bent sharply over or streaked out so they are hardly noticeable in the pale zone at the top of a dish filling (Figures 7.17 and 7.18(b)). Several of the pale areas on dishes that subtly cut through earlier laminations are actually entirely devoid of structure, indicating a severe degree of disruption. However, abrupt offsets are occasionally noticeable at pale zones that are not well developed. Seldom is prior layering in the thicker, coarser-grained beds reported in conjunction with dish structures. In these beds, structureless sand is typically covered by a sequence of flat grading up to concave dishes (Figure 7.18(a), Sequence B). It's possible that the arrangement of dish shapes observed in deep-water sands isn't representative of other facies. Rust observed a slight upward tendency in the dish concavity when studying sand found in a river.

Figure 7.17. Dish structure developed after convolute lamination, Jackfork Group, Oklahoma, U.S.A. Numerous examples of laminae severed or distorted between successive dishes can be seen. Specimen approximately 0.15 m across.

Source: John R. L. Allen. Developments in sedimentology. Elsevier. ISBN: 0-444-41945.

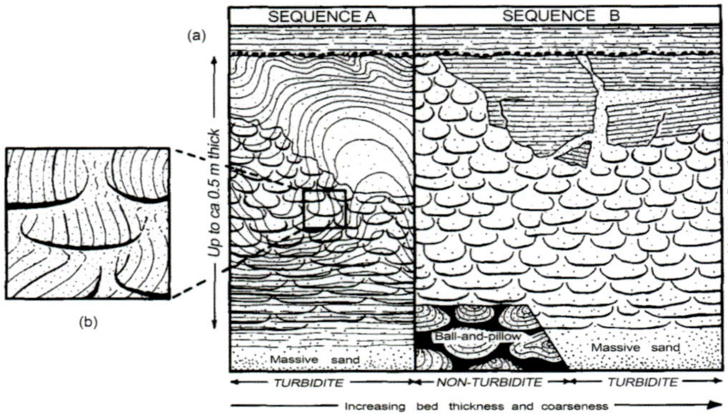

Figure 7.18. Dish structures. (a) Schematic vertical section through beds with dish structures, in which the dish-structured zone has a roof in sediment similar to that involved in the dishes. (b) Some schematic details of dish structures developed in a previously convolute-laminated bed.

Source: John R. L. Allen. Developments in sedimentology. Elsevier. ISBN: 0-444-41945.

Though at least two modes of termination are identifiable, the way that sheets of dish structures are limited upwards receives little attention. As suggested by Klein et al., Mutti, Ricci-Lucchi, and Kruit et al. depict sheets that are truncated by planar erosion surfaces covered in additional sand (Figure 7.18(a), Sequence B, left). In the second mode (Figure 7.18(a), Sequence A, Sequence B, right), the dish-structure zone ends within the affected sedimentation unit at a roof formed by either parallel laminated sand or metadepositional convolute laminated sand. The roof in Figure 7.19 is not level, and irregular veins of sand without structure pierce it; these veins usually cut but occasionally follow the bedding. View the photo from an angle to see how some veins seem to completely enclose asymmetrical blocks of sand, whose lamination has been twisted or tilted in comparison to the unaffected areas above. The laminae at the edge of a sizable block have bent and streaked upward in another example, as reported by Kruit et al.

Figure 7.19. Vertical section in dish-structured sandstone (below) extending up to an irregular roof beneath parallel laminated rocks. The latter are penetrated by veins of destratified (? liquidized) sand. Tertiary deep-sea fan deposits, San Sebastian area, northern Spain. Width of bed 0.45 m.

Source: John R. L. Allen. Developments in sedimentology. Elsevier. ISBN: 0-444-41945.

The earliest scholars to examine dish structures regarded them as primary. Wentworth explained them away as antidunes, but he also believed that the bed's dewatering had a role that was not clear. He also pointed out that some of the hydrodynamic challenges were a result of his assumptions. Stauffer believed that when highly concentrated debris flows, dish structures develop either internally or externally. Much like Wentworth, he wrote that "this trapped water may be the essential factor in the formation of dish structure," giving the pore fluid an unclear role. Corbett extended Stauffer's interpretation by suggesting that dish structures developed during dewatering through the modification of essentially primary laminations that were formed during the mass-shearing of liquidized sediment. Mostly because the features of dish structures don't match the proposed mechanisms, these explanations are hard to believe.

Dish structures are treated by Lowe and LoPiccolo as incidental features resulting from the slow dewatering of quickly consolidated or under-consolidated beds. The escaping water is said to be forced to follow horizontal flow paths beneath semi-permeable laminations or beneath dead water trapped in closure regions, causing the dishes to change.

Although an origin in earlier sediments is not ruled out, the escaping water is thought to have originated primarily in the bed in which the dish structures are developing. Lowe and LoPiccolo's explanation depends on the preexistence of semi-permeable laminations, which do not have to be laterally extensive. These laminae redirect water flowing upward through the bed in a horizontal direction. Under a semi-permeable lamina, the horizontal flow is thought to be consistently strong enough to entice any clay, silt, and organic matter that may be present in the sand, and in certain cases, strong enough to move the sand itself. The pale zones produced by this cleanup procedure were subsequently referred to by Lowe as "horizontal fluidization channels," which is probably a misnomer in the context of fluidization as previously understood. When part of the mostly horizontal flow in the developing pale zone below seeps upward through the semi-permeable lamina, mobilized fines are filtered out and trapped, forming a dish. The lamina simultaneously expands laterally due to deposition at its extremities and thickens upward. However, the escaping flow takes a vertical path that turns upward at the growing dish's edge. A second dish may form if there is another semi-permeable lamina in this path that is at a higher level, and so on, until a sizable section of the initial sedimentation unit has dish structures.

Numerous issues imply that at least one additional model should be taken into account. Semi-permeable laminae do not appear to be required for dishes to be produced. The 179 tuffs that Pedersen and Surlyk found to have two dish-shaped structures are categorically devoid of internal lamination, despite typically having a clearly defined normal grading. Their photos also show no evidence of previous laminations (apart from the dishes) within the two falls that retain the structure. The relative flatness of a dish is understandable where there is undeformed parallel lamination, but difficult to comprehend where the prior laminae are inclined, as in cross-laminated or convolute-laminated zones, or where the sand was apparently structureless. If dishes grow by deposition at their edges, as Lowe and LoPiccolo suggest, why is it that the structures are generally of such a modest concavity? One would expect marginal growth on far steeper upward paths than seem typical because the flow around the dish is compressed at its edge—compare that to that of a sphere. As suggested by Lowe and LoPiccolo, the relative flatness and horizontality of the dishes strongly suggest that production is in fact primarily controlled by gravity and not by the distribution of pore-fluid pressure in an essentially passive granular framework. Lastly, the uneven roofing of sheets by metadepositional parallel- or convolute-laminated sediment (Figure 7.19) implies that dishes did not form until most or all of the bed had already partially consolidated and cohesive, but it also indicates that at least some dish structures are postdepositional. This is further supported by the previous lamination, which is frequently offset between dishes. The fact that the sand masses used to produce dishes were mostly or entirely surrounded by fracture surfaces suggests that they were able to move independently.

Given these challenges, it is worthwhile to consider whether dish structures could arise from a stopping process within a water-saturated sedimentation unit that is marginally cohesive but not fully consolidated. Before stopping, there must be a horizontal cavity in the bed that is filled with water. These cavities, which only need to be very shallow, could form in the bed as a result of sedimentation following the liquefaction of areas where the sand had not consolidated. Crucially, Rust discovered that zones of ball-and-pillow sands were overlain by dish-shaped sands, which may have left water-filled spaces above them as they formed and solidified. However, if there was no cohesive sand left, a cavity might form if a small amount of water from the outside were forcefully injected into the bed; this would only be possible if the bed was below one or more layers of mud that could stretch under tension to make room for and, at the very least, temporarily seal the newly formed fluid layer. Whichever the case, the cavity should show up in the lower, coarser-grained portion of graded sand, where the permeability is greatest and the strength-giving mud content is lowest. The water will likely contain distributed fines in this situation.

A potentially unstable system in the gravity field can even be found in a very shallow water-filled cavity covered in sediment. Sand that is only marginally cohesive will fail and fall into this cavity in large masses rather than grain by grain. Only in cases where prior laminations are essentially horizontal, as in the case of parallel-laminated portions of the bed, will the subhorizontal failure surfaces follow. The failure surfaces in cross-laminated or convoluted portions will define flake-shaped masses instead of sheet-like masses, and they will be largely independent of the bedding. The cavity will rise through the bed as the masses are gradually released from the roof, stopping either where the sand is too dense to allow for more falls or when the cavity is abruptly emptied through a hole in its roof.

As each mass of sand sinks toward the cavity floor, three things ought to happen to it. A portion of the dispersed fines will be filtered out when the cavity fluid seeps upward through the sinking mass. Their rate of deposition will decrease inward along the flow paths and be largest over the mass's underside. Since the flow entering this area is least deflected from the vertical, the majority of fines will be trapped in the center of the mass. As can be seen, the dish-entrapped fines will consequently thin out from the center. Similar to Kuenen's experimental sand balls, the mass's edges will experience the greatest fluid drag, causing them to turn sharply upward. The intervening cavity fluid will be driven outward with rapidly increasing force as the developing dish structure gets closer to the cavity floor, where earlier dish structures lie. The outward flow could mobilize and shear the tops of dish structures already present on the floor, which could be the source of the pale zones and offset laminae observed, as well as replenish the cavity fluid with fines, as demonstrated by lowering a flat pebble a short distance through the water onto sand. In this manner, the floor will be formed to accommodate every dish, and sand will be moved laterally to fill in the gaps created by the dishes sinking closely together. Lastly, the dish's impact on the floor will likely cause the sediment to settle after liquefaction, which will further consolidate the material. Because the roof will rise faster than the floor, the cavity should therefore get deeper as it passes through the bed.

This model explains most of the properties of dish structures, including their form, vertical succession, roofing, and non-complete reliance on semi-permeable laminae. It involves the instability of a two-layer system in which one layer is ephemeral. The main challenge it poses is that, in instances where previous lamination is observed, the offsets between dishes are not as frequent nor as significant as one may anticipate. However, expecting these characteristics following descents of a few millimeters or centimeters may be incorrect (Shrock, 1948; Chakrabarti, 1977; Seilacher, 1984).

7.3. DEFORMATION STRUCTURES RELATED TO NON-UNIFORM CONFINING LOADS

7.3.1. Load-Casted Ripple Marks

Three types are recognized: (1) postdepositional, which involves partial ripple marks; (2) syndepositional, which involves a series of ripple marks; and (3) syndepositional, which involves a single ripple that consistently grows.

In their discussions, Dzulynski and Kotlarczyk, Dzulynski and Walton, and others have addressed load-casted ripple marks originating from post-deposition. In cross-section, the structure (Figure 7.20(b)) resembles a row of coarse silt or sand-embedded, plano-convex or biconvex lenses embedded in mud. The lenses are typically shorter than 0.1 m in length and rarely thicker than 0.01 m, with the thickest region typically located near an edge. The internal lamination of the lens varies, going from nearly horizontal near the thickest area of the lens too steep near its sharp side. As shown in Figure 7.20(a) and (b), Dzulynski and Kotlarczyk propose that the structures are the consequence of differential loading of a hydroplastic substrate. While the mud beneath the ripples is partially loaded according to the local ripple height and sand bulk density, the mud in the troughs between the ripples is essentially unloaded. The ripples would have submerged far deeper and formed sand balls similar to Kuenen's experiments if the mud had liquidized.

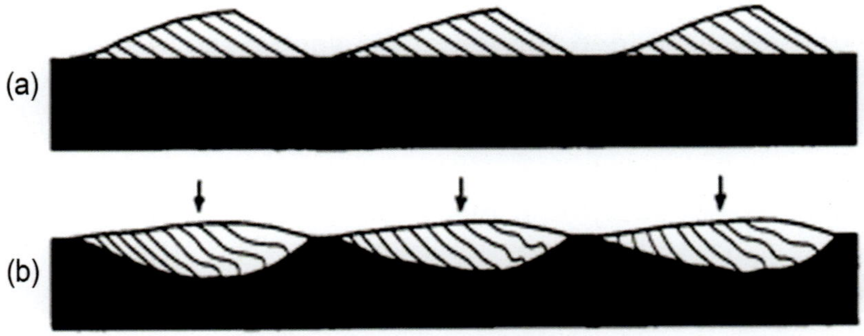

Figure 7.20. Schematic development of a row of load-casted ripple marks.

Source: John R. L. Allen. Developments in sedimentology. Elsevier. ISBN: 0-444-41945.

Dzulynski, and Kotlarczyk, and Dzulynski and Walton have also described clustered load-casted ripple marks of syndepositional origin. These formations are cylindrical masses of coarse silt or sand that are arranged in rows in the bedding plane. A radial internal structure of laminated sand wedges and mud fingers is visible in the cross-section of these structures (Figure 7.21(a)). An inventive explanation for them was put forth by Dzulynski and Kotlarczyk (Figure 7.21(b)). Every mass is thought to be the result of multiple ripples being loaded into weak mud at roughly the same location, each adding to the previous one and pushing it forward and downward until a well-developed radial structure is formed. If multiple adjacent sites experience simultaneous loading, a row of cylinders may be produced. Dzulynski and Kotlarczyk did not address the question of how sand gets to every part of the bed or why the distribution of the cylinders requires that foundering be localized in a certain way. However, after becoming weaker locally, the mud would usually continue to be weaker, allowing for more deformation. If there was a significant amount of sand in suspension, it could reach every area of the bed.

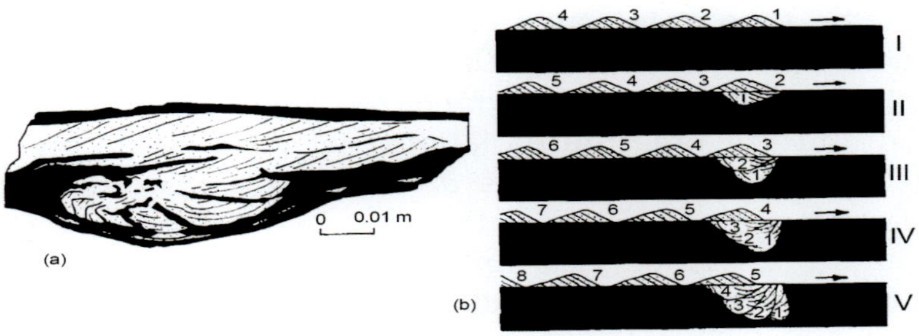

Figure 7.21. Clusters of load-casted ripple marks. (a) A cluster of load-castcd ripple marks, Kliva Sandstone (Oligocene), Polish Carpathan Mountains. (b) Stages in development of cluster of load-casted ripple marks.

Source: John R. L. Allen. Developments in sedimentology. Elsevier. ISBN: 0-444-41945.

Aario describes a second kind of syndepositional load-casted ripple mark from esker silts and sands. Begin describes "asymmetric ripple marks"—a structure that is most likely the same—from interbedded fluvio-deltaic sands and silts.

Figure 7.22 shows the structure that Aario described. It is composed of an extensive row of laterally oriented, inclined pouches in the deposit below that are filled in the troughs of strongly asymmetrical synclinal folds. The pouches are made of coarse-grained cross-laminated sand. From steep and even overturned at the up-current base of each pouch to their normal attitude at the down-current top, the cross-laminae gradually decline in dip. Normally, the sediments below are cross-laminated sands that have been graded, and they rise to a thick layer of laminated silt.

Figure 7.22. Load-casted ripple marks, Kurkiselka Esker, Kiiminki, Finland.

Source: John R. L. Allen. Developments in sedimentology. Elsevier. ISBN: 0-444-41945.

Aario concluded that each trough-filling sand pouch and corresponding anticline developed at the same time as a ripple moved downstream (Figure 7.23). Integrating concepts from Kuenen and Sanders, which were explored in relation to convolute lamination, he proposed that the folds symbolized the joint influence of fluid drag on the undulating bed and the pressure variations between crests and troughs. He also came to the conclusion that a continuous drop in current strength was the cause of the upward and downstream decrease in the pouches' deformation intensity. If Aario's explanation is accurate, it can only be applied to the initial phases of pouch growth because, contrary to what his theory suggests, the ripple troughs in Figure 7.22 now lie above anticlines rather than synclines. It is more likely that different sediment loading caused the deformation to start and continue. The ripples that capped the cross-laminated sand below were draped by the laminated silt beneath the pouches, but they were not completely erased. As a result, coarse sand started to be deposited to the lee of leftover mounds when rippling started up again, filling in the troughs in between. It is reasonable to assume that the silt would have had the lower bulk density due to its fine grade, so a differential loading would have been combined with a latent gravitational instability. The troughs would have been highlighted if liquefaction had taken place by a progressive rotation that was similar to the one proposed by Dzulynski and Kotlarczyk for clustered load-casted ripples. Pressure changes related to flow separation at the ripples may have caused the liquefaction that allowed deformation to occur.

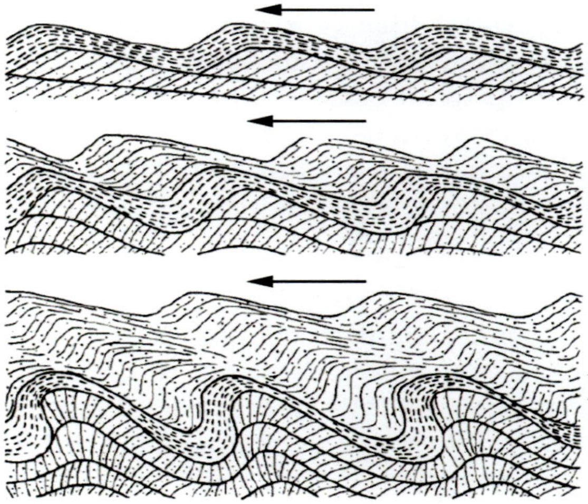

Figure 7.23. Schematic stages in the development of syndepositional load-casted ripple marks, according to Aario (1971).

Source: John R. L. Allen. Developments in sedimentology. Elsevier. ISBN: 0-444-41945.

7.3.2. Folds Related to the Lateral Spread of Sand Mounds

There are numerous examples of layered muds that underwent folding as intricate drifts, sand dunes, and deltas moved over them. Except for a small area where they were liquidized as a direct result of deformation, the mud remained hydroplastic. The features and relationships reported are summarized in Figure 7.24.

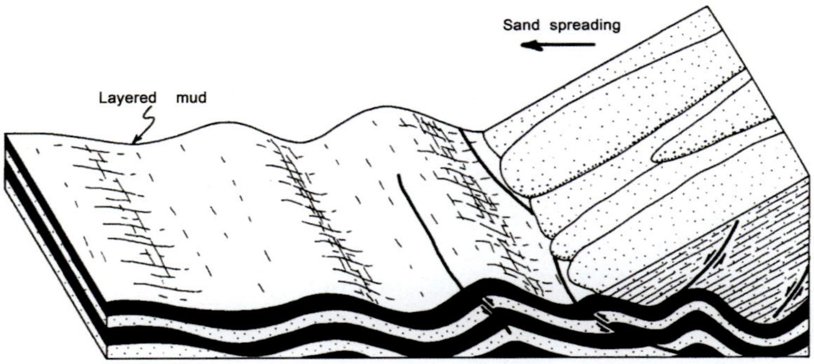

Figure 7.24. Deformation by folding and faulting of layered plastic muds by a migrating sand body.

Source: John R. L. Allen. Developments in sedimentology. Elsevier. ISBN: 0-444-41945.

According to RG Brown and IJ Townsend, the largest and most extensive folds are found at locations in South Australia's Coorong Lagoon where significant aeolian sand-drifts are moving inland from a coastal barrier. Here, folds in thickly layered muds, up to 150 m wide and 1.5 km long, are situated in front of advancing and overriding drifts. The structures are long, exposed ridges (anticlines) that are subparallel to the leading edges of the drifts and approximately one meter high. They are arranged en echelon, with crestal spacing ranging up to 20 meters. The height of the anticlines decreases as they move away from the sand spreads. Folds that are symmetrical to mildly leaning and open to closed are common. Some have enlarged hinge zones, while others resemble boxes. Some anticlines have hinges that are impacted by tiny reverse faults and thrusts. Normal faults are present locally in the sands of the advancing drifts, downsloping toward the exposed mud.

WR Parker describes a second, smaller, contemporary example from a barred beach where underwater dunes pass over thin sheets of mud deposited on the floors of runnels parallel to the shore. Dune crests and fold crests are subparallel. It is typical to have an amplitude of 0.15–0.20 m and a fold wavelength of roughly 0.5 m. The anticlines can be strongly asymmetrical, facing the direction of dune travel, or upright and symmetrical, resembling buckle folds in certain situations. Parker observed that the biggest folds were found in the highest parts of the dunes, and that the folding intensity and size of the dunes decreased together.

An example of the deformation of mud during the spread of a large aeolian dune was described by BG Jones from the Devonian rocks of central Australia. The folds bear similarities to those reported by RG Brown and WR Parker in terms of scale, form, and attitude. Similar to the folds in the Coorong Lagoon, they seem to have been subjected to erosion during their formation, as evidenced by the local appearance of thick intraclast conglomerates in their synclinal troughs.

A mobile mud's differential loading during the spread of a significant sand body is assumed to be the cause of the folding in each of these cases, with some support coming from experimental data presented by Kindle, Rettger, McKee, and Goldberg. These researchers discovered that some of the mud was harmoniously squeezed forward and compressed laterally when a sand body, like a delta, spread quickly across layered mud. This resulted in the formation of folds with outward-declining amplitude that were arranged parallel to the edge of the advancing load. While some folds had thrust-out or thinned limbs, others were asymmetrical and resembled buckles. Rettger replicated what RG Brown saw from the Coorong Lagoon, producing normal faults in the sand body that threw downward toward the side with the least amount of load on the mud (Shrock, 1948; Chakrabarti, 1977; Seilacher, 1984).

7.4. DEFORMATION STRUCTURES RELATED TO A DOWNSLOPE BODY-FORCE COMPONENT

7.4.1. Sheet Slumps Formed from Multilayers

The stratigraphic record is full of sheet-like bodies of deformed sediment that were deposited over a significant distance, seemingly after the collapse and downslope translation of a superficial layered deposit consisting of either mud with carbonate sediment or sand interbedded with mud. At that time, the sediments seem to have been mostly unconsolidated and only occasionally partially lithified. Within many layered igneous intrusions, there have been reports of lumps of still mushy crystal layers from the cooling surfaces that dip more steeply.

Sheet slumps (Figure 7.25) are known to occur in a variety of settings and at various scales. Folded and faulted silts and clays with down-thrust sand masses occur in settings influenced by glaciers. These sheets typically die out over a few tens of meters and seldom exceed 0.5 meters in thickness. Pyroclastic sequences also record thin slump sheets. Larger slumps can sometimes be found in deltaic and fluviatile deposits; these can stretch laterally over 10 km or more and have a thickness of up to 15 m. Sequences from lacustrine and shallow-marine environments have also been found to contain slump sheets with a thickness of roughly 1–10 m. The majority, though, are found in deposits found in deep water basins. The majority of these sheets are about one meter thick, many are around 10 meters thick, and a few are thicker than 100 meters. Although lateral extents are typically unknown, thicker sheets appear to be more extensive. Ballance mentioned thinner sheets that were even less persistent, and he described a slump that was about 1 m thick and passed 200 m laterally into undisturbed strata. However, thick sheets that appeared to be traceable laterally for many kilometers were reported by Gregory and Rupke.

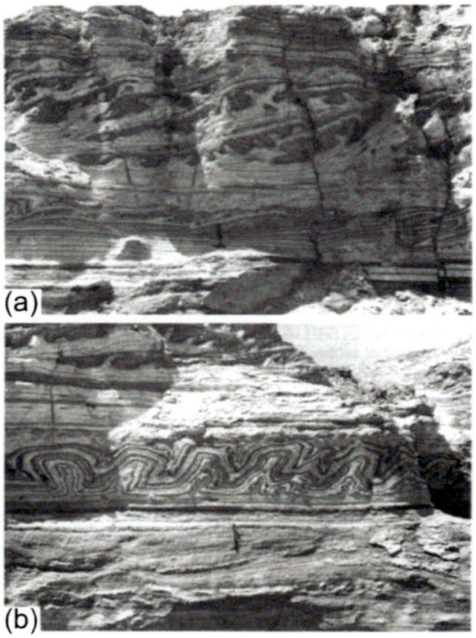

Figure 7.25. Features of slump sheets, lacustrine muds and silts, Dead Sea, Israel. (a) Imbricately thrust folded layers in two zones. Hammer 0.33 m long. (b) Box folds with probable basal dicollernent. Hammer stuck into cliff below folded zone casts vertical shadow about 0.4 m long.

Source: John R. L. Allen. Developments in sedimentology. Elsevier. ISBN: 0-444-41945.

"Distinction between slump structures and small tectonic structures is sometimes difficult, especially where intensive tectonic movements have taken place later or when the structure is only partially exposed," according to the stratigraphic record. The only conclusive large-scale criteria of a slump origin are the inter-bedding of sheets with undisturbed strata and the existence of what OT Jones called a "welded" (i.e., erosional) contact at the top of each sheet, even though many characteristics of slump sheets suggest slumping. These requirements are simpler to determine for thin sheets than for thick ones, although many authors have reported on them. Therefore, it is important to pay attention to small-scale slumping indicators, such as burrowed slump folds, sand volcanoes on top of sheets, and features that show the hydroplasticity of the sediment.

Helwig examined the question of why and how the sedimentary materials that make up a slump sheet deform given the initial geometry of failure. In the thinner sheets, two distinct types of sediment are likely to mix together: (1) cohesionless, water-saturated sands and coarse silts, which, if liquidized, fail by viscous flow or by the formation of slip surfaces; and (2) cohesive muds, which, depending on how much water they contain, deform as either plastic solids or viscous liquids. Deeply layered sequence failures can also involve cemented sand beds, which fail as brittle solids over time. In shallow failures, early formed concretions and early cemented carbonate sediments ought to exhibit comparable characteristics. Gravitational-induced shear is the main stress in slumping, and it can only cause multilayer folding and thrusting if it is deflected from its nominal bedding-parallel sense. According to Helwig, there are five possible reasons why there could be a deflection: (1) rotational movements in the slump's head and tail regions; (2) transverse variations in the slump's local flow speed caused by non-uniform frictional stresses; (3) overriding of bottom irregularities; (4) suddenly decreasing slope or collision with obstacles; and (5) localized liquidization during flow or rise of density-controlled instabilities.

A variety of deformational styles, including the presence of essentially unstrained masses, are the distinguishing feature of slump sheets due to the varied nature of the parent sediments and the localized action of the stresses. Slump sheet folds are usually isolated disharmonic structures or single hinges. The majority of axial surfaces tend to fall into the flat-lying or steeply inclined category (Figure 7.26(a)). It is possible to see box folds, concentrated folds, and other related folds on a single sheet. Most single folds have uneven limbs (Figure 7.26(b)), and some have a spiral shape that looks like a carpet that has been rolled up (Figure 7.26(c)). Numerous faults that split some sheets into identifiable packets internally include low-angle thrusts and bedding-parallel slips. In an imbricate structure, a sequence of thrusts that repeat a bed are commonly seen. A small displacement, low-angle reverse fault is formed by certain thrusts (Figure 7.26(f)). The thinner sheets are most frequently affected by decollement due to box folds. The presence of unwarped blocks surrounded by bedding surfaces and joint-like fractures indicates the brittle response of relatively competent beds. A bed that broke in this way while being stretched and sheared is represented by the strung-out imbricate blocks of Figure 7.26(i).

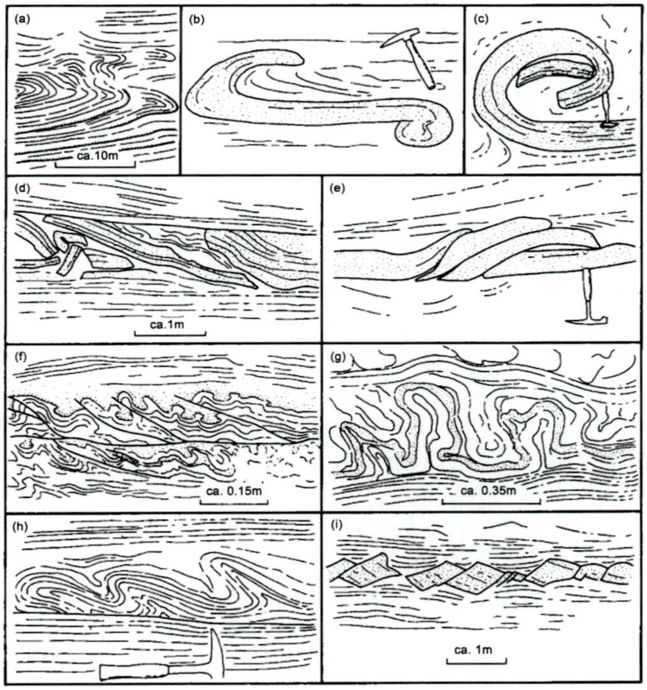

Figure 7.26. Structures (slightly simplified) in slump sheets. Sandstone or marker sandstone beds shown stippled.

Note: (a) A large flat-lying fold, Waitemata Group (Miocene), New Zealand. (b) Unequal folds in turbidite sandstone, Marnosa-arenacea romagnola (Tertiary). Appennines. (c) Spiral fold in turbidite sandstone, Waitemata Group (Miocene), New Zealand. (d) Imbricate structure in turbidite sandstones, Waitemata Group (Miocene), New Zealand. (e) Imbricate structure in fluvial sandstones, Peel Sandstone (?Devonian), Isle of Man. (f) Thrusts and reverse faults in Pleistocene silts and clays, Glasgow, Scotland. (g) Box folds in Pleistocene silts and clays, Finland. (h) Overfolds with decollement, Pleistocene silts and clays, south Quebec, Canada. And (i) Imbricate blocks formed from a single thick turbidite sandstone in tension, Waitemata Group (Miocene), New Zealand.

Source: John R. L. Allen. Developments in sedimentology. Elsevier. ISBN: 0-444-41945.

Fold attitude and slumping direction in slump sheets do not have a straightforward relationship. By discovering that the folds face primarily in the direction of slump movement, have axes primarily aligned perpendicular to this direction and parallel with the strike of the palaoslope, and have axial surfaces that are, on average, gently tilted down toward the slump source, Woodcock has significantly improved upon the early model put forth by OT Jones. Although there is experimental support for the model, Woodcock points out that it lacks a priori validity. Some alternative models could provide a more coherent explanation for the significant variation in fold axial trends found in some slump sheets. Based on an observation of a slumped snow layer, Lajoie's model has fold axes that lie perpendicular to the slope strike at the slump's sides and tail, but turn parallel at the head. The folds are shown as overlapping lobes in Eisbacher's model, with each lobe's curved axis changing from being strike-parallel at the head to strike-perpendicular on the flanks. Another question is how the direction of the slumping relates to the original current path that deposited the beds. Ballance discovered that the slumping and current directions in relation to deep-water deposits coincided, while Murphy and Schlanger, KM Scott, and Corbett established an azimuth difference of roughly 90°. Turbidity currents most likely flowed parallel to the slopes where slumping happened in many basins.

Research conducted in contemporary settings demonstrates that unconsolidated sediments can slump on subaqueous slopes as low as 0.5°. Large amounts of sediment can be involved, as evidenced by the slumps that Walker and Massinghill, Coleman et al., and Prior and Coleman reported from the Mississippi delta and deep-sea fan. Earthquake shocks, pore-fluid pressure accumulation brought on by pore water migration, and oversteepening of slopes through deposition or other processes are the three main causes of slumping. Liquidization likely contributes significantly to most cases of slumping because unconsolidated sediments are involved, but it is not always the most important factor. The most frequent explanations for slump sheets are oversteepening or seismic shocks.

7.4.2. Deformed Cross-Bedding

The third type of deformed cross-bedding that Allen and Banks identified shows a localized blurring or destruction of lamination along with a combination of folding and faulting. Most cases that have been documented come from aeolian sands.

The deformation is therefore metadepositional with respect to foresets rather than to the set itself, having frequently happened during the accumulation of a set of cross-strata. Three features are described: (1) normal faults downthrowing in the direction of foreset dip and occasional small thrusts; (2) small folds ranging from disharmonic and relatively upright, through drag folds, to nappe-like forms; and (3) patches of vaguely laminated or apparently structureless sediment. McKee et al. were able to connect these structures, in part through experimentation, to the mechanisms of gravitationally induced sand failure that accumulate to the lee of dunes; an important controlling factor in this process was the sand's moisture content and wetting and drying history. Sand that had been recently dried out or was wet from rain or dew tended to break easily. Occasionally, the superficial layer produced drag folds in the laminae it passed over as it faulted or broke up into large slabs or a jumble of blocks that moved somewhat downslope. However, wet crusts had a tendency to fold over and fail. Ahlbrandt and Andrews have reported on a type of deformation that arises from the melting of snow that is interbedded with the sand foresets in aeolian dunes in cold climates.

Within a dune on a river point bar, Harms et al. discovered deformed cross-bedding that offers an intriguing analogy from an aquatic environment. Upward and downcurrent deformation increases in the cross-bedding set of the dune. Little disharmonic folds with a broad range of attitudes represent the deformation upcurrent, which is linked to thickened and disrupted laminae. As the set is traced down-current, shear discontinuities among the folds become more significant. It is proposed that downslope mass-movement happened as a result of a rise in porewater pressure within the dune, which led to a fall in river stage in the presence of a silty drape over the bed, rejecting fluid drag as the cause of deformation.

From intertidal dunes in the Bay of Fundy, Dalrymple has described a type of deformed cross-bedding that he can plausibly attribute to the flow and slumping of the loosely packed sand after the waves broke against the bedforms during the tidal ebb, liquefying the sand. Small and localized volumes of sand are involved. Within each affected set, the extent of the deformation increases upward, taking the form of rounded basin-shaped folds that resemble pseudonodule folds, as well as sharp, diaper-like anticlines. This type of deformation lacks faults.

Occasionally, more than one bed is involved in the deformation, which is metadepositional with respect to sets and has a large lateral extent in cases that are nominally of the third kind of Allen and Banks. These cases include the complex folds with faulting and occasionally brecciation from the Casper Sandstone described by Knight and Steidtmann; Triassic sandstones described by Rice; the Hopeman Sandstone described by Peacock; and the Navajo Sandstone described by Kiersch, Sanderson, Freeman, and Visher, and Doe and Dott. These rocks are most likely aeolian, except the Triassic sandstones; however, Visher makes an unpersuasive case for the Navajo Sandstone's shallow-marine origin. There aren't many conclusive answers available for the deformations mentioned. Knight explained folding in the Casper Sandstone by proposing downslope mass-movement into the troughs between dunes. Rice postulated that the Triassic structures were caused by underwater sand bodies collapsing under their own weight. Sanderson argued that the Navajo dunes were saturated with water when they were formed, despite being pushed by wind. Peacock proposed a decision between earth tremors, undercutting at the base of the dune slope, and the downslope movement of partially dry and partially moist sand. Doe and Dott came to the conclusion that "contortion of the sand took place below a water table" in relation to the sandstones they examined, and that "slight changes in the physical setting, such as saturation of the dunes or changes in the level of the groundwater table, might have triggered failure." Some of these theories are easily disproved because they fail to provide a satisfactory explanation for the structures' significant lateral extent (Shrock, 1948; Chakrabarti, 1977; Seilacher, 1984).

ROLE MODEL

THEODOR FUCHS

Theodor Fuchs (15 September 1842 in Eperies – 5 October 1925 in Steinach am Brenner) was an Austrian geologist and paleontologist.

He studied geology and paleontology at the University of Vienna as a pupil of Eduard Suess (doctorate 1863). Following graduation, he worked as an assistant at the *Hofmineralienkabinett* in Vienna, being named its curator in 1880. From 1889 to 1904 he was director of the geologic-paleontological department at the Natural History Museum in Vienna. In 1897 he became an associate professor of paleontology at the university.

His primary research dealt with studies of stratigraphic conditions of the Late Tertiary of the Vienna Basin and investigations of Tertiary deposits in the Mediterranean.

In 1894 he proposed the Chattian age, a chronostratigraphic stage of the Oligocene epoch. In 1895 he was the first to report on soft sediment deformations known today as "load casts" — at the time, Fuchs used the descriptive term *Fließwülste* (flow crests).

Ø **Selected Works:**

- *Die Conchylienfauna der Eocaenbildungen von Kalinowka, im Gouvernement Cherson im südlichen Russland*, 1869 – The conch fauna in the Eocene strata at Kalinovka, in the Kherson Governorate of southern Russia.
- *Beitraege zur Geologie und Paleontologie der libyschen Wüste und der angrenzenden Gebiete von Aegypten*, 1883 (with Karl Alfred von Zittel) – Contributions to geology and paleontology of the Libyan Desert and neighboring areas of Egypt.
- *Studien über Fucoiden und Hieroglyphen*, 1895 – Studies on fucoides and hieroglyphs.
- *Über einen Versuch die problematische Gattung Paleodictyon auf mechanischem Wege künstlich herzustellen*, 1905 – On an attempt to artificially produce tunnels formed by the problematic genus Paleodictyon.
- *Wissenschaft und religion. Eine naturwissenschaftlich-philosophische studie*, 1911 – Science and religion. A scientific and philosophical study.

He was a proponent of Esperanto; in 1912 he translated the introductory chapter of Eduard Suess's Das Antlitz der Erde ("The face of the Earth") as La viza□o de la tero.

CLASS ACTIVITY

Want to integrate the "A" in STEAM? Give your students various craft materials such as buttons, stamps, colored paper, or even different edible items such as beans, noodles, candy, or any small objects to create a collaborative art piece.

Students will work together to use the materials to form various layers stacked on top of each other no larger than an 8.5×11 piece of paper.

Give time for students to share with other teams why they chose the materials and what led them to make their patterns. This activity is a great way to introduce how sedimentary rock forms in layers!

SUMMARY

- Soft-sediment deformation structures develop at deposition or shortly after, during the first stages of the sediment's consolidation. This is because the sediments need to be "liquid-like" or unsolidified for the deformation to occur. These formations have also been put into a category called water-escape structures by Lowe (1975).
- Stratigraphical and sedimentological studies over many years have shown that soft sediments often become deformed non-tectonically. The structures induced take myriad forms and are increasingly called soft-sediment deformations. It is clear from field evidence, and consistent with experiment and theory, they were created either during deposition or shortly after burial started.
- The description hydroplastic is appropriate for the original material, of a significant yield strength and effective viscosity.
- Convolute lamination is typically found in cross-laminated very fine sands or coarse silts that contribute to graded beds in the order of 0.1–1.0 m thick, either turbidites or their fluvial and deltaic analogues.
- Post-Depositional Convolute Lamination arises sometime after the start of burial. The folds have the greatest amplitude in the middle of the convoluted zone and die away upward and downward toward undeformed laminae. They lack other thannormally expectable internal discordances (e.g., between cross-laminated sets).
- Metadepositional convolute lamination arises either just before or immediately after deposition ceases. The folds are truncated either by the unit top, on which sand volcanoes may lie above anticlines, or by an internal erosion surface close beneath.

- Dish structures are thin, dish-shaped formations that normally occur in siltstones and sandstones. The size of each "dish" often ranges from 1 cm to 50 cm in size, and forms as a result of dewatering.
- Slump structures are mainly found in sandy shales and mudstones, but may also be in limestones, sandstones, and evaporites. They are a result of the displacement and movement of unconsolidated sediments, and are found in areas with steep slopes and fast sedimentation rates. These structures often are faulted.
- Seismites are sedimentary beds disturbed by seismic waves from earthquakes. They are commonly used to interpret the seismic history of an area. The term has also been applied to soft sediment deformation structures, including sand volcanos, sand blows, and certain clastic dikes.
- Load-casting is never accompanied by such extreme deformation that the upper sediment layer is pierced by the diapers rising from below. There are other two-layer systems, however, and some multi-layered ones, in which the more dense layers are completely disrupted, an extensive bed or beds of sand having been reduced to a series of isolated masses embedded in more or less continuous mud of a lower density that represents one or a blend of several original strata.

REVIEW QUESTIONS

1. What is an example of a soft-sediment deformation?
2. How does soft-sediment deformation occur?
3. What is the size of convolute laminations?
4. Where are slope structures found?
5. Name the three types of convolute lamination.

REFERENCES

1. Allen, J. R. L. (1982). Sedimentary Structures, Their Character and Physical Basis (Vol. 2). Elsevier Scientific Publishing Company.
2. Anketell, J. M., Cegla, J., & Dzulynski, S. (1970). Annales Societatis Geologorum Poloniae, 40, 3–30.
3. Bachmann, G. H., & Aref, M. A. M. (2005). A seismite in Triassic gypsum deposits (Grabfeld Formation, Ladinian), southwestern Germany. Sedimentary Geology, 180, 75–89. https://doi.org/10.1016/j.sedgeo.2005.06.006.
4. Boggs, S. (2006). Principles of Sedimentology and Stratigraphy (4th ed.). Pearson Prentice Hall.
5. Chakrabarti, A. (1977). Senckenbergiana Maritima, 9, 285–305.
6. Coleman, J. M. (1969). Sedimentology, 3, 129–239.
7. Dzulynski, S., & Kotarba, A. (1979). Zeitschrift für Geomorphologie, 23, 172–191.

8. Elliott, R. E. (1965). Sedimentology, 5, 193–209.
9. Helwig, J. (1970). Journal of Geology, 78, 172–187.
10. Ksiazkiewicz, M. (1958). Annales Societatis Geologorum Poloniae, 28, 123–150.
11. Lowe, D. R. (1975). Sedimentology, 22, 157–204.
12. Rich, J. L. (1950). Bulletin of the American Association of Petroleum Geologists, 34, 717–741.
13. Sanders, J. E. (1960). Geological Magazine, 97, 409–421.
14. Scott, A. M., & Bridgewater, J. (1975). Industrial & Engineering Chemistry Fundamentals, 14, 22–33.
15. Seilacher, A. (1969). Fault-graded beds interpreted as seismites. Sedimentology, 13, 15–159. https://doi.org/10.1111/j.1365-3091.1969.tb00113.x.
16. Seilacher, A. (1984). Sedimentary structures tentatively attributed to seismic events. Marine Geology, 55, 1–12. https://doi.org/10.1016/0025-3227(84)90075-6.
17. Shrock, R. R. (1948). Sequence in Layered Rocks. McGraw-Hill.
18. Ten Haaf, E. (1956). Geologie en Mijnbouw, 18, 188–194.

CHAPTER

8

Biostratigraphy

LEARNING OBJECTIVES

After studying this chapter, you will be able to:

- Discuss on nature of biostratigraphic units;
- Explain the fossils and stratigraphy;
- Understand the evolutionary trends;
- Describe the biozones and zone fossils;
- Define the taxa used in biostratigraphy;
- Discuss the biostratigraphic correlation;
- Focus on spores and pollens;
- Describe the foraminifera.

KEY TERMS FROM THIS CHAPTER

Ammonoids
Assemblage biozones
Cephalopoda
Foraminiferida
Interval biozones
Marine microfossils
Paleoclimatology
Aquatic animals
Biostratigraphy
Echinoderms
Gastropods
Marine macrofossils
Paleozoic rocks
Paleoecology

INTRODUCTION

Biostratigraphy is a branch of stratigraphy that utilizes fossil organisms to date and correlate strata within the geological record. This field relies on the principle of faunal succession, which posits that fossil assemblages succeed one another in a recognizable and predictable order through geological time. By identifying and analyzing the presence of particular fossil species, biostratigraphers can establish relative ages of rock layers, enabling the construction of a chronological framework that is crucial for understanding Earth's history. Key to this process are index fossils, which are species that were widespread, rapidly evolving, and limited to a short span of geological time. These characteristics make index fossils particularly useful for pinpointing specific intervals in the geologic record. For example, the presence of ammonites, which thrived during the Mesozoic era, can precisely indicate Mesozoic-aged rocks. The precision of biostratigraphic correlation allows for detailed regional and global stratigraphic correlations, facilitating the reconstruction of past environments and the interpretation of sedimentary basins. Additionally, biostratigraphy plays a vital role in the petroleum industry, where it aids in the exploration and development of hydrocarbon resources by providing age constraints for sedimentary sequences and helping to predict the locations of reservoir rocks. Through detailed biostratigraphic studies, geologists can better understand sedimentation rates, depositional environments, and the timing of geological events such as transgressions and regressions of the sea. Biostratigraphy also intersects with other disciplines like paleoecology and paleoclimatology, offering insights into past climate changes and ecological shifts. Modern advancements, such as high-resolution stratigraphy and biostratigraphic zonation techniques, have further refined the accuracy and resolution of this field. Techniques like quantitative biostratigraphy, which employs statistical methods to analyze fossil data, and biochronology, which integrates biostratigraphic data with absolute dating methods, have revolutionized the precision of biostratigraphic correlations. Despite its many advantages, biostratigraphy does have limitations, including the potential for diachronous occurrences of species and the dependence on well-preserved fossil records, which may not always be available. Nevertheless, it remains a cornerstone of geological sciences, providing a crucial tool for unraveling the complexities of Earth's stratigraphic record and contributing to our understanding of geological time and processes.

8.1. NATURE OF BIOSTRATIGRAPHIC UNITS

The bodies of strata that make up biostratigraphic units, or biozones, are classified or defined according to the fossils they contain. Only in cases where the specific diagnostic attribute or feature upon which they are based has been determined do biostratigraphic units exist. Thus, objective units based on the identification of fossil taxa are biostratigraphic units.

The identification of either their defining or characterizing attributes is necessary for their recognition. When more information is gathered, biostratigraphic units may be expanded to encompass a larger portion of the stratigraphic record in terms of both vertical and geographical dimensions. A single taxon combinations, relative abundances, particular morphological characteristics, or variations in any of the numerous other characteristics connected to the distribution and content of fossils in strata can all serve as the basis for a biostratigraphic unit. Different zones may be assigned to the same interval of strata based on the selected fossil group or diagnostic criteria. As a result, different types of biostratigraphic units with gaps or overlaps in their vertical and horizontal ranges may exist within the same interval of strata (Blatt, Berry, & Brande, 1991; Doyle, 1996; McGowran, 2005).

8.2. FOSSILS AND STRATIGRAPHY

In addition to the value of fossils as markers of deposition processes and environments, fossil research has yielded essential knowledge regarding the evolution of life on Earth. Fossilized plant or animal remains, such as skeletons and shells, provide unequivocal proof that the makeup of life on Earth has evolved over time. While some of these fossils are unlike anything we are familiar with, others clearly relate to modern life forms and resemble plants or animals that exist today. The more striking of these fossils often evoke memories of historical periods when, for instance, dinosaurs inhabited ecological niches in the sea, on land, and even in the air. The remains of aquatic animals like ammonites and pieces of plants that are unlike anything we see living around us today can be found even when fossil hunting is done casually. The first hints regarding the passage of geological time were discovered during the 18th and 19th centuries when a catalog of fossils discovered in sedimentary rocks was created. Early naturalists and scientists noticed that distinct rock units either contained assemblages of fossils that were quite different from one another or similar fossil remains. Furthermore, it was occasionally possible to trace the units that contained the same fossils laterally and demonstrate that they belonged to the same layer. According to general stratigraphic principles, individuals with distinct fossils could be younger or older. A given fossil type was frequently found in rocks with the same lithology, but occasionally it was discovered in a different rock type, which was important for the formation of stratigraphy. It became clear that there were patterns in the distribution of fossils as paleontology advanced. It was discovered that some organism types predominated in specific strata groups. As a result, the "ancient life," or Paleozoic, "middle life," or Mesozoic, and "recent life," or Cenozoic, were established as the three geological eras into which deposits were formed. When these concepts were first introduced in the 19th century, the actual time periods that these represented were pure speculation, and the numerical ages for these eras were unknown until the development of techniques for radiometric dating. At this point, the presence of particular fossil types in specific stratigraphic units was only an observation; once theories regarding the evolution of life were developed, an explanation for the distribution of fossils in the stratigraphic record could be found.

8.2.1. Classification of Organisms

The systematic framework for arranging the enormous diversity of life on Earth is provided by the classification of organisms or taxonomy. This hierarchical system makes it easier for scientists to identify, investigate, and comprehend organisms by grouping them according to shared traits and evolutionary relationships. The system begins with broad categories and narrows down to more specific ones, typically involving several levels:

8.2.1.1. Species

The classification of organisms is based on the concept of species, which was initially defined as groups of interbreeding organisms that are reproductively isolated from other such groups. The genetic traits of organisms can also be taken into account when defining a species thanks to the additional information provided by modern biological analyses. Discrepancies or similarities in an organism's genetic makeup enable precise species definitions and the identification of their relationships. Therefore, genetics provides the foundation for a hierarchical classification system; however, the Linnaean System predates our understanding of the nature of genes by a long margin. According to the Linnaean scheme, similar genera belong in families, closely related species belong in genera, and so on up to the Kingdom, the largest classification unit (Figure 8.1). A taxon (plural taxa) is the general term for any one of the ranks specified by the Linnaean System, and the species is the fundamental taxon rank.

Although the system was created for living things, paleontology also makes use of the same classification scheme and nomenclature. However, it is difficult to apply the definitions of a species to fossils because genetic material can typically only be extracted from relatively recent fossil material and because DNA is typically too deteriorated in fossils to be demonstrated in most cases. Because soft parts of organisms are only preserved under exceptional circumstances, paleontologists must focus on similarities or differences in morphology to define species. As a result, the hard parts of organisms are primarily used. In addition, some species exhibit significant form variations, such as dogs, and there are numerous examples of similar-looking organisms that are genetically distinct, particularly among birds, indicating that the hard-part morphology is not always a reliable method of defining species. Because of this, it is never certain that a fossil's skeletal resemblance provides enough evidence to conclude that the two species are related. It has been customary to define a fossil species by establishing a holotype, or a single representative specimen that can be used to compare other possible representatives of the species. These days, morphometrics—a statistical measure of an organism's size and form—as well as details about the paleoenvironmental habitat, related fauna, and associated information can all be utilized to further define a species.

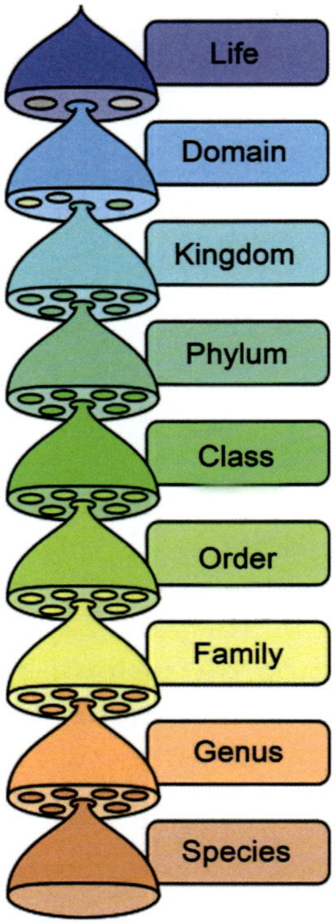

Figure 8.1. The Linnaean hierarchical system for the taxonomy of organisms.

Source: https://en.wikipedia.org/wiki/File:Biological_classification_L_Pengo_vflip.svg.

8.2.1.2. Other Ranks in Taxonomy

Subspecies and races are separate groups that differ from one another but share traits that make them unique, yet they belong to the same species. The geographical separation of the sets, which causes the development of distinct characteristics, is frequently the cause of the variations. Paleontology uses the concept of subspecies, even though it is typically impossible to determine the underlying genetic basis for this. When an organism is named, it is assigned both a genus and a species. For instance, the Linnaean classification name for the human species is Homo sapiens. A genus, also known as plural genera, is a collection of closely related species. In paleontology, it is typically acceptable to identify and classify a fossil only at the generic level; species-level identification is typically only necessary for biostratigraphic purposes. For instance, fossil oysters discovered in limestone may just be referred to as Ostraea since identification up to this point offers enough paleoenvironmental details, negating the need to identify specific Ostraea species.

According to scale, the higher positions in the hierarchy are family, order, class, phylum, and kingdom (Figure 8.1). The principal phyla (Arthropoda, Mollusca, etc., Figure 8.2) have persisted throughout the Phanerozoic, and fossils and contemporary representatives of these subgroups of the major kingdoms (plant and animal) can be compared. Nonetheless, a great deal of families, orders, and classes have been identified as fossils that have no modern counterparts. For example, from the Ordovician to the Cretaceous, the ammonites formed a very large and diverse order. There are no modern counterparts to the ammonites; instead, the only creatures that resemble them are nautiloids, which are members of the same class in the phylum Mollusca called the Cephalopoda. There are no living members of the class of graptolites, which is typically found in Paleozoic rocks. The difficulties in classifying organisms increase with decreasing resemblance to contemporary organisms, as the importance of morphological distinctions becomes less clear. Because new fossil discoveries provide light on the likely relationships between fossil organisms, the Linnaean hierarchy used to classify fossils is constantly changing (Blatt, Berry, & Brande, 1991; Doyle, 1996; McGowran, 2005).

Figure 8.2. An example of a macrofossil.

Source: By SNP at English Wikipedia, CC BY-SA 3.0, https://commons.wikimedia.org/w/index.php?curid=17271119.

8.3. EVOLUTIONARY TRENDS

The biosphere that surrounds us today is the culmination of a general trend of lifeforms becoming more sophisticated and complex, beginning with those found in older rocks. This trend is accompanied by evidence of organism diversification and emergence, as well as indications that some groups are declining from abundance to insignificance or even extinction. When we examine specific groups, we find that many of them exhibit morphological trends over the course of the stratigraphic record. These trends are linked to the evolutionary history of that particular group of organisms. Evolutionary theory's central tenet is that changes are one-way processes: once an organism has evolved a new trait to become more "advanced," subsequent changes do not lead to a reversion to the more "primitive" form. Thus, the idea of evolutionary trends gives us a means of understanding how biological changes have occurred over time based on the fossil content of rocks. This makes it possible to correlate rocks and establish the relative ages of those rocks based on the fossils they contain.

8.3.1. Population Fragmentation and Phyletic Transformation

Numerous contemporary plant and animal species exhibit form variation based on region. Charles Darwin proved that a population's geographical isolation can cause some members to diverge significantly from the rest of the species in terms of traits (Darwin, 1859). These alterations are the outcome of mutations that benefit the isolated population by improving their adaptation to the surrounding environment. Examples could include adaptations that improve the organisms' suitability for varying temperatures, different salinities of the water, distinct features of the sediment, or decreased susceptibility to local predators. This can result in speciation, the process by which an isolated group of animals develops traits that set them apart from the rest of the population and qualify them as a new species. If the process of population fragmentation weren't generally counterbalanced by extinction, the number of species would explode over time. This process increases the number of species. When a population of a species reaches a terminal decline in size due to competition from other organisms or because it is no longer well adapted to changing environmental conditions, it becomes extinct. A different mechanism through which a new species could emerge is the population's overall change over time. The "interbreeding test" is speculative because the ancestral and descendant species are separated in time, but they do have different morphological and, in cases where it can be tested, genetic traits. Because it results in a change in form between the ancestral and descendant members of the evolutionary lineage (or phylogeny) and does not cause the number of species to increase, this phenomenon is known as phyletic transformation. The term "extinction" in this instance refers to "pseudoextinction," or the disappearance of the ancestral species, rather than the death of the entire population. Nevertheless, "real" extinction is still possible and can result in the extinction of an entire lineage's branch.

8.3.2. Phyletic Gradualism and Punctuated Equilibrium

The basic tenet of the theory of evolution is that organisms undergo genetic changes in response to environmental stimuli and that these changes ultimately lead to the emergence of sufficiently distinct organisms to be classified as separate species.

This would suggest that evolution proceeds steadily and gradually—a gradualism known as phyletic gradualism. An alternate theory holds that a lineage evolves episodically rather than gradually, with brief intervals of rapid change during which new species emerge and periods of stasis during which a species does not alter its genetic makeup or form. Eldredge & Gould (1972) refer to this process as a punctuated equilibria pattern of evolution. One would think that the stratigraphic record would reveal which of these processes has predominated. If the theory of punctuated equilibria is true, then a continuous succession of strata should theoretically contain fossils that show either a pattern of gradual change in morphology through time, up the succession, or the form would be the same through the beds and abruptly change to a different morphology at a certain horizon. In actuality, the stratigraphic record is unable to offer definitive responses (Blatt et al., 1991). First, there may be long stretches of time during which there is no deposition of sediment, and as a result, no fossils, except some deep marine and lake environments. Additionally, erosion may remove portions of the sedimentary record, creating a gap in the record. Second, a species' populations may shift due to environmental changes, possibly returning to their original location only after speciation has taken place. In this case, it will seem as though the population underwent a gradual change while speciation at the original site happened all at once. Third, the stratigraphic record only retains a very small fraction of the total number of organisms that existed, and only a very small portion of a population is ever fossilized.

8.3.3. Speciation and Biostratigraphy

Based on the identification of distinct species at various stratigraphic horizons, biostratigraphy is created. Strata can be defined in terms of clearly different fossils if evolution is a punctuated process with periods of stasis followed by rapid change. Because speciation would occur too quickly for intermediate forms to be preserved, a zonation scheme based on the appearance and disappearance of species can be developed (Doyle et al., 1994; Bennett & Doyle, 1998). A new species would be defined at any random point in the lineage if morphology evolved gradually; as a result, zonation would be based on these points. In many cases, the lack of intermediate forms gives the impression that speciation occurred suddenly in the geological record, even though the process was actually gradual. When a phyletic gradualism is evident, it becomes imperative to utilize statistical methods to characterize species based on morphological variability, like the ratio between a shell's length and width (Blatt, Berry, & Brande, 1991; Doyle, 1996; McGowran, 2005).

8.4. BIOZONES AND ZONE FOSSILS

A rock body's composition of fossils defines it as a biostratigraphic unit. As a result, it differs significantly from a lithostratigraphic unit, which is determined by the lithological characteristics of the rock. The biozone is the basic unit of biostratigraphy. The zone fossils that define a biozone are typically species or subspecies. Biozones are stratigraphic units. Although environmental factors are frequently taken into account when defining and interpreting biozones, in theory, they are independent of lithology. Every biozone needs to have a type section that is identified as a stratotype and described, just as formations in lithostratigraphy must be defined from a type section. Their names derive from the trait or common taxon (or sometimes taxa) that characterizes the biozone. According to the zone fossils they contain, biozones can be categorized in a number of ways (Figure 8.3).

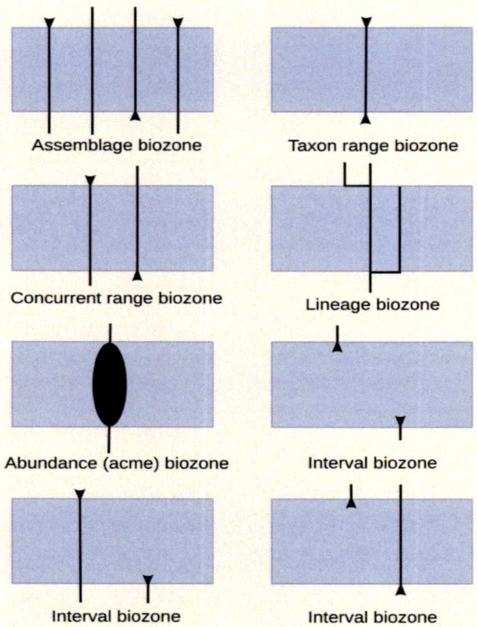

Figure 8.3. Zonation schemes used in biostratigraphic correlation.

Source: By PePeEfe – This file was derived from: Tipos de biozonas.svg:, CC BY-SA 4.0, https://commons.wikimedia.org/w/index.php?curid=66858760.

1. Interval Biozones: They are characterized by their occurrences in a series of one or two taxa. A taxon-range biozone is one in which a single taxon's initial appearance and eventual extinction serve as the definition. A biozone with two taxa that overlap in range is called a concurrent range biozone; it is characterized by the appearance of one taxon at the base and the disappearance of the other taxon at the top. A third option is a partial range biozone, which is founded on two taxa with non-overlapping ranges; here, the top is determined by the disappearance of a taxon, and the base is determined by the appearance of one taxon. A taxon's defined biozone is referred to as a lineage biozone (also called a consecutive range biozone) when it can be shown that it belongs to a phyletic lineage and that it preceded a third taxon.
2. Assemblage Biozones: In this instance, at least three distinct taxa, which may or may not be related, define the biozone. A stratigraphic interval is defined by these taxa's presence, absence, and appearance/vanishing behavior. When appropriate taxa are unavailable to define interval biozones, assemblage biozones are used instead, and they may represent shorter time periods than those based on one or two taxa.
3. Acme Biozones: A biozone can be defined by an interval that contains a statistically significant proportion of a given taxon, even though the abundance of that taxon may change over time. Because the relative abundance is influenced by local environmental factors, this approach may not always be accurate. An organism that was abundant, could live in all depositional environments on Earth, had easily preserved hard parts, and belonged to an evolutionary lineage that regularly produced new, distinct species would be the ideal zone fossil. Naturally, there has never been a fossil taxon like that, and the selection of fossils for biostratigraphy depends on a variety of factors.

8.4.1. Rate of Speciation

The resolution that can be used in biostratigraphy depends on how frequently new species emerge and displace older species within the same lineage. Certain organisms appear to have undergone minimal evolution; for example, the brachiopod Lingula bears little biological significance because it resembles fossils found in Lower Paleozoic rocks almost exactly. Vertebrate groups seem to exhibit the highest rates of speciation; on average, new species of fish, mammals, and reptiles are created every one to 3 million years (Stanley, 1985). In contrast to marine mollusks, which are far more common as fossils but have slower average speciation rates (about 10 million years), vertebrates have a poor stratigraphic record. Certain groups seem to have evolved new forms on a regular and frequent basis: during the Jurassic and Cretaceous periods, new ammonite species seem to have evolved roughly every million years, and in some Cambrian regions, certain trilobite lineages seem to have evolved new species roughly every million years (Stanley, 1985). Biozones can typically be defined for time periods of up to a million years by utilizing multiple species to define them; higher resolution can be achieved in some areas of the stratigraphic record, particularly in younger strata.

8.4.2. Depositional Environment Controls

It is not possible for a single species, genus, or family to exist in every depositional environment due to the extreme differences in conditions between them. The organisms that inhabit these environments must differ due to the adaptations needed to survive in a desert as opposed to a swamp or a sandy shoreline as opposed to a deep ocean. The distribution of taxa is heavily regulated by the environment today, and it stands to reason that the distribution of fossil groups was also heavily impacted by the environment. Body fossil preservation is more likely in some settings than in others. For instance, the potential for preservation is lower in a low-energy lagoon than it is in a high-energy beach. Since very few animals or plants are found in both continental and marine environments, there is a fundamental issue with correlation between them. Planktonic (free-floating) and nektonic (free-swimming) animals are the most common types of organisms in marine environments. The creatures and plants that reside on the sea floor, known as benthonic or benthic, are typically restricted to a specific range of water depths and are therefore not as helpful.

The distribution and preservation of stratigraphically significant fossils are also influenced by the rates of sedimentation in various depositional settings. Since the remains of the organisms are exposed on the sea floor or land surface and are vulnerable to biogenic degradation, slow sedimentation rates frequently lead to poor preservation. However, the higher concentration of fossils brought about by the decreased supply of sediment can facilitate the collection of biostratigraphically valuable material because the rate of accumulation will be slower in an environment where organic material has a higher chance of preservation (such as an anoxic environment). Additionally, hundreds of meters of strata may lie within a single biozone if sediment accumulation rates are high, increasing the likelihood that a first or last appearance datum will be discernible in a single outcrop section.

8.4.3. Mobility of Organisms

An organism's lifestyle influences not only how widely it is found in depositional environments but also how quickly it migrates from one place to another. A new species' value as a zone fossil in a regional or global context will depend on how quickly it migrates to occupy ecological niches elsewhere after evolving in one particular geographic location. Again, due to their relatively fast movement, planktonic, and nektonic organisms are typically the most valuable in biostratigraphy. Because their larval stage allows them to swim freely, some benthic organisms can spread fairly quickly throughout oceans. For biostratigraphic purposes, organisms with a sessile lifestyle—those with little movement—generally yield poor fossils.

8.4.4. Geographical Distribution of Organisms

Even though two environments have nearly identical physical characteristics, very different species of plants and animals may live in them if they are on different sides of the globe. The differences are most pronounced in continental settings where tectonic plate movements have caused geographical isolation of communities, giving rise to remarkably distinct families and orders. Australia's mammal population is a remarkable illustration of how geographical isolation has led to the evolution of a group of animals that are very different from those found in similar environments in Europe or Asia. Provincialism is the term for this geographical isolation of groups of organisms; it also affects marine life, especially benthic forms, which have a difficult time moving across oceans. Even though the depositional environments may have been similar, the oceans of today and the past have been sufficiently apart to support the development of localized communities. Different biostratigraphic schemes need to be developed in different parts of the world due to this faunal provincialism (Blatt, Berry, & Brande, 1991; Doyle, 1996; McGowran, 2005).

8.4.5. Abundance and Size of Fossils

A species needs to be sufficiently abundant to be easily found in sedimentary rocks to be useful as a zone fossil. It must be feasible for the geologist to locate representatives of the relevant taxon without needing to search for too long. Additionally, there is a trade-off between abundance and size. Smaller organisms are typically more numerous, and as a result, their fossils are typically the most prevalent. The issue with tiny fossils is that they could be hard to locate and identify. More use is being made of microfossils, which are fossils too small to be recognized in hand specimens but which may be abundant and easily identified under the microscope (or electron microscope in some cases) because biostratigraphic schemes must be applicable to subsurface data from boreholes. Microfossil-based schemes have been developed concurrently with macrofossil schemes. Finding an entire ammonite in the core of a borehole is unlikely, despite the fact that an ammonite-based scheme might perform admirably in the field. In biostratigraphy, where drilling only uncovers fragments of the lithologies in the drilling mud rather than recovering the core, microfossils are the only material that is practical for use.

8.4.6. Preservation Potential

Since very few species or individuals are ever preserved as fossils, it is impossible to estimate the number of people or species that have existed on Earth over geological time. For a variety of reasons, the fossil record only makes up a very small portion of the planet's biological history. First of all, some organisms lack the hard parts necessary to withstand burial in sediments; as a result, we are unsure of the exact number of different kinds of worms that may have existed in the past. Lagerstatten, or places where the soft parts of fossils have been remarkably preserved, offer tantalizing hints about the diversity of lifeforms about which we know very little (Whittington & Conway-Morris, 1985; Clarkson, 1993). Second, the environment in which the remains are deposited may not be favorable for their preservation; in the dry, oxidizing conditions of deserts, only the hardest-wearing bone fragments survive, and nearly all other material is destroyed. Since all living things are a part of a food chain, some sort of predator or scavenger will typically consume their bodies. Consequently, for the majority of animals and plants, preservation is the exception. Lastly, only a small portion of the environmental niches that have existed are preserved in sedimentary rocks, making the stratigraphic record extremely incomplete. The material available for biostratigraphic purposes is severely limited by the low preservation potential, limiting it to taxa with hard parts that existed in suitable depositional environments (Blatt, Berry, & Brande, 1991; Doyle, 1996; McGowran, 2005).

8.5. TAXA USED IN BIOSTRATIGRAPHY

Numerous taxonomic groups have been employed to define biozones through the stratigraphic record, as no single group of organisms satisfies all requirements for the ideal zone fossil (Clarkson, 1993). Certain species, like the graptolites found in the Ordovician and Silurian, are utilized for global correlation, while others, like corals found in the Carboniferous of northwest Europe, are only used in specific facies within a specific succession. Below is a summary of some taxonomic groups that are commonly used in biostratigraphy.

8.5.1. Marine Macrofossils

Throughout the Phanerozoic, sedimentary rocks deposited in marine environments were frequently found to contain the hard parts of invertebrates (Clarkson, 1993). The stratigraphic column was divided into Systems, Series, and Stages in the 18th and 19th centuries based on these fossils. Fossils of creatures like echinoderms, arthropods, and mollusks, among others, are comparatively simple to distinguish in hand specimens, giving the field geologist a way to determine the age of rocks to the appropriate period or possibly epoch. Based on the expert paleontological analysis of marine macrofossils, the rocks are divided into stages.

8.5.1.1. Trilobites

The primary group utilized in the Cambrian zonation is these Paleozoic arthropods. It is believed that the majority of trilobites were benthic organisms that lived on and in the sediment in shallow marine waters. They exhibit a great range of morphologies and seem to have developed quickly into taxa with recognizable traits. As fossils, they are only plentiful in the area.

8.5.1.2. Graptolites

These strange and somewhat mysterious creatures are thought to be individual colonial groups joined by a skeleton. They are widely distributed in Ordovician and Silurian mudrocks and seem to have had a planktonic lifestyle. Typically, a thin layer of compressed organic material on the bedding planes of sedimentary rocks with fine grains serves as preservation. Examined under a microscope or with a hand lens, the shapes of the skeletons and the "teeth" where individuals in the colony were located are unique. The Ordovician and Silurian systems now have high-resolution biostratigraphy thanks to the traceage of lineages that show rapid evolution. The primary disadvantage of using graptolites is their poor preservation in rocks with coarser grains, like sandstones.

8.5.1.3. Brachiopods

Shelly, sessile organisms such as brachiopods generally make poor zone fossils but in shallow marine, high-energy environments where graptolites were not preserved, brachiopods are used for regional correlation purposes in Silurian rocks and in later Paleozoic strata.

8.5.1.4. Ammonoids

Goniatites from Paleozoic rocks and the more well-known ammonites from the Mesozoic are both members of this taxonomic group of cephalopods (phylum Mollusca). The closest living group to one another is the nautiloids. These cephalopods are a great group to study biostratigraphy because of their large size and free-swimming behavior. Fossils are relatively robust, abundant, and found in many fully marine environments. The "suture line," a remnant of the bounding walls that once separated the chambers of the coiled cephalopod, and the external shape of the organisms underwent morphological changes over time. The primary zone fossils in Mesozoic rocks are ammonites and other ammonoids, while goniotites have been used to correlate Devonian and Carboniferous rocks. The ammonoids went extinct at the end of the Cretaceous period.

8.5.1.5. Gastropods

These are also members of the Mollusca family and are common fossils in Cenozoic rocks, resembling marine "snails." In the deposits of practically all shallow marine environments, they are extremely prevalent. There are many different taxa in this group, and identification is made easy by the distinctive shapes and ornamentation on the calcareous shells.

8.5.1.6. Echinoderms

Sea lilies and sea urchins belong to the phylum Echinoids. Despite their abundance in certain Paleozoic limestones, most crinoids likely lived attached to their substrate, which makes them relatively poor zone fossils. Benthic organisms, echinoids inhabit soft sediments. Their morphology is characterized by subtle yet distinct changes, and their relatively robust form has made them useful for regional and global correlation in some Cretaceous regions.

8.5.1.7. Corals

Corals abound in the vast outcrops of shallow marine limestones found in Lower Carboniferous (Mississippian) and Devonian rocks found in some regions of the world. Because of the extremely limited depositional environments in which they occur, this group is therefore used for zonation and correlation within these strata, even though they are generally not suitable for biostratigraphic purposes.

8.5.2. Marine Microfossils

Taxa that leave behind fossil remnants too small to view well through a hand lens or the naked eye are known as microfossils. Optical microscopes are typically used to examine them, though scanning electron microscopes are the only tools necessary for a detailed analysis of certain forms. Foraminifers, radiolaria, and calcareous algae (nanofossils) are the three main groups used in biostratigraphy; additional microfossils used in biostratigraphy include ostracods, diatoms, and conodonts.

8.5.2.1. Foraminifera

The Protozoa Subkingdom is home to single-celled marine organisms known as "forams," which is the common abbreviation for foraminifers. Although they did not become well established until the Devonian, forms with hard calcareous shells, or "tests," have been discovered as fossils in strata dating back as far as the Cambrian. Calcareous forams are widespread in many Mesozoic and Cenozoic marine strata and generally increased in abundance during the Phanerozoic. Although larger benthic forms, which can measure more than a centimeter in diameter, do occur during some periods, especially the Paleogene, the calcareous tests of planktonic forams are usually a millimeter or less across. Since planktonic forams are common, abundant, and seem to have evolved quickly, they make excellent zone fossils. Because microfossils can be easily recovered from boreholes and because both regional and global zonation schemes are used, schemes that use forams for correlation in the Mesozoic and Cenozoic are widely used in the hydrocarbon industry.

8.5.2.2. Radiolaria

These creatures, which belong to the subclass of planktonic protozoans, have been preserved as fossils in deep marine layers for the entirety of the Phanerozoic. Radiolaria are roughly spherical, frequently spiny organisms that are less than a millimeter across. They typically have silica skeletons. The fact that the skeletons survive in siliceous oozes deposited at depths below the CCD makes them significant for deep-marine deposit dating calculations. The fossil assemblages found in these deposits, which are preserved in the stratigraphic record as radiolarian cherts, usually contain a large number of taxa, allowing for the use of fairly high resolution biozonation schemes. They are significant for Paleozoic strata dating because their stratigraphic range is likewise larger than that of forams.

8.5.2.3. Calcareous Nanofossils

Nanofossils are defined as fossils that are too small to be seen with the unaided eye and that are barely visible under a high-power optical microscope. The best way to examine them is with a scanning electron microscope; they range in size from microns to tens of microns. The spherical calcareous cysts of marine algae known as coccoliths are the most prevalent type of nanofossil. Coccoliths are the primary component of some fine-grained limestones, such as the Upper Cretaceous Chalk of northwest Europe, and can be found in enormous quantities in certain sediments. They are found on the shelf or in any depths above the CCD, below which they are not typically preserved. They are found in fine-grained marine sediments. They are employed in Mesozoic and Cenozoic strata according to biostratigraphy.

8.5.2.4. Other Microfossils

The closest relatives of ostracods, which are crustaceans with a two-valve calcareous carapace, are crabs and lobsters. Their distribution and abundance are erratic, but they have a long history and can be found in a very wide range of freshwater and marine depositional environments. Only locally is ostracod zonation used in marine and non-marine environments. Diatoms are chrysophyte algae that can be found in large quantities in freshwater and shallow marine environments. They have a siliceous frustule, or skeleton. Less than a millimeter in diameter, the diatom frustules can comprise the majority of the sediment in certain lacustrine environments, resulting in the formation of a diatomite deposit. They are utilized in biostratigraphy very infrequently. Conodonts are phosphate-based, tooth-like structures that are found in Paleozoic strata. They are somewhat mysterious. They are valuable stratigraphic microfossils in the older Phanerozoic rocks, which typically contain few other microfossils, despite the uncertainty surrounding their origins. Acritarchs are tiny, spiny structures found in both Proterozoic and Paleozoic rocks. They are composed of organic material. They are helpful as a biostratigraphic tool in rocks of this age because of their occurrence in Precambrian strata. They are not clearly related to dinoflagellates, which are primitive organisms that have been present since the Phanerozoic era and that also produce microscopic cysts, even though they are most likely the cysts of planktonic algae. Particularly in non-calcareous strata of Mesozoic and Cenozoic ages, dinoflagellate zoning is highly significant locally; however, the schemes utilized are typically geographically local and have restricted stratigraphic ranges.

8.5.3. Terrestrial Fossil Groups Used in Biostratigraphy

Due to the lower ability of most materials to be preserved in a subaerial environment, correlation in the deposits of continental environments is always more challenging. In the majority of continental deposits, only the hardest-wearing materials—such as the organophosphates found in vertebrate teeth and the coatings found on pollen, spores, and plant seeds—have managed to fossilize. Using the teeth of small mammals and reptiles, stratigraphic schemes have been established to correlate continental deposits from the Neogene era. Much more frequently used are palynomorphs, which are seeds, spores, and pollen. They can be dissolved out of siliceous sedimentary rocks using hydrofluoric acid because they are composed of organic material that is extremely resistant to chemical attack. Pollen, spores, and certain seeds are examples of airborne particles that can spread widely. The presence of these aeolian palynomorphs in marine strata enables the correlation of marine and continental successions. Even so, palynomorphs seldom offer the same high resolution as marine fossils when used as zone fossils. After strong acids are used to chemically separate the palynomorphs from the host sediment, identification is done under an optical or electron microscope (Blatt, Berry, & Brande, 1991; Doyle, 1996; McGowran, 2005).

8.6. BIOSTRATIGRAPHIC CORRELATION

A high-resolution basis for strata division and, consequently, a way to correlate disparate successions can be found in biostratigraphy. Nonetheless, the strategy needs to meet a few requirements to be effective. The first and most obvious requirement is that the rocks must contain the right fossils. The presence or absence of these fossils will mostly depend on the deposition environment, which may not have been ideal for the important taxa. The diagenetic history is also important because physical or chemical processes, like compaction, can change or remove fossil material entirely. Examples of these processes include mineral dissolution and replacement. Another important factor is the relative rate of speciation events and sedimentation in the successions. When there is infrequent speciation and rapid sediment accumulation, two thick successions may be shown to lie within the same biozone, but no further subdivision or correlation is possible.

8.6.1. Correlating Different Environments

It frequently happens that the rocks under study contain valuable fossils from the biostratigraphic record, but that the taxa represented in the global biostratigraphic zonation scheme for that specific section of the stratigraphic record are absent from the fossils. Provincealism, the propensity of populations to reside solely in a small geographic area, or the depositional environment could be the cause of this. Since different climatic conditions exist at different latitudes, the fossils found in the deposits of contrasting environments—such as muds deposited in an offshore setting compared to a sandy foreshore—are probably different or more numerous. Variations in fossil content resulting from provincialism are caused by the geographic isolation of evolutionary lineages rather than environmental factors. A more indirect method of correlating using fossils may be necessary in these situations, involving the establishment of a local or regional zonation scheme based on the taxa that are present in the area. Next, by locating a succession elsewhere where taxa from both the local and global schemes are preserved, the strata containing the local scheme's flora or fauna must be correlated with the general scheme.

Zone fossils may appear or disappear because of environmental changes rather than stratigraphic significance. If the depositional environment remains unchanged, the emergence of a taxon could be attributed to a speciation event, which would carry stratigraphic significance. On the other hand, it's also possible that the species migrated to the section under study after having lived for a while in a different region. If the depositional environment has not altered, the extinction of a species from the stratigraphic succession is most likely to have occurred because a population is not likely to shift from a favorable environment. The appearance and disappearance of taxa within a succession may therefore be caused by changes in sea level rather than speciation and extinction events. Relative sea level is one of the factors that affect depositional environment and hence fossil content. The most versatile and valuable organisms as zone fossils are those that can adapt to a variety of environments. Corals and other highly sensitive taxa are only beneficial in situations where the environment of deposition has been stable.

8.6.2. Graphical Correlation Schemes

The rate of sediment accumulation during the time period represented by the biozone determines the thickness of a biostratigraphic unit at any given location. As a reference section, a succession that is thought to have been a site of constant, continuous sedimentation is selected, and the locations of biozone markers—the appearance and disappearance of taxa—are noted within it. Then, this reference section can be used to compare another vertical succession of strata with the same biozone markers (Carney & Pierce, 1995). The biostratigraphic data is used to establish tie-points, and a graphic correlation between intermediate levels can be produced (Figure 8.4).

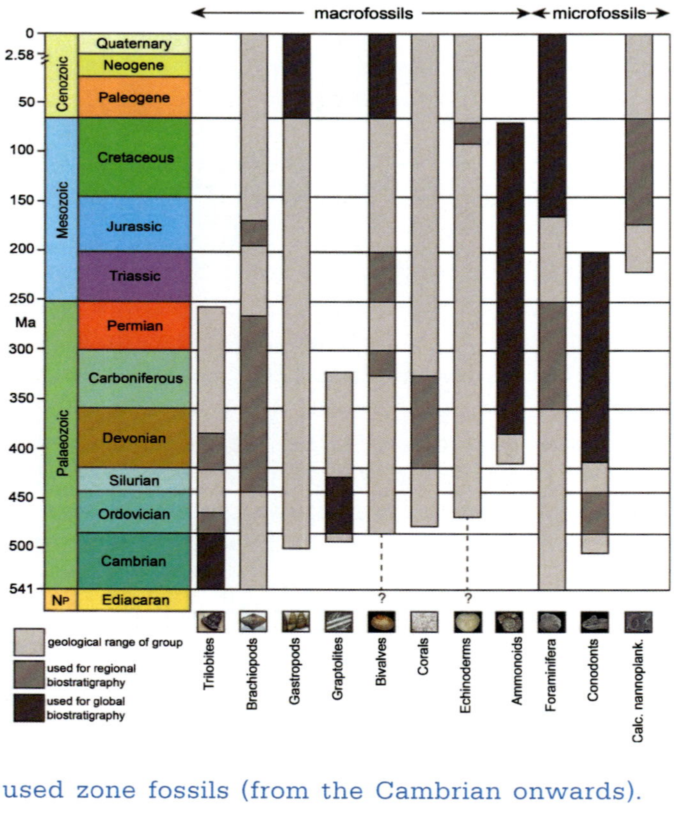

Figure 8.4. Commonly used zone fossils (from the Cambrian onwards).

Source: By David Bond – Own work, CC BY-SA 4.0, https://commons.wikimedia.org/w/index.php?curid=53499176.

This method works especially well for detecting variations in sedimentation rates and determining whether a hiatus—a time of erosion or nondeposition—has occurred in a succession (Figure 8.4). Depositional hiatuses have been widely employed in subsurface correlation and are crucial to sequence stratigraphic analysis of successions.

8.6.3. Biostratigraphy in Relation to Other Stratigraphic Techniques

There are inherent limitations to strata correlation based on lithology (lithostratigraphy). Above all, it is unable to offer any foundation for correlation across great distances, particularly between continents. The principles of evolutionary biology, however, require that if fossils from different locations belong to the same species, the rocks in which they are found must also have been deposited during the period when the species was extant. This is especially true if fossils are utilized as a correlation tool. As a result, biostratigraphy and lithostratigraphy are essentially independent, though some factors are related because the depositional environment determines both the facies of the sediment and the kinds of organisms that may be present. Biozones could be argued to be chronostratigraphic units in the event that a speciation event occurs quickly enough to be regarded as an "instant" in geological time. An isochronous horizon, or surface that represents a specific point in time, can be thought of as the base of a biozone if the new species disperses very quickly—that is, geologically instantaneously. Thus, biozones can be thought of as chronostratigraphic units within the resolution that is available if the idea of punctuated equilibria is accepted and biozones are defined by the first instances of free-swimming or floating organisms. Additionally, some taxa or taxonomic groups may go extinct in a relatively short amount of geological time; in this case, the upper boundaries determined by these events can also be thought of as roughly corresponding to isochronous surfaces.

Subsurface analysis relies heavily on biostratigraphy, even though microfossils are the primary source of information. A borehole's strata can be dated at various levels using forams, radiolarians, and calcareous nanofossils that are recovered from drill cuttings and core. These fossils also serve as a foundation for correlating strata with those of other boreholes and with successions that are exposed at the surface. Compared to the much older biostratigraphic approaches, the techniques utilized in the sequence stratigraphic analysis of successions are relatively new. However, biostratigraphy is still necessary to provide a broad temporal framework for the sequence stratigraphic analysis, so even though the identification of depositional sequences offers a new method of correlation, it does not replace it. Furthermore, graphical correlation schemes offer data on sedimentation rates and evidence of hiatuses, two critical components of sequence stratigraphic analysis (Blatt, Berry, & Brande, 1991; Doyle, 1996; McGowran, 2005).

8.7. SPORES AND POLLENS

Plants produce pollen and spores as part of their life cycle. Gymnosperms and angiosperms produce pollen, while bryophytes and ferns produce spores. The walls of both varieties of grain are extraordinarily resilient to microbial invasion as well as the effects of pressure and temperature changes following burial. These minuscule grains are produced in large quantities and have the ability to move quickly and far in the wind or water, landing on the bottom of ponds, lakes, rivers, and seas. They are useful for biostratigraphy because of these characteristics, especially when relating continental and nearshore marine deposits that are Silurian or younger in age. Spores and pollen can be used for paleoecological and paleoenvironmental studies when the ecology of the parent plant is known.

1. Spore Morphology: It can be characterized by their size, shape, apertures, and wall structure. The type of meiotic divisions that occur in the spore mother cell greatly influences the shape of the resulting spore. The mother cell divides into a tetrad, or group of four smaller cells, during simultaneous meiosis. Each of the four spores in a tetrahedral tetrad is in contact with its three neighbors on the proximal face. A Y-mark or trilete mark centered on the proximal pole defines three contact areas that make up the proximal face. The trilete mark's arms can resemble raised ridges or surface fissures called laesurae, and they can reach the equator. The distal polar face is the spore's external surface in the tetrad. The mother cell divides into two cells in the initial stages of meiosis, and these cells further divide along a single plane perpendicular to the initial division or along two planes perpendicular to each other. Here, the tetrads are tetragonal, with a shape akin to orange segments; each spore has two contact areas, a single scar, and is only in contact with two of its neighbors. These spores frequently have a bean shape. In fossilized material, spores are most frequently compressed proximo-distally. We refer to the equatorial contour as the amb.

Vascular plant spores are distinguished by regular, well-formed germinal aperture placement. These facilitate prompt germination of the prothallus and adapt to variations in size resulting from variations in humidity. When characterizing and categorizing fossil spores (and pollen), the shape and location of these apertures are crucial. Three laesurae, extending 120° from the proximal pole, are present in triploid spores. Trilete spores have heteropolar, or differently formed polar faces, despite having radial symmetry. Although they may be abundant in Paleogene-Recent assemblages, monolete spores are generally less common. The contact areas are separated by a single proximal laesura, known as the monolete mark. As a result, monolete spores have bilateral and heteropolar symmetry. Certain spores with tetrad scars but no laesurae have a hilum. In many bryophytes, this develops on the proximal or distal faces and serves as the germinal exit. Aletes are spores that don't appear to have any dehiscence structures.

The exospore is either a single layer or multi-layered and mostly composed of sporopollenin, whereas the inner cellulose layer, or endospore, rarely survives fossilization. More electron-dense sporopollenin material makes up the perispore, which is located outside of the exospore. Numerous fossil spores (sporoderms) have a single exine layer on their wall. When there are two layers, they can either be completely separated (cavate) or in contact (acavate). The cavum typically develops in an equatorial or distal orientation. Both homogeneous and finely lamellate layers are possible. The layers can have varying thicknesses or be the same thickness throughout. A continuous equatorial flange is called a zona, and a continuous equatorial thickening is called a cingulum. Cingulizonate spores are those that possess composite equatorial features. The smooth valvae and the ear-like, frequently fluted auriculae are examples of discontinuous equatorial features that typically develop in the radial areas. Moreover, flanges, coronae, or kyrtomes may form in the interradial regions (Figure 8.5).

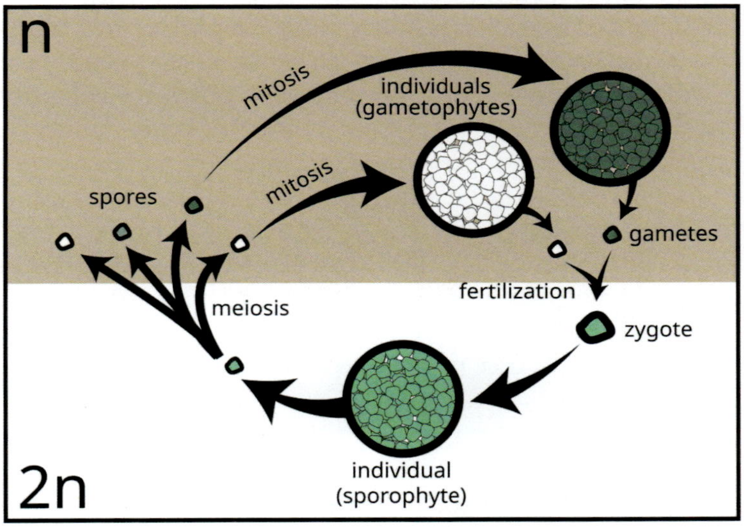

Figure 8.5. Meiosis and the production of bilaterally or radially symmetrical spores.

2. Pollen Morphology: Gymnosperm pollen can range in size from large bisaccate and ornamented grains and polyplicate forms to small, simple, spherical, and inaperturate grains. Gymnosperms are known for producing saccate pollen, which can be found in grains in one, two, or very rarely three sacs (bisaccate or trisaccate). The outer, extremely resilient exine and the inner, cytoplasm-enclosing intine make up the pollen grain wall.

8.7.1. Distribution and Ecology

Spores and pollen generally mirror the environment of the parent plants. However, a plant's leaves, wood, seeds, and spores are rarely preserved together due to size sorting in sediments. Even so, it is possible to infer the habitat and ecology of the plants that produce pollen and spores; however, this requires a prior knowledge of sedimentation and dispersal.

8.7.2. Dispersal and Sedimentation

Pollen and spores that travel through the air are largely determined by their size, weight, sculpture, and the state of the atmosphere. During the day, they are usually found 350–650 meters above the ground, but many of them sink to the surface at night or are carried below by rain. Pollen grains have been known to drift at least 1,750 kilometers when favorable conditions are met, but 99% of them usually settle within 1 km of the source. By aerial dispersal, very little of it ever makes it to the oceans. Following their landing, pollen grains or spores may become embedded in the fossil record through direct fall into bogs, swamps, or lakes, or through erosion into them and into rivers, estuaries, and oceans. The pollen record may now go through a similar filtering process in water since it has already been filtered by differential dispersal in the air at this point. For instance, there may be size sorting across the continental shelf; small miospores and pollen grains may settle out in outer shelf and oceanic conditions, whereas large miospores, pollen grains, and megaspores will typically settle out in rivers, estuaries, deltas, or shallow shelf areas. Those that are not buried in reducing sediments are more likely to oxidize and possibly even die. The fossil record may become somewhat confused as a result of spores and pollen going through multiple cycles of reworking and redeposition. Skilled palynologists can identify these altered forms through variations in preservation (e.g., color, corrosion, abrasion, and fragmentation), discrepancies in the ecology or stratigraphy, and related proof that needs to be reworked.

8.7.3. Geological History

Cryptospore monads, dyads, triads, and tetrads can be found in sediments from deltaic and lacustrine deposits that date from the Mid-Ordovician to the Early Silurian period. Palynological preparations from this era may also include cuticle tubes and sheets, which could be remnants of the earliest subaerial plants. The earliest Cooksonian macroplant remains have been discovered in Late Silurian deposits. From this point on, there is a sharp increase in the variety of spore types and macroplant fossils discovered, which indicates a significant diversification of early plants. These early spores' parent plants seem to have been distributed throughout the world.

The Devonian is most likely the peak period for pteridophytic plants. Progymnosperms, which by the Late Devonian produced pollen grains and actual seeds, joined these in the Early Devonian. These early pollen grains are known as pre-pollen because at first, they were identical to trilete miospores. Different equatorial-low latitude (North American-Eurasian), Australian, and southern Gondwana floras resulted from growing provinciality during the Devonian. The Devonian continents' wider latitudinal spread or the global cooling brought on by the start of glacial conditions could have contributed to this rise in provincialism. True megaspores appear to have become less significant after the Carboniferous, but they may reappear as common in non-marine deposits during the Jurassic and especially the Cretaceous. Large coal deposits are a major factor in the high level of knowledge about carboniferous floras. Lycopsids, seed-fern trees and shrubs, sphenopsid trees and shrubs, and shrub cordaitaleans, such as cordaites, which were composed of primitive conifers, were the characteristic vegetation of carboniferous coal swamps. Carboniferous coal swamps have been analogized to tropical deltas. There are numerous known spore-plant relationships for the Carboniferous. Several spore types were produced by some plants in a single microsporangium.

After the early evolution of the angiosperms, pollen grains gradually replaced spores in Mesozoic palynological assemblages, especially from the mid-Cretaceous onward. By the Permian, the gymnosperms' seed and pollen habits had taken over as the dominant life cycles. The earliest plants to produce pollen were pteridosperms, or seed ferns. Though the precise nature of this event is unknown, they evolved from pteridophytes; heterosporous pteridophytes were most likely an intermediary stage in their emergence. The Late Devonian (Famenian) is when the earliest known pollen, known as pre-pollen, originated. The earliest known locations of gymnosperm pollen with distal germination are Upper Carboniferous deposits. In the later Paleozoic, a great variety of gymnosperm pollen types evolved. The easiest to identify among these are saccate pollen grains, which are found in many different groups, such as the now-extinct pteridosperms, conifers, and cordaitaleans. During the Carboniferous, Early Permian, and Late Triassic periods, monosaccate grains were more prevalent than bisaccate grains.

Although the exact relationships are debatable, angiosperms are thought to have originated from a group of advanced gymnosperms. Several angiosperm pollen grains are triaperturate, and non-laminate endoexine and fully differentiated ektexine are features of angiosperm pollen. According to the palynological record, the Early Cretaceous is when the angiosperms first appeared. One of the earliest angioperm pollen grains is Clavatipollenites hughesii (Barremian, Lower Cretaceous); it is monosulcate and has a columellate, tectate exine. This geographic spread may have been regulated by alterations in paleoclimate and paleogeography, or by rapid evolution driving plants into lower latitudes. One of the key evolutionary innovations that led to the success of the earliest angiosperms was the emergence of tricolpate pollen, along with a carpel-protected seed. By the end of the Cenomanian, all of the structural characteristics of modern pollen grains had evolved. During the Late Cretaceous, angiosperm diversification led to a more regional dispersion. The extinction of extinct Cretaceous and Paleogene species was the primary cause of the gradual emergence of the modern flora from the Neogene onward. The Asteraceae (Compositae) and Poaceae (grasses) are two new, modern families that spread widely in the mid-Tertiary. With a large number of surviving species, they are the most successful group among modern organisms, having emerged as a result of climatic deterioration. Because the composites are entomophilous and the grasses are anemophilous, their pollen has a very different morphology. Hay fever is mostly caused by the simple spheroidal, monoporate pollen of grasses. The structure of contemporary plant communities has evolved since the end of the last ice age, and some communities have only been established in the past 200 years as a result of human influence.

8.7.4. Applications of Fossil Spores and Pollen

A continuous record of the vascular plants' evolutionary past is provided by spores and pollen. After being economically employed for the first time in coal-seam correlation and biostratigraphy, spores are now widely used in paleoenvironmental, paleoecological, and phytogeographical studies as well as source rock provenance. In hydrocarbon exploration, spores and pollen grains are frequently used in palynofacies analysis and thermal maturity studies (Thermal Alteration Index (TAI) and equivalents). Paleoshorelines can be defined by the combination of spores and pollen with other palynomorphs (Blatt, Berry, & Brande, 1991; Doyle, 1996; McGowran, 2005).

8.8. FORAMINIFERA

An important order of single-celled protozoa called the Foraminiferida can be found living among marine plankton or on the sea floor. The foraminiferid cell's soft tissue, or cytoplasm, is primarily encased in a test or shell that can be made of agglutinated particles or calcite. This test consists of one chamber (unilocular) or several chambers (multilocular), all of which are primarily smaller than one millimeter in diameter and joined by one or more foramens (foramina). The group, which derives its name from these foramina, has been known since the Early Cambrian period and peaked in the Cenozoic. Foraminiferid species can be found in large quantities; in the contemporary ocean, they account for more than 90% of deep-sea biomass and more than 55% of Arctic biomass. Foraminiferid counts in marine sediments usually range from a few individuals per kilogram to Globigerina ooze and Nummulitic limestone, which form rocks. Because foraminifera are plentiful, diverse, and simple to study, they are valuable as biostratigraphic indicators in marine rocks from the Late Paleozoic, Mesozoic, and Cenozoic eras. The widespread distribution and fast evolutionary history of planktic foraminifera have contributed significantly to the inter-regional correlation of Cretaceous, Paleogene, and Neogene strata. Since they are more prevalent, smaller benthic foraminifera are frequently employed in regional stratigraphy. The biostratigraphy of Tethyan and other tropical limestone can be aided by studying the intricate internal structures of larger benthic foraminifera, which are usually larger than 2 mm in diameter and 3 mm^3 in volume. Among them are the biggest known single-celled organisms, with a maximum diameter of 180 mm. Foraminiferid life history and developmental stages are well-suited to evolutionary studies because they are preserved in the test. Foraminifera can be found in a broad variety of environments, including deep sea, tropical, and terrestrial settings. The group is especially helpful in studies of recent and ancient environmental conditions because of its ecological sensitivity. Variations in the makeup of foraminiferal assemblages can be used to monitor shifts in seawater depth and water mass circulation. Their $CaCO_3$ tests' isotopes, which record variations in ocean chemistry and temperature, make them particularly valuable for research on the climate history of the Mesozoic to Quaternary eras.

8.8.1. The Test

The test is supposed to lessen stress on the body, mind, and spirit. Biological pressures include infections, predators, and the possibility of inadvertent ingestion. Abrasion, turbulence in the water, and UV radiation from the Sun are examples of physical stresses. As a result, test strength probably matters. Chemical stressors in water include changes in salinity, pH, CO_2, O_2, and toxins. In each of these scenarios, the cytoplasm may retreat into the inner chambers, leaving the outer ones as "lobbies" for protection, or the aperture may be sealed off by a detrital plug. The acidity of environments or digestive tracts that are rich in organic matter but low in oxygen may also be mitigated by $CaCO_3$ shells. A subset of organisms specifically adapted to a benthic lifestyle benefit from the test's negative buoyancy, which is one of its additional advantages. In addition to improving adhesion, strengthening the test against crushing, and facilitating ectoplasmic flow to and from the apertures, pores, and umbilicus, surface sculpture may also help planktic forms (such as spines and keels) achieve positive buoyancy.

8.8.2. Foraminiferal Ecology

8.8.2.1. Smaller Benthics

There are approximately 5,000 species of smaller living benthic foraminifera known. Because they have inhabited marine habitats ranging from the most extreme tidal marshes to the deepest ocean trenches, they are particularly significant as environmental indicators. Test morphology adaptations reflect resource exploitation across this broad range of habitats.

1. Light: The Sun's rays' incident angle and water clarity have an impact on the photic zone, which is the area where light penetrates the oceans. As a result, the photic zone is deeper (less than 200 meters) in tropical waters and gets shallower as one approaches the poles, where it also exhibits strong seasonal variations. Foraminifera, particularly the Miliolina, find this zone to be attractive due to the primary production of planktic and benthic protozoa, as well as the protection and substrates offered by algae and sea grasses. Quinqueloculina and other miliolines have porcelaneous walls.
2. Food: Because they are micro-omnivores, foraminifera are important members of marine ecosystems, i.e., they consume invertebrates, protozoa, and tiny bacteria as food. Because epifaunal organisms in the photic zone primarily eat diatoms, their populations may vary according to the seasonal cycle. Tests on these are frequently flattened on one or both sides. It is known that some smaller benthic forms can support photosymbiont culture. Others graze on bacteria or live infaunally in the sediment or beneath the photic zone, where they eat dead organic matter. Active forms typically have lenticular or elongated tests. To collect the seasonal rain of phytodetritus, organisms inhabiting the abyssal plains may extend their pseudopodia into the water column. These forms typically have fixed-to-the-substratum erect, tubular, frequently branched tests. It is believed that some hyaline foraminifera, which have degenerate unilocular tests and may have a parasitic lifestyle, shield the cytoplasm in shallow equatorial waters from harm by dispersing short-wavelength ultraviolet light.
3. Substrate: Foraminifera that favor hard substrates, such as rocks, shells, sea grasses, and algae, are typically affixed, either permanently or temporarily, by a concave or flat lower surface. Hydrodynamically stable growth forms include irregular, concavo-convex, planoconvex, discoidal, and dendritic

shapes. Compared to sediment-dwelling and planktic forms, adherent forms typically develop a relatively thin test and show more morphological variability. The bulk of foraminifera are found in the top 10 mm or at the surface, though they have been found to exist up to 200 mm below the sediment surface. From these rougher substrates, foraminifera typically take the form of adherent forms or free-living, thick-shelled, highly decorated lenticular or globular forms. Low-energy environments, such as the mid-shelf to bathyal slope and lagoons with their silty, muddy substrates, are frequently rich in organic debris, and their tiny pore spaces tend to promote bacterial blooms. Because of this, these substrates are appealing to free-living foraminifera and can sustain sizable but sporadic populations of them. A large number of infaunal species have delicate, long, thin shells.

4. Salinity: Since most foraminifera are acclimated to normal marine salinities of approximately 35%, these conditions support the highest diversity assemblages. Certain hyaline forms and low-diversity assemblages of agglutinated foraminifera are favoured by the low salinity of brackish lagoons and marshes. The porcelaneous Miliolina appears to benefit from the high carbonate ion concentrations found in hypersaline waters (salinities over 40%), while most other groups seem to be discouraged. It appears that the endoplasm is better shielded from the stressful osmotic gradients of high salinity by the imperforate tests of Textulariina and Miliolina. Thus, triangular plots representing the relative abundances of Textulariina, Miliolina, and hyaline forms have shown promise as paleosalinity indices.

5. Nutrients and Oxygen: The rates of primary productivity in seas and oceans are significantly influenced by the biolimiting nutrients phosphate and nitrate. The rates of primary productivity in seas and oceans are significantly influenced by nutrients and oxygen. The biolimiting nutrients phosphate and nitrate are particularly important in regulating low rates of food supply, such as those found in the deep sea. Foraminiferal densities ($<10/10$ cm^2) are typically low in low-food supply environments, such as the deep sea, but diversity can be high. Foraminiferal diversities typically decrease in upwelling zones with high rates of nutrient supply to the surface for many reasons. High nutrient flux rates tend to inhibit photosymbiosis, which inhibits the growth of larger benthic foraminifera and planktic organisms that cultivate symbionts and other

oligotrophic species. High rates of primary production at the surface also cause anaerobic bacterial blooms on the sea floor and in the oxygen minimum zone of the midwater. Foraminifera may be rare in anaerobic environments, but eutrophic benthic foraminifera with densities greater than 1,000/10 cm^2 may predominate in dysaerobic environments. Smaller organisms like foraminifera are not completely eliminated by oxygen deficiency, most likely due to their low oxygen demand and high diffusion rates brought on by a high surface area–volume ratio.

6. Temperature: Every species has evolved to survive in a specific range of temperatures, the most important of which is the range for successful reproduction. For low latitude faunas acclimated to steady, tropical climates, this range is typically narrowest. The lower layers of water, however, are progressively colder due to ocean stratification. This is seen, for instance, in tropical waters, where the surface may average 28°C while the abyssal plains' bottom waters may only average 4°C. Cool-water benthic assemblages, which are normally found at shallower depths closer to the Poles, are what distinguish these colder, deeper waters.
7. Water Mass History: Until the 1970s it was widely thought that certain smaller, hyaline, benthic foraminiferal species were adapted to specific water depths, largely controlled by temperature, and could therefore be used to estimate ancient water depth (paleobathymetry). Research has since shown that these species are closely tied to specific water masses. For example, Epistominella is typical of North East Atlantic Deep Water, Fontbotia of North Atlantic Deep Water and Nutallides of Antarctic Bottom Water. This means that the ancient distribution of such benthic species can be used to reconstruct the history of a specific water mass in relation to changes in global climate or in basin geometry.
8. Diversity: The quantity of taxa in an assemblage is indicated by this. One species is typically found to be more abundant than all others in living assemblages; this species is referred to as dominant. Lower dominance is typically found with higher diversity. Species dominance is commonly expressed as a percentage of the population. Compared to normal marine and deep sea habitats, the diversity of contemporary benthic foraminiferal assemblages from marginal marine habitats is lower. Increased diversity in the latter could indicate a more equitable distribution of resources among species. On the other hand, foraminiferid blooms with higher abundance

but lower diversity are produced by fluctuations in environmental stability, which are present in marshes and lagoons.

8.8.2.2. Larger Benthic Species

The majority of larger benthic foraminifera inhabit oligotrophic reef and carbonate shoal environments with minimal effects from the land and the seasons. They cultivate rhodophytes, chlorophytes, dinoflagellates, and endosymbiotic diatoms. High rates of $CaCO_3$ precipitation during test growth are made possible by these endosymbionts, which release photosynthates to their hosts and also absorb respiratory CO_2 during photosynthesis. Larger foraminifera are therefore extremely sensitive to light levels. The light wave lengths needed by their symbionts are also closely correlated with the depth distribution of larger living benthic foraminiferal taxa. Therefore, it seems that co-evolution with endosymbionts is how fossilized larger benthic foraminifera, which have evolved repeatedly since the Carboniferous, acquired their enormous size (up to 180 mm in the Oligocene) and skeletal complexity. Due to their adaptation to living in mobile carbonate sands, many larger foraminifera have robust, fusiform, conical, or biconvex tests. Large, discoidal-shaped specimens are typically found resting on sediments in the lowermost portion of the photic zone. Forms with strong spines for anchoring or small, flat sizes are typical of those adapted for adhesion to seagrass or algae fronds. Larger benthic foraminifera, which can produce up to 2,800 g of $CaCO_3/m^2$ annually in contemporary tropical oligotrophic environments, are important contributors to the carbonate sedimentation of today's environments due to their large test sizes and quick rates of growth. Larger fossil foraminifera have also colonized and occasionally built up vast areas of carbonate ramp environments, particularly during the Carboniferous, Permian, and Tertiary periods. Particularly significant as hydrocarbon reservoirs are the nummulitic sands of the Middle East, which could contain up to 60% of the world's petroleum reserves.

8.8.3. Planktic Foram Ecology

Because temperature and salinity are the only significant ecological factors affecting planktic foraminifera, their environmental controls are far better understood than those affecting benthics. Large latitudinal provinces exhibiting some bipolar distribution of species are found, with temperature acting as the main controlling factor. Using the fossil record of living species, this trait has proven extremely valuable in estimating Quaternary sea-surface temperatures.

8.8.4. Temperature and Latitude

Five biogeographic provinces can be distinguished from contemporary assemblages: (1) Polar (which encompasses the Arctic and Antarctic); (2) Subarctic (which encompasses Subantarctic and Subarctic); (3) transitional (4); subtropical; and (5) tropical. The term used to describe it in planktic foraminifera is latitudinal provincialism. Here, several trends are worth mentioning. A few species are typical of both northern and southern subtropical waters due to the bipolar distributions. Diversity rises as one moves closer to the tropics due to an increase in endemic forms. In waters that are colder than 5°C, keeled forms are not found at higher latitudes. In line with the lower density of warmer water, the test porosity of shallow and intermediate species likewise rises toward the equator. Neogloboquadrina pachyderma populations classified as subpolar or polar based on the predominant left-handed (sinistral) or right-handed (dextral) coiling patterns. There is a strong correlation between the distribution of these assemblages and the pattern of surface circulation. Thus, the distribution of planktic foraminifera preserved in deep sea cores can be used to infer the history of Quaternary oceanic and temperature fluctuations.

Almost two-thirds of the world's oceans are covered by the warm-water provinces. The boundary between the warm subtropical and colder transitional province is marked by the annual isotherm of 18°C, which corresponds approximately to the latitude of the balanced radiative heat budget. Most extant species are cosmopolitan within their preferred bioprovince, although three Indo-Pacific (Globigerinella adamsi, Globoquadrina conglomerata, Globorotaloides hexagona) and one Atlantic tropical species (G. ruber pink) are endemic.

The distribution and abundance of plankticforaminifera species is strongly linked to surfacewater properties. Sea-surface temperature (SST) appears to be the single most important factor controlling assemblage composition, diversity, and shell size. Planktic foraminifera species survive under a considerable range of SST, but that their optimum ranges, defined by highest relative and absolute abundances, are typically narrow and distinct. At present, the polar waters of both hemispheres are dominated by a single small species (N. pachyderma), while the highest diversity and largest sizes are found in the oligotrophic subtropical gyres. The increase in SST toward the equator is accompanied by a proportional increase in surfacewater stratification. The strength of vertical gradients in the ocean determines the number of physical niches available for passively floating plankton, and may thus control their diversity and morphological disparity.

After nutrient entrainment, these opportunists can respond quickly to phytoplankton blooms and the redistribution of organic particles. The seasonal SST cycle and periodic pulses in primary productivity lead to predictable successions of planktic foraminifera species, which respond to shifting environmental conditions in accordance with their ecological preferences. Numerous sediment-trap studies have documented these successions, and understanding them is crucial for geochemical proxies. Planktic foraminiferal shell flux rates typically follow primary productivity cycles, albeit a few weeks behind. In temperate oceans, the spring flux maximum is frequently followed by a smaller autumn peak; in polar oceans, however, the flux peak is observed during the summer. All year long, foraminiferal shells fall from the sky in tropical and subtropical oceans.

Within the range of normal marine conditions (33–36‰), salinity does not appear to exert any significant influence on planktic foraminifera. Laboratory experiments indicate that some species can tolerate a remarkable range of salinities (G. ruber: 22–49‰) and that salinity tolerances differ among species. In nature, no planktic foraminifera are known to live under hyposaline conditions. N. pachyderma is known to avoid low salinity (<32%) surface layers. Low salinity waters are also inhabited by a distinct assemblage dominated by G. ruber pink and Neogloboquadrina dutertrei. At the other end of the spectrum, planktic foraminifera inhabiting the Red Sea live at salinities in excess of 40‰ and the Antarctic N. pachyderma live in sea-ice where brine salinities exceed 80‰. The influence of ecological factors other than temperature, salinity, and fertility is difficult to disentangle, because most surface-water properties are strongly inter-correlated.

In the vicinity of oceanic gyre margins, where upwelling and mixing occur and nutrient levels are high, planktic foraminiferal densities can be extremely high. In lower latitudes, where seasonal disturbances occur due to monsoonal upwelling, an ecological succession of species is observed. The ratio of planktic to benthic tests within the total foraminiferal assemblage usually increases in a transect that runs from the inner shelf to the bathyal slope. This occurs partly because the amount of food that reaches the sea floor tends to decrease with increasing water depth and partly because an increase in depth increases the biomass of plankton above a specific area of the sea floor. However, because local conditions can vary the test production rate of either planktic or benthic foraminifera, the ratio is only a rough indicator of paleobathymetry.

8.8.4.1. Globigerina Ooze

Together with coccoliths, planktic foraminifera play a significant role in deep sea sedimentation and are responsible for over 80% of the carbonate deposition that occurs in seas and oceans today. Currently, foraminifera make up a greater contribution than coccolithophores, a situation that did not exist during the Mesozoic. The deposition of Globigerina ooze is influenced by three key factors (i.e., ooze in which Globigerina planktic foraminifera shells make up more than 30% of the sediment): (i) climate, (ii) lysocline depth, and (iii) terrigenous sediment supply.

Plankton productivity is influenced by the position and strength of currents, particularly diverging and upwelling currents, which are highly influenced by climate. When the empty foraminiferal tests approach the lysocline, which is typically located between 3,000 and 5,000 m depth, they settle more quickly and are less prone to dissolution than coccoliths, which do not have organic outer layers. It is now understood that variations in the calcite compensation depth throughout the Mesozoic and Cenozoic triggered cycles of deposition and dissolution, which excluded some of the smaller or more delicate forms and left the deep sea fossil record lacking. Globigerina oozes are rarely found on continental shelves because, even in favorable conditions, they cannot accumulate where terrigenous clastics are influxing. These oozes are currently primarily building up between 50°N and 50°S at depths of 200–5,000 m, particularly near the midoceanic ridges. However, they are frequently diluted with the siliceous remnants of radiolarians and diatoms (Blatt, Berry, & Brande, 1991; Doyle, 1996; McGowran, 2005).

8.8.5. Calcite Compensation Depth (CCD)

In comparison to cool waters, warm waters have less solubility of $CaCO_3$. This helps explain why foraminiferid limestones and oozes are more common at low latitudes and why the tests are thicker. The vertical shift in $CaCO_3$ solubility, which likewise rises with increased pressure and, consequently, with increased ocean depth, is more significant. Because there is no photosynthesis below the photic zone, but bacteria and animals can still breathe there, the partial pressure of CO_2 also rises with depth. These elements caused the pH to drop with depth, from roughly 8.2 to as low as 7.0. The calcium carbonate compensation depth, or CCD, is the point in the water column where the supply and demand of $CaCO_3$ are equal. The idea of the lysocline, or the level of maximum change in the rate of solution of foraminiferal test calcite, is widely used because it is impractical to locate this in the geological record. Benthic agglutinated foraminifera predominate populations from abyssal depths as a result of the overall decrease in calcareous organisms with depth—few exist below 3,000 m.

8.8.6. Geological History of Foraminifera

Shelled protozoa and invertebrates appeared at the same time, as evidenced by the earliest fossil foraminifera, which are simple, agglutinated tubes from the earliest Cambrian that resemble the modern genus Bathysiphon. Although agapitinated foraminifera started to proliferate in the Ordovician, true multichambered forms did not emerge until the Devonian, when the Fusulinina started to flourish and gave rise to the intricately designed tests of the Fusulinacea in the Late Carboniferous and Permian eras. The end of the Paleozoic era saw the extinction of this super-family. It was during the Early Carboniferous that Miliolina and Lagenina first emerged. The emergence and radiation of the Rotaliina, Miliolina, and complex Textulariina in the Jurassic, which was shortly followed by the appearance of the first unquestionably planktic foraminifera, are significant Mesozoic events. Larger miliolines and rotaliines blossomed in the Cretaceous tropical regions, and a thriving planktic population was supported by the newly opened Atlantic Ocean and the extensive chalk seas.

The end of the Cretaceous marked the extinction of the planktic Globotruncanidae. About 75% of the species in the Tethys Ocean's low latitude vanished at or close to the K–T boundary. Only ecological generalists survived extinction because it was a very selective process. The observed mass extinction pattern is accompanied by marked variations in temperature, salinity, oxygen, and nutrients across the boundary. These variations are caused by a combination of long-term environmental changes, such as sea level, volcanism, and climate change, as well as short-term effects, like the impact of proposed bolides. The emergence of the planktic Globigerinidae and Globorotalidae in the Paleocene and the development of orbitoids and soritids in the Neocene—which eventually spread nearly everywhere—and nummulites and soritids in the Old World and Old World, respectively, marked the beginning of a comparatively quick radiation. Since the extinction of orbitoids during the Miocene, the distribution and diversity of larger foraminiferal stocks have gradually decreased, primarily due to deteriorating climate. Since the Late Cretaceous, the diversity of planktics has also decreased.

8.8.7. Applications of Foraminifera

Because they are tiny, numerous, widely dispersed, and frequently incredibly diverse, foraminifera are in many ways the perfect index fossils for marine rocks. Additionally, a lot of them have complex morphologies that make it easy to track evolutionary changes. Planktic foraminifera serve as the foundation for significant intercontinental correlation schemes between Mesozoic (particularly upper Cretaceous) and Cenozoic rocks. Although their distribution is typically more constrained, benthic foraminifera offer helpful schemes for local correlation. Comparisons with the many studies of contemporary ecology serve as the foundation for environmental interpretations that make use of fossil foraminifera. For instance, research on the foraminifera of late glacial and postglacial raised beaches and beach deposits has revealed significant shifts in depth, salinity, and climate (Bates et al., 2000; Roe et al., 2002). The established depth distribution of contemporary foraminifera has served as the foundation for the value of benthic foraminifera as indicators of the depth of deposition. To plot changes in depth, trends in species diversity, planktic–benthic ratios, shell type ratios, and morphology have been used. Broadly speaking, both species diversity and the planktic–benthic ratio rise offshore to the continental slope. Planktic life assemblages exhibit depth stratification, resulting in the formation of more diverse death assemblages in deeper waters relative to those in shallower waters. It is also possible to identify assemblages connected to benthic depth in Cretaceous sediments.

Jurassic and younger rocks can be interpreted using the planktic–benthic ratio. The proportions of agglutinated, porcelaneous, and haline in different test types seem to have varied throughout the Paleogene. The salinity of modern marginal marine species is a major determinant. Many researchers have identified unique foraminiferal assemblages in the deep sea, upper slope, and inner and outer continental shelves in water with a typical marine salinity. Numerous studies of the upper portions of DSDP and ODP cores are starting to reveal patterns and distribution of deep sea benthic foraminifera. The distribution of water masses and ocean currents are related to the biogeography of modern foraminifera. Benthic and planktic foraminifera's paleobiogeographical patterns are therefore essential for deducing paleoceanography. The foraminiferal assemblage and the oxygen minimum zone are highly correlated at bathyal depths. In regions where upwelling occurs, benthic foraminifera are also useful markers of productivity. Changes in the advection of North Atlantic Deep Water during the Quaternary are correlated with the relative abundances of Bulimina alazanensis and Cibicides wuellerstorfi. The distribution and stable isotope chemistry of foraminifera have also been used to reconstruct Cretaceous current patterns and ocean stratification. Price & Hart (2002) documented shifting oceanic temperature gradients in the Pacific Ocean during the Early-Middle Albian by utilizing both $\delta^{13}C$ and $\delta^{18}O$ values in benthic and planktic foraminifera. A reduction in the poleward heat flux is suggested by an increase in ocean temperature (and/or a decrease in salinity) in the Cenomanian, which encourages the accumulation of limited polar ice.

Stable isotopes in planktic and benthic foraminifera also indicate a 100–500 kyr long period of instability in oceanic bottom-water temperature and sea level before the K-T boundary transition at Stevns Klint, Denmark. During the latest Maastrichtian, bottom water temperatures gradually cooled by about 1.5°C as surface water temperatures remained constant, perhaps consistent with the initiation of a thermohaline circulation and the formation of some polar ice. The narrow temperature ranges of living planktic species have become useful tools in paleoclimatology especially of Quaternary sediments. Neogloboquadrina pachyderma and Globigerina bulloides have been extensively used as a paleotemperature proxy from the Late Miocene through the Quaternary, exploiting the modern polar affinity of the sinistrally coiled forms. However, the relationship between sea surface temperature and coiling direction is not a simple one. Pliocene and Pleistocene sinistral forms of N. pachyderma are morphologically and ecologically different and the modern sinistral form only appeared ~1 Myr. This suggests that N. pachyderma (sinistral) should not be used for calibrated paleoceanographic reconstructions before the Middle Pleistocene. N. pachyderma (sinistral) may have evolved in response to the onset of the 100 kyr climate regime in the Middle Pleistocene. The proportions of sinistral and dextral forms of N. pachyderma and G. bulloides have also been shown to change in response to the vigour of oceanic upwelling.

One of the main instruments used in paleoceanographic and paleoclimatic research is the analysis of the oxygen isotope ratios of calcareous foraminiferid shells. To distinguish between variations in global temperature and ice volume during the Cenozoic and even to refute theories linking ice advance to orbital forcing, paired Mg/Ca and $^{18}O/^{16}O$ measurements are being performed on both benthic foraminifera. Surface water stratification can be inferred from variations in the $\delta^{18}O$ and $\delta^{13}C$ of planktic foraminifera that live in shallow and deep waters, respectively.

When used with other paleoceanographic proxies, foraminifera are especially helpful in paleoecology and paleo-oceanography. The most well-studied global cooling events occurred during the Quaternary, the Miocene, and the Eocene–Oligocene boundary, and are documented in a large number of case studies from the Cenozoic. Foraminifera have been used to document the onset and orbital forcing of the Messinian Salinity Crisis, and the $\delta^{13}C$ profile in Globigerinoides ruber has been used to infer changes in solar irradiance over the last 2000 years. The tectonic history of ocean gateway openings and closings has been reconstructed using planktic foraminifera. Hydrocarbon exploration has made extensive use of foraminifera. For biosteering, paleoenvironment analysis, and biostratigraphy, they are crucial instruments (Blatt, Berry, & Brande, 1991; Doyle, 1996; McGowran, 2005).

CASE STUDY

USING ROCKWORKS FOR STRATIGRAPHIC ANALYSIS:

RockWorks is a versatile geological software used for stratigraphic analysis, borehole visualization, and geological modeling. This case study demonstrates the application of RockWorks for stratigraphic analysis in a hypothetical geological project.

- **Project Background:** A geotechnical engineering firm is conducting a study on a proposed construction site to assess its subsurface geological conditions. The objective is to create a detailed stratigraphic model to identify different soil and rock layers, their thicknesses, and their spatial distribution.
- **Data Collection:** The first step in the project involves collecting borehole data from various locations across the site. Each borehole log includes information on:
  - o Depth intervals;
  - o Lithology descriptions;
  - o Grain size;
  - o Color;
  - o Moisture content;
  - o Presence of groundwater.
- **Steps in RockWorks Stratigraphic Analysis:**

1. Data Entry:
   - o Input borehole data into RockWorks. Each borehole log is entered with detailed depth intervals and corresponding lithological descriptions.
2. Creating Borehole Logs:
   - o Generate detailed borehole logs using the Borehole Manager in RockWorks. These logs provide a visual representation of the subsurface conditions at each drilling location.
3. Lithology Correlation:
   - o Use the lithology correlation tool to correlate similar lithological units across different boreholes. This step is crucial for identifying continuous strata and understanding their spatial distribution.
4. Stratigraphic Modeling:
   - o Develop a stratigraphic model by defining stratigraphic units based on correlated lithological data. Assign distinct colors and patterns to different stratigraphic units for better visualization.
   - o Use the Stratigraphy Picker tool to manually adjust stratigraphic boundaries where necessary.

5. 3D Visualization:
   - o Create 3D visualizations of the subsurface stratigraphy. RockWorks allows for the generation of 3D block models, fence diagrams, and cross-sections to better understand the spatial relationships between different stratigraphic units.
6. Isopach Maps:
   - o Generate isopach maps to show the thickness variations of different stratigraphic units across the site. Isopach maps are valuable for identifying areas with thick or thin sedimentary layers, which may impact construction decisions.
7. Hydrostratigraphic Analysis:
   - o Integrate hydrogeological data to create hydrostratigraphic models. This involves mapping aquifers and aquitards to understand groundwater flow and storage within the subsurface.
8. Report Generation:
   - o Compile the results into a comprehensive report. RockWorks provides tools for creating detailed reports, including tables, charts, and graphical representations of the stratigraphic model.

- **Results:** The stratigraphic analysis using RockWorks revealed several key findings:
  - o Identification of five distinct stratigraphic units: Topsoil, Sand, Clay, Gravel, and Bedrock.
  - o The spatial distribution of these units showed variability across the site, with the sand layer being thicker in the northern part and thinning towards the south.
  - o The gravel layer, which is significant for foundation stability, was found to be continuous across the site but varied in thickness.
  - o Groundwater was encountered primarily within the sand layer, suggesting it acts as a major aquifer.
- **Conclusion:** The use of RockWorks for stratigraphic analysis provided a detailed understanding of the subsurface conditions at the proposed construction site. The ability to visualize and model the stratigraphy in 3D was particularly valuable for identifying areas of concern and making informed decisions about construction planning and design. This case study demonstrates the effectiveness of RockWorks as a tool for geotechnical and geological projects, offering comprehensive data analysis and visualization capabilities.

CLASS ACTIVITY

Students form teams to sort through specimen sets to identify the species of their sample (noted on the group's range sheet). Using the range sheet, they determine the possible age of their sample set. The individual teams then come together to discuss the results as a whole class. The individual age ranges based on each team are compared to further constrain the age of the sample. Together, all groups determine the final age range of their sample.

SUMMARY

- Biostratigraphic units (biozones) are bodies of strata that are defined or characterized on the basis of their contained fossils. Biostratigraphic units exist only where the particular diagnostic feature or attribute on which they are based has been identified.
- The importance of fossils as indicators of processes and environments of deposition, but the study of fossils has also provided fundamental information about the evolution of life on Earth.
- Modern biological analyses provide additional information that also allows the genetic characteristics of organisms to be considered when defining a species: similarities or differences in genetic make-up make it possible to rigorously define species and determine the relationships between them.
- A biostratigraphic unit is a body of rock defined by its fossil content. It is therefore fundamentally different from a lithostratigraphic unit that is defined by the lithological properties of the rock.
- The abundance of a particular taxon may vary through time, in which case an interval containing a statistically high proportion of this taxon may be used to define a biozone. This approach can be unreliable because the relative abundance is due to local environmental factors.
- These Paleozoic arthropods are the main group used in the zonation of the Cambrian. Most trilobites are thought to have been benthic forms living on and in the sediment of shallow marine waters.
- This taxonomic group of cephalopods (phylum Mollusca) includes goniatites from Paleozoic rocks as well as the more familiar ammonites of the Mesozoic. The nautiloids are the most closely related living group.
- This phylum includes crinoids (sea lilies) and echinoids (sea urchins). Most crinoids probably lived attached to substrate and this sessile characteristic

makes them rather poor zone fossils, despite their abundance in some Paleozoic limestones.

- These organisms form a subclass of planktonic protozoans and are found as fossils in deep marine strata throughout the Phanerozoic. Radiolaria commonly have silica skeletons and are roughly spherical, often spiny organisms less than a millimeter across.
- Fossils that cannot be seen with the naked eye and are only just discernible using a high-power optical microscope are referred to as nanofossils. They are microns to tens of microns across and are best examined using a scanning electron microscope.
- Ostracods are crustaceans with a two-valve calcareous carapace and their closest relatives are crabs and lobsters. They occur in a very wide range of depositional environments, both freshwater and marine, and they have a long history, although their abundance and distribution are sporadic.
- Spores and pollen are produced during the life cycle of plants. Spores are produced by the bryophytes and ferns, and pollen by the gymnosperms and angiosperms.
- Sediments from deltaic and lacustrine deposits of Mid-Ordovician to Early Silurian age yield cryptospore monads, dyads, triads, and tetrads. Palynological preparations of this age can also contain tubes and sheets of cuticle that may represent debris from the first subaerial plants.
- The Foraminiferida are an important order of single-celled protozoa that live either on the sea floor or amongst the marine plankton. The soft tissue (cytoplasm) of the foraminiferid cell is largely enclosed within a shell or test, variously composed of calcite, or of agglutinated particles.
- Planktic foraminifera are important contributors to deep sea sedimentation and, with coccoliths, account for more than 80% of modern carbonate deposition in seas and oceans.
- Foraminifera are in many respects ideal index fossils for marine rocks, being small, abundant, widely distributed and often extremely diverse. Many also have an intricate morphology in which evolutionary changes can be readily traced.

REVIEW QUESTIONS

1. What is biostratigraphy and how is it used to date and correlate rock strata?
2. Define population fragmentation and explain its role in evolutionary processes.
3. Compare and contrast phyletic gradualism and punctuated equilibrium.
4. Explain how biostratigraphy can be used to infer speciation events.
5. Describe the methods used to estimate the rate of speciation from fossil data.
6. How does the depositional environment affect the preservation of fossils?
7. Explain the significance of the geographical distribution of organisms in biostratigraphy.

8. Define marine microfossils and describe their role in biostratigraphy.
9. Describe the morphological features of spores that are important for biostratigraphic analysis.
10. Explain the processes of dispersal and sedimentation of fossils.

REFERENCES

1. Blatt, H., Berry, W. B., & Brande, S. (1991). Principles of Stratigraphic Analysis. Blackwell Scientific Publications.
2. Cantalejo, B., Pickering, K. T., McNiocaill, C., Bown, P., Johansen, K., & Grant, M. (2020). A revised age-model for the Eocene deep-marine siliciclastic systems, Aínsa Basin, Spanish Pyrenees. Journal of the Geological Society, 178(1), jgs2019–131.
3. Clarkson, E. N. K. (1993). Invertebrate Paleontology and Evolution (3rd ed.). Allen & Unwin.
4. Doyle, P. (1996). Understanding Fossils: An Introduction to Invertebrate Paleontology. Wiley.
5. Gard, G., Crux, J., Lee, B., & Belgarde, C. (2015). Remote monitoring of well site biostratigraphy. APPEA Journal, 55(2), 469.
6. Loutit, S., Romine, K. K., & Foster, C. B. (1997). Sequence biostratigraphy, petroleum exploration and A. cinctum. APPEA Journal, 37(1), 272.
7. McGowran, B. (2005). Biostratigraphy: Microfossils and Geological Time. Cambridge University Press.
8. Neal, J. E., Stein, J. A., & Gamber, J. H. (1994). Graphic correlation and sequence stratigraphy in the Paleogene of NW Europe. Journal of Micropaleontology, 13(1), 55–80.
9. Scherer, R. P., Gladenkov, A. Y., & Barron, J. A. (2007). Methods and applications of Cenozoic marine diatom biostratigraphy. Paleontological Society Papers, 13, 61–83.

CHAPTER

9 Dating and Correlation Techniques for Stratigraphy

LEARNING OBJECTIVES

After studying this chapter, you will be able to:

- Understand the concept of radiometric dating;
- Describe other isotopic and chemical techniques for dating;
- Define magnetostratigraphy;
- Explain dating in the quaternary

KEY TERMS FROM THIS CHAPTER

Amino acid racimization
Biostratigraphic correlation
Chemostratigraphy
Magnetic material
Radiogenic decay
Sedimentary rock
Thermochronometry
Authigenic minerals
Carbonate minerals
Closure temperature
Metamorphism
Radiometric dating
Thermochronology

INTRODUCTION

Scientists employ dating techniques to ascertain the age of fossils, rocks, and artifacts. Absolute dating techniques give an approximate date in years, whereas relative dating techniques only indicate whether one sample is older or younger than another. The latter have primarily only been accessible since 1947. The process of radioactive decay, in which an element's radioactive form gradually transforms into a non-radioactive product, is a useful tool for many absolute dating methods. Some rely on chemical alterations in the organic or inorganic composition of a sample, such as cation-ratio dating and amino acid racemization. A few of these approaches have been the subject of intense scrutiny in recent years as researchers work to create the most precise dating methods available.

Since the early 20th century, when radioactivity was discovered, geologists have had access to radiometric dating techniques, which have allowed them to pinpoint specific dates throughout Earth history on an absolute time scale of millions or even billions of years. Although there are several distinct radioactive decay series, the main application for them is in the dating of igneous rocks since they all give an age for the formation of a mineral. Dating a grain in a sandstone bed typically yields the age at which the mineral first formed in a sedimentary rock, such as granite, but it does not reveal much about the sedimentary rock's age. Magnetostratigraphy and chemostratigraphy are two techniques that are frequently used in conjunction with other approaches. These methods can be carried out using the magnetic and chemical properties of rocks. In reality, the dating and correlating of rocks is a process that requires the fusion of data from several sources and methodologies. Carbon-14 dating, which can be used to date organic material formed in the last few tens of thousands of years, is one of the many techniques used in the specialized field of Quaternary dating.

The age of the rock is not determined by lithostratigraphic techniques, except for indicating which rock units are older or younger. Moreover, stratigraphy lacks a temporal framework based on the correlation of lithological characteristics. Fossils can be used for both local and regional correlation, allowing for the ordering of strata on a large and small scale through biostratigraphic correlation. Although these are still the most popular and oldest methods in stratigraphy, newer methods that make use of the technological advancements of the last century have been added to the mix. The most significant of these is radiometric dating since it has given rise to a temporal framework with a year-based measurement. New methods utilizing various isotope systems are constantly being developed, and improved instrumentation enables more precise measurements to be made on smaller material samples.

9.1. RADIOMETRIC DATING

The early 1900s saw the discovery of radioactivity and the radiogenic decay of isotopes, which made it possible to date rocks using an absolute method as opposed to a relative one. Until now, theories regarding the rates of evolution, deposition, the Earth's and the Sun's thermal behavior, or the interpretation of sacred texts have been used to determine the age of the planet. Radiometric dating determines the age of a rock by measuring the proportions of different isotopes present today, assuming the proportions of different isotopes present when the mineral formed, and knowing the rate of decay of the elements present in minerals. The primary application of this dating technique is to ascertain the age of the formation of igneous rocks, such as volcanic units found inside sedimentary strata. It can also be applied to authigenic minerals found in some sedimentary rocks, like glauconite. The age of the metamorphism is typically indicated by radiometric dating of the minerals in metamorphic rocks.

When materials like carbon or rocks were formed, trace radioactive impurities were selectively incorporated into them. This process is known as radiometric dating. Using a known constant decay rate, the method measures the abundance of a naturally occurring radioactive isotope in the material and compares it to the abundance of its decay products. Since Bertram Boltwood first described the use of radiometric dating in 1907, it has been the primary method for determining the exact age of rocks and other geological features, as well as the age of the Earth and fossilized life forms. It can also be used to date a variety of natural and artificial materials. In geochronology, radiometric dating techniques are employed in conjunction with stratigraphic principles to determine the geologic time scale. The most well-known methods are uranium-lead dating, potassium-argon dating, and radiocarbon dating. It offers an important source of information about the ages of fossils and the estimated rates of evolutionary change by enabling the establishment of geological timescales. Archaeological materials, such as ancient artifacts, are also dated using radiometric dating.

The range of materials that can be used and the time periods over which various radiometric dating techniques are accurate are different.

9.1.1. Fundamentals

9.1.1.1. Radioactive Decay

Each chemical element combination that makes up ordinary matter has an atomic number that represents the number of protons in the atomic nucleus. Furthermore, elements can exist in distinct isotopes, with the number of neutrons in the nucleus varying amongst isotopes of an element. A nuclide is a specific isotope of a specific element. Certain nuclides are unstable by nature. That is, an atom of such a nuclide will eventually experience radioactive decay and change into a different nuclide on its own. There are several ways to achieve this transformation, such as alpha decay, which involves the release of alpha particles, and beta decay, which involves the release of electrons, positrons, or electrons through capture. Spontaneous fission into two or more nuclides is another possibility.

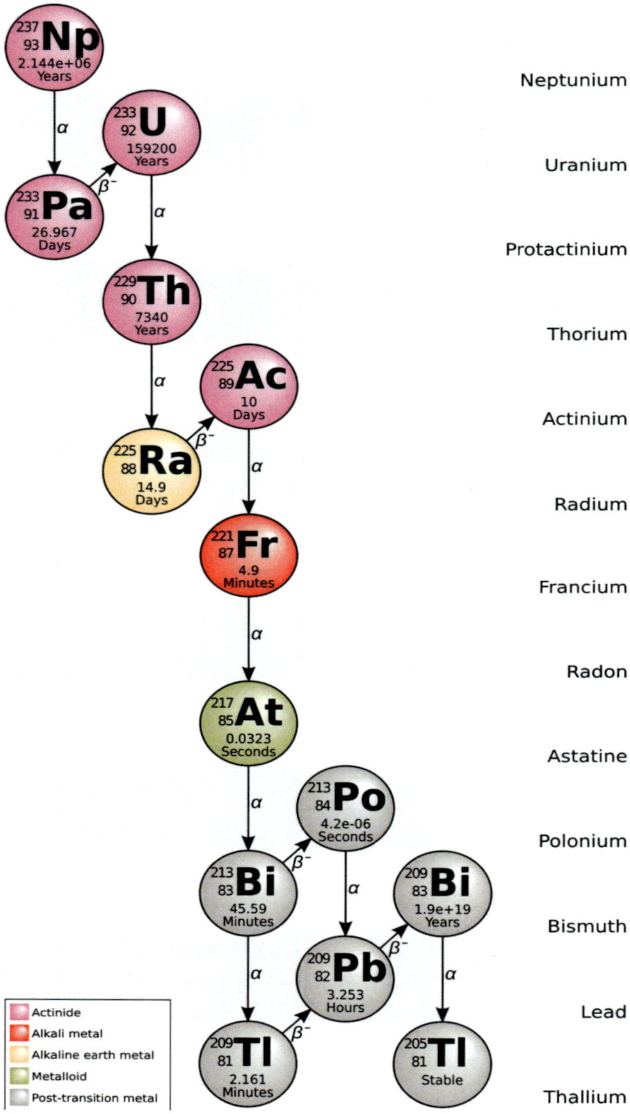

Figure 9.1. Radioactive decay results in the formation of a new 'daughter' isotope from the 'parent' isotope.

Source: By Original: Tosaka~commonswiki Vector: BatesIsBack (2003). "The NUBASE Evaluation of Nuclear and Decay Properties." Nuclear Physics A., 729, 3–128. doi: 10.1016/j.nuclphysa.2003.11.001., CC BY 3.0, https://commons.wikimedia.org/w/index.php?curid=14865030.

The decay of a collection of atoms of a radioactive nuclide follows an exponential pattern at a rate determined by the half-life, often measured in years for dating purposes. After one half-life, half of the atoms will have decayed into a daughter nuclide. In some cases, the "daughter" nuclide is also radioactive, leading to a decay chain that ends with a stable daughter nuclide. Each step in the chain has its own distinct half-life. The longest half-life in the chain is usually the most important for radiometric dating, as it determines the overall transformation of the radioactive nuclide into its stable form. Isotopic systems used in radiometric dating have half-lives ranging from a few years to billions of years (Bernard-Griffiths & Groan, 1989) (Figure 9.1).

The half-life is essentially constant and depends only on nuclear properties for the majority of radioactive nuclides. This is known because the ages of the same materials are consistent across methods and decay constants measured using various techniques yield consistent values within analytical errors. External elements like pressure, temperature, the presence of an electric or magnetic field, or the chemical environment have no effect on it. The only exceptions are nuclides like strontium-85, zirconium-89, and beryllium-7 that decay by electron capture; the rate of decay of these nuclides is dependent on the local electron density. When the original nuclide decays over time, the ratio of the original nuclide to its decay products for all other nuclides varies predictably. Because of their predictability, related nuclide relative abundances can be used as a clock to determine how long it has been since the original nuclides were incorporated into a material (Pommé et al., 2018).

9.1.1.2. Decay Constant Determination

The basis for the widely used method of measuring radioactivity is the radioactive decay constant, which is the likelihood that an atom will decay annually. A nuclide's half-life and age can only be determined with accuracy and precision based on the measurement of the decay constant. One technique for determining a system's decay constant is the in-growth method, which entails the accumulation of daughter nuclides. Unfortunately, it takes a lot of time (decades) for nuclides with high decay constants—which help date extremely old samples—to accumulate enough decay products in a single sample so that measurements of them can be made with accuracy. A quicker approach is to measure alpha, beta, or gamma activity with particle counters and divide the result by the total number of radioactive nuclides. Accurately calculating the quantity of radioactive nuclides is difficult and costly. An alternative method for determining decay constants is to compare isotope data for known-age rocks. Using this method, at least one of the isotope systems—like the Pb–Pb system—needs to be extremely precisely calibrated.

9.1.1.3. Accuracy of Radiometric Dating

The fundamental formula for radiometric dating stipulates that following material formation, neither the parent nuclide nor the daughter product may enter or exit the substance. It is necessary to take into account the potential confounding effects of parent and daughter isotope contamination as well as the effects of any loss or gain of these isotopes since the sample was produced. Thus, it's critical to gather as much information as you can about the material being dated and to look for any indications of tampering. Measuring multiple samples from various locations within the rock body improves precision. On the other hand, an isochron should form if multiple minerals that can be dated from the same sample are thought to have formed from the same event and were in equilibrium with the reservoir at the time of formation. This may lessen the contamination issue. The Concordia diagram is employed in uranium-lead dating, which also lessens the issue of nuclide loss. Lastly, to verify the age of a sample, correlation between various isotopic dating techniques might be necessary. For instance, age estimates of 3.60 ± 0.05 Ga (billion years ago) for the Amitsoq gneisses in western Greenland and 3.56 ± 0.10 Ga (billion years ago) for the same rocks using lead–lead dating are in agreement with one another.

The parent must typically have a long enough half-life to be present in significant amounts at the time of measurement, the parent's half-life must be precisely known, and sufficient daughter product must be produced to allow for accurate measurement and differentiation from the initial amount of the daughter present in the material. Precise and accurate methods must be employed in the isolation and analysis of the parent and daughter nuclides. Usually, isotope-ratio mass spectrometry is used for this.

The half-life of the radioactive isotope in question affects a dating method's accuracy in part. For example, the half-life of carbon-14 is 5,730 years. There is so little carbon-14 remaining in an organism after it has died for 60,000 years that precise dating is impossible. Conversely, the decrease in carbon-14 concentration is so great that the age of relatively young remains can be accurately ascertained to within a few decades.

9.1.1.4. Closure Temperature

The temperature below which the mineral is a closed system for the isotopes under study is known as the closure temperature, also called the blocking temperature. Any daughter nuclides that have accumulated over time will be lost through diffusion if a material that rejects the daughter nuclide selectively is heated above this temperature, which will reset the isotopic "clock" to zero. Isotope diffusion becomes more difficult as the mineral cools and the crystal structure starts to form. The crystal structure has sufficiently formed at a given temperature to stop isotope diffusion. Therefore, until it cools below the closure temperature, a slowly cooling igneous or metamorphic rock or melt does not show any detectable radioactive decay. Thus, the time at which the rock or mineral cooled to closure temperature is the age that can be determined using radiometric dating. Every mineral and isotopic system has a different temperature, which is why a system may be closed for one mineral but open for another. The thermal history of a rock can be tracked over time through dating of different minerals and/or isotope systems within the same rock that have different closure temperatures. This allows for a detailed understanding of the history of metamorphic events. Using a high-temperature furnace to artificially reset sample minerals, these temperatures are measured experimentally in the lab. Thermochronology or thermochronometry is the name of this field.

9.1.1.5. The age Equation

The mathematical expression that relates radioactive decay to geologic time is:

$$D^* = D_0 + N(t)\,(e^{\lambda t} - 1)$$

where; t is the age of the sample; D* is the number of atoms of the radiogenic daughter isotope in the sample; D_0 is the number of atoms of the daughter isotope in the original or initial composition; N(t) is the number of atoms of the parent isotope in the sample at time t (the present), given by $N(t) = N_0 e^{-\lambda t}$; and λ is the decay constant of the parent isotope, equal to the inverse of the radioactive half-life of the parent isotope times the natural logarithm of 2.

Rather than using the constant initial value No, it is more convenient to express the equation in terms of the measured quantity N(t). It is assumed that the system is closed (neither parent nor daughter isotopes have been lost from the system), that D_0 must be negligible or accurately estimated, that λ is known with high precision, and that D* and N(t) can be measured with accuracy and precision to determine the age.

The parent and daughter isotope compositions at the moment the material under test cooled below its closure temperature are used in the above equation. This is generally known for isotopic systems. Nevertheless, an isochron can be created with just the current ratios of the parent and daughter isotopes to a standard isotope, without the need for knowledge of the original compositions. The age of the sample and its original composition are determined graphically by solving the age equation using an isochron plot.

9.1.2. Modern Dating Methods

Since its invention by Ernest Rutherford in 1905 as a means of determining the age of the Earth, radiometric dating has been used. Since then, a century has passed, and the techniques have significantly expanded and improved. A mass spectrometer can now be used to date materials as small as a nanogram. In the 1950s, radiometric dating started to use the mass spectrometer, which had been developed in the 1940s. It works by using the sample being tested to produce an ionized atom beam. After passing through a magnetic field, the ions are directed into various Faraday cups sampling sensors based on their mass and degree of ionization. The ions create a very weak current in the cups upon impact, which can be used to measure the rate of impacts and the relative concentrations of various atoms in the beams.

9.1.2.1. Uranium–Lead Dating Method

Uranium-235 or uranium-238 are used in uranium-lead radiometric dating to determine the precise age of a material. The error margin in the dates of rocks can be as low as less than 2 million years in 2.5 billion years thanks to the refinement of this scheme. It has been possible to obtain an error margin of 2–5% on younger Mesozoic rocks.

The mineral zircon ($ZrSiO_4$) is frequently the subject of uranium-lead dating, however, it can also be applied to other materials like baddeleyite and monazite (see: monazite geochronology). While lead is strongly rejected, uranium atoms are incorporated into the crystalline structures of zircon and baddeleyite to replace zirconium. Zircon is incredibly chemically inert, has a very high closure temperature, and is resistant to mechanical weathering. During metamorphic events, zircon also forms multiple crystal layers, each of which may record an isotopic age for the event. It is possible to perform in situ micro-beam analysis using SIMS or laser ICP-MS methods (Mattinson, 2005). One of its greatest benefits is that every sample has two clocks: one based on the half-life of uranium-235, which decays to lead-207, and another based on the half-life of uranium-238, which decays to lead-206, which has a half-life of approximately 4.5 billion years. This built-in cross-check enables precise age determination of the sample even in cases where some lead has been lost. The concordia diagram illustrates this, with the samples plotting along an errorchron (straight line) that intersects the concordia curve at the sample age.

9.1.2.2. Samarium–Neodymium Dating

A radiometric dating technique called samarium-neodymium dating is helpful in establishing the ages of rocks and meteorites. It is based on the alpha decay of the stable radiogenic neodymium isotope (^{143}Nd) from the long-lived samarium isotope (^{147}Sm). The age and source of igneous melts can be determined using samarium-neodymium ratios and neodymium isotope ratios. It is sometimes believed that the neodymium isotope ratio solely depends on the time of the event that causes crustal material to form from the mantle; however, it then evolves according to the new ratio of samarium to neodymium in the crustal material, which will differ from the ratio in the mantle material. We can ascertain the formation date of the crustal material using samarium-neodymium dating.

Because these two elements are rare earth elements and are therefore, in theory, not particularly susceptible to partitioning during sedimentation and diagenesis, Sm–Nd dating is useful. The Sm/Nd ratio of the resulting materials is altered by the fractional crystallization of felsic minerals. This in turn affects how quickly radiogenic ^{143}Nd is produced, which raises the ^{143}Nd/^{144}Nd ratio. In many cases, Sm–Nd and Rb–Sr isotope data are used together (McCulloch & Wasserburg, 1978).

9.1.2.3. Sm–Nd Radiometric Dating

Neodymium and samarium both have seven naturally occurring isotopes. The two elements are connected in a parent-daughter relationship through the alpha decay of parent radionuclide ^{147}Sm, which produces radiogenic daughter ^{143}Nd with a half-life of 1.06×10^{11} years, and the alpha decay of nearly extinct radionuclide ^{146}Sm, which produces radiogenic daughter ^{142}Nd with a half-life of $1.03(5)\times10^{8}$ years. The technique known as isochron dating can be used to determine the exact date that a rock (or group of rocks) formed. The ratio of radiogenic ^{143}Nd to non-radiogenic ^{144}Nd is plotted against the ratio of the parent isotope, ^{147}Sm, to the non-radiogenic isotope, ^{144}Nd, on the Sm-Nd isochron. Since ^{144}Nd is a stable and reasonably abundant neodymium isotope, it is used to normalize the radiogenic isotope in the isochron.

The Sm-Nd isochron is defined by the following equation:

$$\left(\frac{^{143}\mathrm{Nd}}{^{144}\mathrm{Nd}}\right)_{present} = \left(\frac{^{143}\mathrm{Nd}}{^{144}\mathrm{Nd}}\right)_{initial} + \left(\frac{^{147}\mathrm{Sm}}{^{144}\mathrm{Nd}}\right)\cdot(e^{\lambda t}-1),$$

where; t is the age of the sample; ⊠ is the decay constant of ^{147}Sm; (e$^{⊠t}$−1) is the slope of the isochron which defines the age of the system. On the other hand, it is possible to consider that the substance originated from mantle material that was undergoing a similar evolutionary process in terms of these ratios as chondrites. By doing so, the exact time of formation can be determined.

9.1.2.3.1. Sm and Nd geochemistry

According to Brown's reaction series, the concentration of Sm and Nd in silicate minerals increases with the order in which they crystallize from magma. Since samarium is easier to incorporate into mafic minerals, neodymium will be more concentrated in the melt phase of a mafic rock that crystallizes mafic minerals than samarium. Therefore, both the abundance and ratio of Sm and Nd vary when a melt fractionally crystallizes from a mafic to a more felsic composition. Consequently, ultramicrous rocks have high Sm and low Nd, resulting in high Sm/Nd ratios. Low Sm and high Nd concentrations result in low Sm/Nd ratios in felsic rocks (komatiite, for instance, has 1.14 ppm Sm and 3.59 ppm Nd, compared to 4.65 ppm Sm and 21.6 ppm Nd in rhyolite). The importance of this process is apparent in modeling the age of continental crust formation.

9.1.2.3.2. The CHUR Model

By examining the isotopic compositions of neodymium, DePaolo and Wasserburg found that terrestrial igneous rocks, when they were formed from melts, closely mirrored the chondritic uniform reservoir or CHUR line – the trend of the ^{143}Nd:^{144}Nd ratio increasing with time in chondrites. Chondritic meteorites are believed to represent the primitive material that existed in the Solar system before the formation of planets. Because chondrites have consistent trace-element signatures, studying their isotopic evolution can help model the evolution of the entire Solar system and the bulk Earth. After plotting the ages and initial $^{143}Nd/^{144}Nd$ ratios of terrestrial igneous rocks on a diagram showing Nd evolution over time, DePaolo and Wasserburg concluded that Archean rocks had initial Nd isotope ratios similar to those defined by the CHUR evolution line.

9.1.2.3.3. The Depleted-Mantle Model

DePaolo and Wasserburg noted that the bulk of young oceanic volcanics (Mid Ocean Ridge basalts and Island Arc basalts) lay +7 to +12 ⊠ units above the CHUR line, despite the good fit of Archean plutons to the CHUR Nd isotope evolution line. This resulted in the discovery that Archean continental igneous rocks that plotted within the CHUR line's error may actually lie on a line representing the evolution of depleted mantle that is characterized by gradually increasing Sm/Nd and $^{143}Nd/^{144}Nd$ ratios. A study on the Proterozoic metamorphic basement of the Colorado Front Ranges (the Idaho Springs Formation) was carried out to examine this discrepancy between the Archean CHUR data and the young volcanic samples in more detail. Plotting the initial $^{143}Nd/^{144}Nd$ ratios of the analyzed samples on a ⊠Nd versus time diagram is depicted in Figure 9.2. DePaolo represented the neodymium isotope evolution of a depleted reservoir by fitting a quadratic curve to the Idaho Springs and average ⊠Nd for the modern oceanic island arc data.

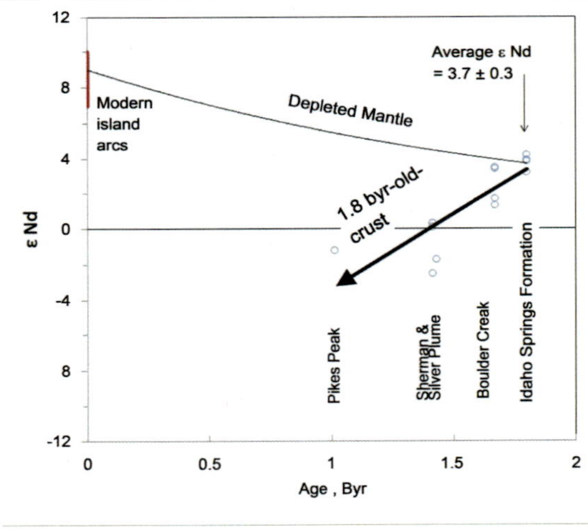

Figure 9.2. A graph showing the depleted-mantle model of DePaolo.

Source: https://upload.wikimedia.org/wikipedia/commons/c/cb/Depleted_mantle_model.jpg.

9.1.2.4. Potassium–Argon Dating Method

The radiometric dating technique known as potassium-argon dating, or K-Ar dating, is employed in archaeology and geochronology. Its foundation is the measurement of the product of an potassium (K) isotope's radioactive decay into argon (Ar). Potassium is a common element present in a wide range of materials, including evaporites, feldspars, micas, clay minerals, and tephra. These materials allow the decay product ^{40}Ar to escape the liquid (molten) rock, but as the rock solidifies (recrystallizes), it begins to accumulate. The purity of the sample, the makeup of the mother material, and several other variables determine how much argon sublimation happens. Due to these factors, the final age determination depends on the conditions during formation, melting, and exposure to low pressure or open air. These factors also introduce error limits on the upper and lower bounds of dating. The ratio of the amount of ^{40}Ar accumulated to the remaining ^{40}K is used to calculate the time since recrystallization. The technique can be used to determine the exact age of samples older than a few 1,000 years due to its long half-life of ^{40}K (Figure 9.3).

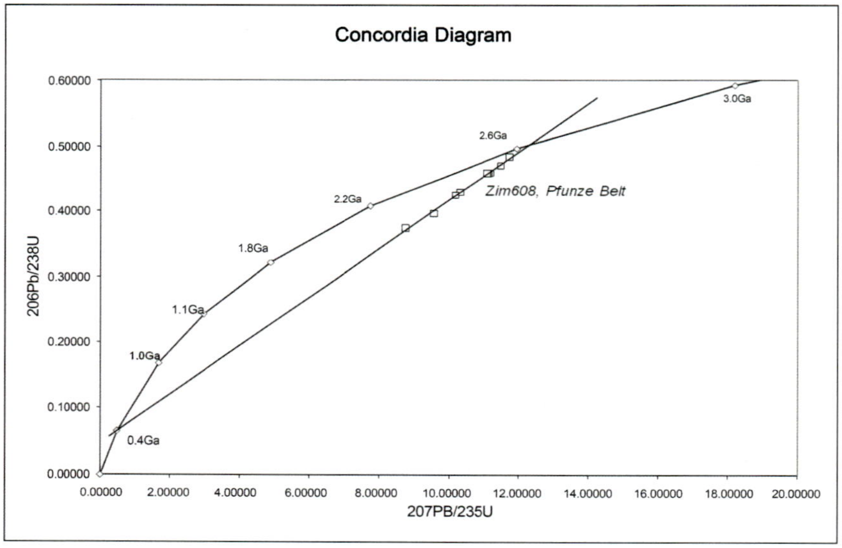

Figure 9.3. A Concordia diagram as used in uranium–lead dating, with data from the Pfunze Belt, Zimbabwe. All the samples show loss of lead isotopes, but the intercept of the errorchron (straight line through the sample points) and the Concordia (curve) shows the correct age of the rock.

Source: By Babakathy – Own work, Public Domain, https://commons.wikimedia.org/w/index.php?curid=5027761.

9.1.2.4.1. Decay Series

Three isotopes of potassium are found in nature: ^{39}K (93.2581%), ^{40}K (0.0117%), and ^{41}K (6.7302%). ^{41}K and ^{39}K have stability. The radioactive isotope ^{40}K decays to ^{40}Ca and ^{40}Ar over a half-life of 1.248×10^9 years. In 89.3% of decay events, electron emission (beta decay) results in the conversion to stable ^{40}Ca. The remaining 10.7% of decay events involve electron capture as the means of conversion to stable ^{40}Ar.

Since argon is an noble gas and does not bind with other atoms in a crystal lattice, it is a minor component of the majority of rock samples of geological interest. Because it is larger than the gaps between the other atoms in a mineral crystal, the atom usually stays trapped within the lattice when ^{40}K decays to ^{40}Ar. However, when the correct circumstances—such as variations in temperature or pressure—are met, it can escape into the surrounding area. As most crystals have melted and the atoms are no longer trapped, ^{40}Ar atoms can diffuse through and escape from molten magma. When magma cools down to form solid rock once more, entrained argon—diffuse argon that is unable to escape from the magma—may once more become trapped in crystals. More ^{40}K will decay and ^{40}Ar will reaccumulate along with the entrained argon atoms trapped in the mineral crystals following the recrystallization of magma. To determine how long it has been since a rock sample has solidified, the number of ^{40}Ar atoms is measured.

Although ^{40}Ca is the most abundant isotope and the favored daughter nuclide, its use in dating is rare due to the abundance of calcium in the crust. Therefore, it is unknown how much calcium was initially present, and this variation can cause problems when measuring the slight increases brought on by radioactive decay.

9.1.2.4.2. Formula

The following equation relates the ratio of ^{40}Ar to ^{40}K to the amount of time that has passed since the rock was sufficiently cool to trap the Ar:

$$t = \frac{t_{\frac{1}{2}}}{\ln(2)} \ln\left(\frac{K_f + \frac{Ar_f}{0.109}}{K_f}\right),$$

where; t is the time elapsed; $t_{1/2}$ is the half-life of ^{40}K; K_f is the amount of ^{40}K remaining in the sample; and Ar_f is the amount of ^{40}Ar found in the sample.

The scale factor of 0.109 adjusts for the portion of ^{40}K that decayed into ^{40}Ca but wasn't measured. Adding together the measured ^{40}K and the adjusted amount of ^{40}Ar reveals the initial amount of ^{40}K at the start of the time period. Typically, these values can be represented as a fraction of the total potassium present, since only relative amounts are needed, not exact quantities.

9.1.2.5. Rubidium–Strontium Dating Method

Scientists use the rubidium-strontium dating method (Rb-Sr), a radiometric dating technique, to ascertain the age of rocks and minerals based on the concentration of particular isotopes of strontium (^{87}Sr, ^{86}Sr) and rubidium (^{87}Rb). ^{87}Rb, one of the two naturally occurring isotopes of rubidium, has a half-life of 49.23 billion years before decaying to ^{87}Sr. Of the four naturally occurring strontium isotopes, only ^{87}Sr, the radiogenic daughter produced in this decay process, was not produced solely by stellar nucleosynthesis before the formation of the Solar System. While the quantity of other Sr isotopes stays constant, the amount of radiogenic ^{87}Sr increases with time due to the decay of ^{87}Rb.

Mass spectrometers are useful tools for accurately measuring the ratio $^{87}Sr/^{86}Sr$ in mineral samples. Assuming that the sample's formation-related Sr and Rb isotope contents can be ascertained, the age can be computed using the rise in $^{87}Sr/^{86}Sr$. The initial $^{87}Sr/^{86}Sr$ of various minerals that crystallized from the same silicic melt will be the same as that of the parent melt. Because Rb replaces K in minerals, the minerals will have had different initial Rb/Sr ratios, and the final $^{87}Sr/^{86}Sr$ ratio will not have increased as much in the minerals lower in Rb. This is because the minerals have different K/Ca ratios. In the order plagioclase, hornblende, K-feldspar, biotite, and muscovite, Rb/Sr generally increases. Given enough time for radiogenic ^{87}Sr to produce significantly (in growth), the measured $^{87}Sr/^{86}Sr$ values in the minerals will therefore differ and increase in the same order. Thus, by comparing various minerals in a rock sample, scientists can determine the age of the rock and infer its original $^{87}Sr/^{86}Sr$ ratio.

Furthermore, Rb is an extremely incompatible element that would rather join the magmatic melt than stay in the mantle minerals during partial melting of the mantle. Consequently, $^{87}Sr/^{86}Sr$ is higher in crustal rock than in mantle rock, and Rb is enriched in crustal rocks relative to the mantle. This enables scientists to differentiate between magma formed from the melting of mantle rock and magma formed from the melting of crust rock, even in cases where the general chemistry produced by subsequent magma differentiation is similar.

It is also possible for scientists to infer the formation date of crust rock from $^{87}Sr/^{86}Sr$ dating of magma that was taken from the mantle, even if the rock is later metamorphosed or even melts and recrystallizes. This offers hints regarding the age of the continents on Earth.

German chemists Otto Hahn and Fritz Strassmann contributed to the development of this process; Strassmann would go on to discover nuclear fission in December 1938.

9.1.2.6. Uranium–Thorium Dating Method

The radiometric dating method known as uranium–thorium dating, sometimes referred to as thorium–230 dating, uranium-series disequilibrium dating, or uranium-series dating, was developed in the 1960s and has been used since the 1970s to ascertain the age of calcium carbonate materials like coral and speleothem. In contrast to other widely used radiometric dating methods like uranium-lead or rubidium-strontium dating, the uranium-thorium method does not track the build-up of a stable end-member decay product. Rather, it determines an age based on the extent to which the radioactive isotope thorium-230 and its radioactive parent uranium-234 within a sample have returned to secular equilibrium.

9.1.2.7. Fission Track Dating Method

To ascertain the density of track markings left in a polished slice of a material by the spontaneous fission of uranium-238 impurities, is done. It is necessary to ascertain the sample's uranium content, which can be done by covering the polished portion of the material with a plastic film and subjecting it to a slow neutron bombardment. In contrast to the spontaneous fission of ^{238}U, this results in the induced fission of ^{235}U. The plastic film records the fission tracks created by this process. The number of tracks and the neutron flux can then be used to determine the material's uranium content.

There is a broad range of geologic dates over which this scheme is applicable. The best materials to use are meteorites, micas, and tektites—glass shards from volcanic eruptions—for dates up to a few million years old. Zircon, apatite, titanite, epidote, and garnet are among the materials with varying uranium contents that can be used to date older materials. The technique has advantages and limitations because temperatures above 200°C cause the fission tracks to heal. The method may be used to provide information about a deposit's thermal history.

9.1.2.8. Chlorine-36 Dating Method

Between 1952 and 1958, irradiation of seawater during atmospheric detonations of nuclear weapons produced large amounts of otherwise rare ^{36}Cl (half-life ~300ky). ^{36}Cl has a one-week residence period in the atmosphere. Thus, ^{36}Cl is also helpful for dating waters that were formed less than 50 years ago, as it serves as an event marker of 1950s water in the soil and groundwater. Other applications of ^{36}Cl in geological sciences include ice and sediment dating.

9.1.2.9. Luminescence Dating Methods

Since luminosity dating techniques do not depend on isotope abundances to determine age, they differ from radiometric dating techniques. Rather, they are a result of background radiation's effect on specific minerals. Mineral grains found in sediments and archaeological materials, such as quartz and potassium feldspar, gradually absorb ionizing radiation. In structurally unstable electron traps, the radiation keeps charge inside the grains. These charges are released when the sample is exposed to heat or sunlight, effectively bleaching it and restarting the timer. The amount of background radiation at the site where the sample was buried affects how quickly the trapped charge builds up over time. The intensity of the luminescence signal that is produced when these mineral grains are stimulated by heat (thermoluminescence dating) or light (optically stimulated luminescence or infrared stimulated luminescence dating) varies based on the amount of radiation absorbed during burial and particular properties of the mineral.

These techniques can be used to determine how old a sediment layer is because sediment deposited on top would shield the grains from sunlight's bleaching and reset. Ceramic fragments can be used to determine when they were last exposed to a lot of heat, usually during a kiln firing.

9.1.3. Dating with Decay Products of Short-Lived Extinct Radionuclides

A quantifiable portion of the parent nucleus must still be present in the sample rock for absolute radiometric dating to work. In the case of rocks that date back to the early solar system, this necessitates extraordinarily long-lived parent isotopes, which leads to imprecise age determination. Short-lived isotopes that are no longer present in the rock can be used to determine the relative ages of rocks from such old material and to obtain a better time resolution than that provided by long-lived isotopes.

In the early days of the solar system, there were various short-lived radionuclides like ^{26}Al, ^{60}Fe, ^{53}Mn, and ^{129}I present in the solar nebula. These radionuclides, believed to have originated from a supernova explosion, no longer exist today, but their decay products can be found in ancient materials like meteorites. By analyzing the decay products of extinct radionuclides using a mass spectrometer and isochron plots, scientists can determine the relative ages of different events in the early history of the solar system. Dating methods based on extinct radionuclides can be cross-referenced with the U-Pb method to provide absolute ages.

This allows for both an approximate age and detailed time resolution to be obtained, with a shorter half-life providing a higher time resolution but sacrificing overall timescale accuracy.

9.1.3.1. The ^{129}I–^{129}Xe Chronometer

^{129}I beta decays with a half-life of 16 million years to produce ^{129}Xe. One isochron method is the iodine-xenon chronometer. Neutrons are introduced to samples within a nuclear reactor. This process uses neutron capture and beta decay (of ^{128}I) to change the only stable isotope of iodine (^{127}I) into ^{128}Xe. Following radiation, samples are heated in stages, and each step's evolved gas's xenon isotopic signature is examined. The sample stopped losing xenon at a certain temperature, which can be inferred from a consistent $^{129}Xe/^{128}Xe$ ratio over several consecutive temperature steps.

Typically, samples of a meteorite known as Shallowater are exposed to radiation to track the efficiency of the conversion from ^{127}I to ^{128}Xe. The variations in the ratios of $^{129}I/^{127}I$ at the time when each stopped losing xenon are then explained by the differences in the measured $^{129}Xe/^{128}Xe$ ratios of Shallowater and the sample. This in turn translates into a variation in the early solar system's closure age.

9.1.3.2. The ^{26}Al–^{26}Mg Chronometer

The ^{26}Al–^{26}Mg chronometer is another example of a short-lived extinct radionuclide dating method that can be used to determine the relative ages of chondrules. ^{26}Al has a half-life of 7,20,000 years before decaying to ^{26}Mg. Finding the difference between the ratio of the stable isotopes, $^{27}Al/^{24}Mg$, and the natural abundance of ^{26}Mg, the product of ^{26}Al decay is all that is required for dating.

The excess of ^{26}Mg (often designated $^{26}Mg^*$) is found by comparing the $^{26}Mg/^{27}Mg$ ratio to that of other Solar System materials.

The ^{26}Al–^{26}Mg chronometer gives an estimate of the time period for the formation of primitive meteorites of only a few million years (1.4 million years for Chondrule formation).

9.2. OTHER ISOTOPIC AND CHEMICAL TECHNIQUES

9.2.1. Strontium Isotopes

Strontium is present in small amounts in many limestones and shares chemical similarities with calcium. Strontium has two common isotopes: ^{86}Sr and ^{87}Sr. Carbonate analysis throughout the Phanerozoic has shown that the ratio of these two isotopes in seawater has varied over time. Using data from carbonate minerals formed from seawater and not subsequently changed or recrystallized, a strontium isotope curve has been created. The age of the sample can be ascertained by comparing the $^{86}Sr/^{87}Sr$ ratio in calcite from an unknown age sample with the established curve. It should be noted that, despite using isotopes, this method differs from absolute rubidium-strontium dating in that it is not an absolute dating technique (Wilken et al., 1987).

When using strontium isotope ratios for dating, there are two crucial factors to take into account. First off, a specific $^{86}Sr/^{87}Sr$ ratio is not exclusive to a date; rather, it could have existed multiple times. To limit the portion of the curve that can be used for comparison, additional controls on the rock's age are needed. Second, the only carbonate minerals that are suitable for use are those that have formed from seawater and haven't been recrystallized yet. Many organisms produce aragonite shells, but these are useless because aragonite eventually recrystallizes into calcite. The only organisms that can be utilized are those that precipitate calcite in their skeletons or shells, and they also need to be unaltered (Figure 9.4).

When dating using strontium isotope ratios, it is important to consider two factors. Firstly, a specific $^{86}Sr/^{87}Sr$ ratio is not unique to a single date, as this ratio may have occurred multiple times. Therefore, another method is needed to establish the age of the rock to accurately compare it. Secondly, only carbonate minerals formed from seawater and not recrystallized can be used for dating. While many organisms produce shells of aragonite, these cannot be used as aragonite recrystallizes into calcite over time. Only organisms that produce calcite in their shells or skeletons, and have not been altered, can be used for dating purposes.

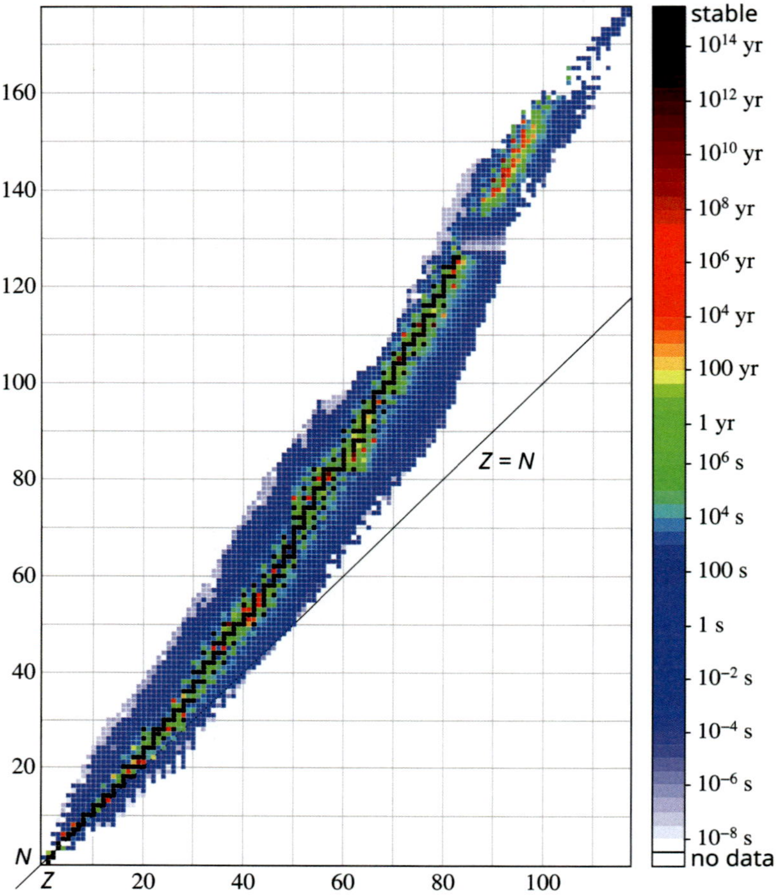

Figure 9.4. Isotope half-lives. The darker more stable isotope region departs from the line of protons (Z) = neutrons (N), as the element number Z becomes larger.

Source: By BenRG – Own work, Public Domain, https://commons.wikimedia.org/w/index.php?curid=7900237.

9.2.2. Thermochronological Techniques

Thermochronology is a method used to determine the temperature at which a rock was at a specific point in time, with its primary application being to calculate the geological history of uplift and erosion of a rock formation. The K–Ar and Ar–Ar dating techniques are commonly utilized in thermochronology because the decay series 'clock' is reset at different temperature ranges for various minerals, typically between 350°C and 700°C. Fission track analysis can also be employed using zircon grains to identify temperatures of 300°C or higher, and with apatite grains to detect temperatures exceeding 110°C. Additionally, lower temperatures can be determined using the U/Th-–He (uranium/thorium–helium) isotopic analysis technique.

9.2.3. Chemostratigraphy

The composition of sediments can vary, making differences in the mineralogical and chemical properties of sedimentary rocks a potential method for distinguishing and correlating deposits. This technique, called chemostratigraphy, is effective in situations where there is diversity in the mineralogy and chemistry of sediment being delivered to the sedimentation area, indicating changes in provenance. Clastic sediments contain detrital grains sourced from high ground areas surrounding the sedimentation basin that are undergoing erosion.

This is known as the provenance of the detrital material. Over time, different rock units in the source areas are exposed to erosion, resulting in changes in the types of clastic material supplied to the basin. The presence of new clast types in sediment successions signifies the time when a new source area lithology was exposed and eroded. A shift in clast types leads to a change in the overall chemistry of the deposit.

Three primary techniques are utilized in sediment analysis: (a) bulk chemical analysis, which is quick and simple but can be challenging to interpret; (b) heavy mineral correlation, a more time-consuming method but highly effective; and (c) clay mineral analysis for specific situations. While bulk chemical analysis is appealing due to its simplicity, its limited applicability is a drawback. Changes in sediment provenance can be subtle, influenced by factors such as grain size and diagenetic processes, making interpretation challenging. In contrast, variations in heavy mineral assemblages in sandstones often provide clearer insights into changes in provenance over time. Chemostratigraphy has proven useful in analyzing fossil-free continental deposits and in correlating sandstone beds in subsurface hydrocarbon reservoirs at a finer scale.

9.3. MAGNETOSTRATIGRAPHY

The Earth's magnetic field changes between normal magnetic polarity, where the field is in its present orientation, and reversed magnetic polarity, where the field is flipped and a magnetic compass would point south instead of north. Evidence from studying past magnetic fields (paleomagnetism) shows that these polarity reversals happen irregularly over geological time. Some reversals are short, occurring every tens of thousands of years, while there was a 30-million-year period in the Cretaceous where the field remained the same. During most of the Cenozoic era, polarity reversals happened every few 100,000 to a few million years. The time it takes for a reversal to occur seems to be quick in geological terms.

9.3.1. The Magnetic Record in Rocks

When a magnetic material cools below the Curie Point, its magnetic dipoles become mobile and unrestricted in their ability to reorient, absorbing the polarity of the surrounding magnetic field. When a material is below the Curie Point, it maintains its magnetic field regardless of its position or changes in the surrounding magnetic field. Numerous distinct magnetic minerals, each with a unique Curie Point temperature, can be found in rocks. Additionally, the rock may undergo changes that result in the formation of new minerals that are sensitive to the surrounding magnetic field. As a result, the residual magnetism or preserved magnetization found in rocks may consist of a complex combination of various field orientations found in various minerals.

The orientation of the Earth's magnetic field at the time of rock formation can be determined by measuring the remnant magnetism in a rock sample. In extrusive igneous rocks, the remnant magnetism in minerals like magnetite and haematite as they cool below their Curie Point can be easily detected by a magnetometer. However, for stratigraphers, the weaker remnant magnetism found in sedimentary rocks is more valuable. Fine magnetized particles in sedimentary rocks, such as grains containing iron minerals like haematite, tend to align themselves parallel to the Earth's magnetic field as they settle out of water. Although not all particles will align perfectly, a statistically significant pattern in their orientation can indicate the sediment's remnant polarity. This phenomenon is most pronounced in fine-grained sediments with a high iron mineral content, while coarser-grained sediments may not exhibit detectable remnant magnetism due to orientation by sediment flow. The remnant magnetism in a rock sample can be reset by heating the minerals above their Curie Point during metamorphism, or by alterations from diagenesis or weathering.

9.3.2. Practical Magnetostratigraphy

The main goal of a magnetostratigraphic study is to identify periods of normal and reversed magnetic polarity found in a series of rock layers. Fieldwork typically involves drilling small cores from rock beds at the outcrop, with the cores' orientation and bedding attitude being carefully measured in three dimensions. Multiple cores are usually extracted from a single bed to ensure enough samples for a reliable analysis of the remnant magnetism at that particular site. The distance between sampling sites in the rock layers is determined by the sediment accumulation rates and the frequency of magnetic field reversals during that time period. In slower sedimentation rates, samples need to be taken more frequently to guarantee all polarity reversals are detected, while faster rates allow for wider spacing of sample sites. To pinpoint the exact location of a reversal within the rock layers, closer sampling intervals between sites with opposite magnetic field directions may be necessary.

A magnetometer is used in the lab to measure the residual magnetism in the samples. Magnetic fields in samples that are several orders of magnitude weaker than the Earth's magnetic field can be found and measured by modern instruments. By placing the sample in an area protected from the current magnetic field and either heating it up or exposing it to an alternating current field, the effects of the current magnetic field are eliminated. The orientation of the residual magnetism that still exists will be remnants from a previous period in its history, most likely the formation of the rock. To confirm the statistical significance of the outcome, remnant magnetism measured in different samples taken at the same location is compared.

9.3.3. Magnetostratigraphic Correlation

An account of the normal and reversed stratigraphic periods can be created by measuring the remnant paleomagnetism across a series of beds. Traditionally, these are displayed as intervals with normal polarity indicated in black and reversed polarity indicated in white. For a large portion of the Phanerozoic, the pattern of reversals in the Earth's magnetic field over time has been determined, and many reversal events have been precisely dated.

Having some kind of tie-point to the geological time scale is crucial for connecting the pattern measured in a single succession to the established polarity stratigraphy. Biostratigraphic data that can be utilized to connect a point in the succession to the time scale or absolute dating of a unit, like lava, within the successions may be used to provide this. One significant caveat is that any gaps in the record that the succession provides must be acknowledged and taken into consideration. For example, if sedimentation is absent or if erosion has removed the deposits from a period of normal or reversed polarity, then that period may not be represented. Correlation within a sedimentary basin can be achieved by matching the reversal patterns at other localities, again accounting for any evidence of breaks in the sedimentary record, once a reversal stratigraphy has been established in a portion of the basin.

In continental successions that lack age-diagnostic flora or fauna and that cannot be biostratigraphically dated, magnetostratigraphy is frequently used. Normally, the technique is only used when other (biostratigraphic) methods cannot be used or a high-resolution stratigraphy is required.

9.3.4. Technique

Magnetostratigraphic units, also known as magnetozones, are a group of related but distinct stratigraphic units that arise from stratigraphical variations in the measurable magnetic properties of rocks. The shift in the direction of the rocks' remanent magnetization, brought on by reversals in the Earth's magnetic field's polarity, is the magnetic property most helpful in stratigraphic research. The stratigraphic sequence can be divided into units based on their magnetic polarity, with the direction of the remnant magnetic polarity being the foundation for this division. These units are known as chrons, or magnetostratigraphic polarity units.

The strata maintain a normal polarity if the ancient magnetic field was oriented similarly to the current field, that is, with the North Magnetic Pole close to the Geographic North Pole. The strata show reversed polarity if the data suggests that the North Magnetic Pole was close to the Geographic South Pole.

9.3.4.1. Polarity Chron

An Earth's magnetic field polarity chron, also known as an in context chron, is the period of time between polarity reversals. It is the period of time that a magnetostratigraphic polarity unit represents. It represents a certain time period in geologic history where the Earth's magnetic field was in predominantly a "normal" or "reversed" position. Chrons are numbered in order starting from today and increasing in number into the past. As well as a number, each chron is divided into two parts, labelled "n" and "r," thereby showing the position of the field's polarity. Chrons are also referred by a capital letter of a reference sequence such as "C." A chron is the time equivalent to a chronozone or a polarity zone.

When the interval is less than 2,00,000 years, it was referred to as a polarity subchron; however, in 2020, the term was redefined to refer to an approximate duration of 10,000 to 1,00,000 years, and to an approximate duration of 1,00,000 to a million years, as a polarity chron. For a duration of 10^8–10^9 years, Megachron is another term used; for a duration of 10^7–10^8 years, Superchron is another term used; and for a duration of less than 3×10^4 years, Crytochron is the term used.

9.3.4.1.1. Chron Nomenclature

Naming the sequence of polarity intervals can be difficult, especially when changes occur briefly or are not consistent due to the complexity of the Earth's magnetic field. The "C-sequence" and "M-sequence" are two standardized marine magnetic anomalies sequences that range from the Middle Jurassic to the present day. The C polarity chrons series starts with C1n, known as Brunhes, and goes back to C1r, known as Matuyama, marking the Brunhes–Matuyama reversal around 0.773 Ma. The C sequence ends with the Cretaceous Normal Superchron, C34n, which lasted from 120.964 Ma to 83.650 Ma, defining the Santonian geologic age. The M series starts with M0, labeled M0r, at 121.400 Ma and goes up to M44n.2r before 171.533 Ma in the Aalenian period.

There is particular nomenclature for subdivisions in the sequences, so C8n. Of the normal-polarity Chron C8n, 2n is the second-oldest subchron with normal polarity, and C1n–1 is the name of the Emperor cryptochron, which is the youngest. Some terms used in the literature to describe a hypothetical brief reversal at roughly 118 Ma, such as M-1r, are provisional.

9.3.4.2. Sampling Procedures

Using a rock core drill or by hand, oriented paleomagnetic samples are gathered in the field (chunks broken off the rock face). At least three samples are taken from each sample site to average out sampling errors. The stratigraphic section's age and deposition rate determine how the sample sites are spaced within it. Because their magnetic grains are finer and more likely to align with the surrounding field during deposition, mudstones, claystones, and very fine-grained siltstones are the preferred lithologies in sedimentary layers.

9.3.4.3. Analytical Procedures

To determine the samples' natural remanent magnetization (NRM), they are first examined in their natural state. The stable magnetic component is then exposed by gradually removing the NRM using thermal or alternating field demagnetization techniques.

After comparing the magnetic orientations of every sample collected at a site, directional statistics—most frequently Fisher statistics or bootstrapping—are used to calculate the average magnetic polarity of the samples. Every average's statistical significance is assessed. Plotting the Virtual Geomagnetic Pole latitudes against the stratigraphic level at which they were collected is done for those sites where the latitudes were found to be statistically significant. The standard black and white magnetostratigraphic columns—where black denotes normal polarity and white denotes reversed polarity—are then abstracted from these data.

9.3.4.4. Correlation and Ages

The thickness of a particular polarity zone can differ from one place to another due to differences in the rate at which the sediment accumulated because a stratum's polarity can only be normal or reversed. This raises the issue of how to relate similar polarity zones across stratigraphic sections. It is necessary to gather at least one isotopic age from each section to prevent confusion. This is frequently extracted from volcanic ash layers found in sediments. If that doesn't work, a polarity can be linked to a biostratigraphic event that has previously been associated with an isotopic age. The Global Magnetic Polarity Time Scale (GMPTS) and the local magnetostratigraphic column are correlated with the help of the independent isotopic age or ages.

Numerous time lines are established through the stratigraphic section by the correlation because the age of every reversal displayed on the GMPTS is reasonably well known. For features in the rocks like fossils, alterations in the composition of sedimentary rocks, shifts in the depositional environment, etc., these ages offer reasonably accurate dates. Additionally, they limit the ages of cross-cutting elements like unconformities, dikes, and faults.

9.4. DATING IN THE QUATERNARY

Various techniques are employed to date and correlate Quaternary deposits, but these methods are only applicable to the most recent tens or hundreds of thousands of years and are not suitable for older stratigraphic units.

9.4.1. Carbon-14 Dating

Since carbon-14, also known as radiocarbon dating, is so often used in archaeology to date materials like bone and charcoal, it is most likely the most well-known of all radiometric dating techniques. The idea behind this method is that carbon, including the isotope ^{14}C, is constantly taken up by living things. The atmosphere's constant cosmic bombardment of ^{14}N produces this radioactive isotope. The ^{14}C in it radioactively decays at a rate that is determined by its half-life of 5,730 years when the organism dies. The amount of time that has passed since a sample of once-living tissue died and ceased to absorb ^{14}C from the atmosphere can be ascertained by measuring the percentage of ^{14}C present in the sample. The primary constraint on this method from a geological perspective is that, due to its brief half-life, the amounts of ^{14}C in a sample become too low to accurately detect in materials older than 50,000 years, though more recent analytical methods are extending this to 70,000 years or more. The assumption that the environment's ^{14}C levels had remained constant during this time also introduced error; it is now known that these levels have fluctuated over time, necessitating some adjustments to the dates determined.

Cross-referencing carbon-14 dating with other dating techniques yields consistent results, suggesting that the rate of creation of carbon-14 is roughly constant. However, localized carbon-14 concentrations can be lowered and dates can be erroneous by volcanic eruptions or other events that release copious amounts of carbon dioxide. In contrast, above-ground nuclear bomb tests carried out until the early 1960s increased the amount of carbon-14. The releases of carbon dioxide into the biosphere as a result of industrialization have also decreased the proportion of carbon-14 by a few percent. Additionally, the amount of carbon-14 created in the atmosphere would decrease if the solar wind or Earth's magnetic field increased beyond their current levels.

9.4.2. Uranium-Series Dating

Utilizing uranium-series dating methods, it is possible to partially overcome the age constraints for dating the Pleistocene using carbon-14 dating. Another isotope of uranium, 234U, is one of the byproducts of the radioactive decay series of ^{238}U. Seawater contains both isotopes, but due to differences in their reactivity, there is some fractionation and a higher concentration of ^{234}U than would be predicted during decay. The ratios of the two isotopes gradually return to normal as the ^{234}U decays if the fractionated uranium is absorbed from the seawater into a mineral, such as the carbonate of a marine organism. By calculating the ratio of ^{234}U to ^{238}U, marine carbonates that are hundreds of thousands of years old can be dated. This process takes place over a period of approximately 1 million years. Since the fractionation in non-marine waters varies, this uranium series dating method is not applicable to carbonates formed in fresh water. It is most effective when applied to corals.

9.4.3. Oxygen Isotope Stratigraphy

The most prevalent isotope of oxygen is ^{16}O, while ^{18}O makes up approximately 0.2% of naturally occurring oxygen. Both isotopes are stable, meaning they do not decay radioactively. Because of the process of isotopic fractionation, which causes the lighter isotope to evaporate more readily into the atmosphere, atmospheric water has a comparatively higher concentration of ^{16}O. A larger amount of the world's water is retained in ice caps during glacial periods, and these caps are nourished by snow-melt water from the atmosphere. Thus, the oceans are relatively enriched in ^{18}O during glacial periods when there is more low-^{18}O ice. An oxygen isotope stratigraphy, which was established for global ocean waters back to around 750 ka and now spans the entire Quaternary, is based on this variation of the $^{18}O/^{16}O$ ratio in the ocean waters with temperature. This was produced by dating the deposits in which the carbonate minerals occur by other methods and measuring the $^{18}O/^{16}O$ ratio in carbonate minerals that are directly formed from seawater by organisms. When comparing a sample's oxygen isotope ratio to a standard, it is expressed as $\delta^{18}O$, where positive values are enriched in ^{18}O and negative values are depleted in ^{18}O. Oxygen isotope stages, also known as marine isotope stages, are a common way to describe the marine record of Quaternary deposition. These stages represent times when the climate was relatively warm and cool.

9.4.4. Luminescence and Electron Spin Resonance Dating

Ionizing radiation from naturally occurring radioactivity in elements like potassium, thorium, and uranium exposes sedimentary materials to a flux of radiation. The electrical charge in mineral crystals is redistributed by the radiation; while the majority of this charge swiftly returns to its initial state, some of it becomes trapped in lattice imperfections at higher energy states. The duration of the dosage determines how much additional energy the crystal retains. This energy can be released by heat, and when it does, the material will luminesce. This phenomenon is known as thermoluminescence (TL). Optical stimulating luminescence (OSL), a technique that uses a burst of light to illuminate the sample, is an alternative to heating.

Light exposure of the sediment at the surface results in "bleaching," or the release of stored energy; hence, luminescence cannot occur until the sediment is buried. Thus, the duration of sediment burial can be ascertained by measuring the amount of luminescence generated by heating or optically stimulating a sample. The methods are limited to materials that have been completely bleached upon deposit, such as slowly deposited aeolian and fluvial sediment. With an accuracy of roughly 10%, the TL and OSL techniques can be used to date the sediment's burial back to 150 ka. With roughly twice the age range, they can also be used with comparable accuracy on stalagmites found in caves. Similar to luminescence techniques, electron spin resonance (ESR) dating relies on the accumulation of energy within crystal lattices following burial. The atoms' magnetic field, or spin, is altered by the movement of electrons within the lattice. Since the magnetic field gradually changes with time, dating sediments can be done by measuring the electron spin resonance of this effect. Samples can be dated more than once since the ESR method does not destroy the sample, unlike TL and OSL dating. Quartz sand and biogenic calcium carbonate can be dated using electron spin resonance.

9.4.5. Cosmogenic Isotopes

As elements in minerals exposed at the surface react with cosmic rays (primarily high-energy protons) bombarding the surface, cosmogenic isotopes are formed. Thus, the amount of these unstable isotopes created by this bombardment can be utilized to determine the length of time a rock has been exposed. This method makes use of the isotopes ^{26}Al, ^{10}Be, and ^{36}Cl. The isotopes of aluminum and beryllium are typically combined in quartz, where ^{26}Al is produced from silicon and ^{10}Be is derived from oxygen. The majority of production takes place in the upper 50 cm of the rock surface, and this system can be used to measure exposure times spanning hundreds of thousands of years. The calcium and potassium isotopes, which are both prevalent in a variety of rock types, are bombarded to produce the chlorine isotope, ^{36}Cl. Rock exposure periods spanning tens of thousands of years can be ascertained through ^{36}Cl measurements. When dating features like river terraces and other depositional surfaces in Quaternary continental successions, cosmogenic isotope techniques are especially helpful. The findings limit the geomorphological past of a land surface by revealing details about occurrences like uplift and river incision (Figure 9.5).

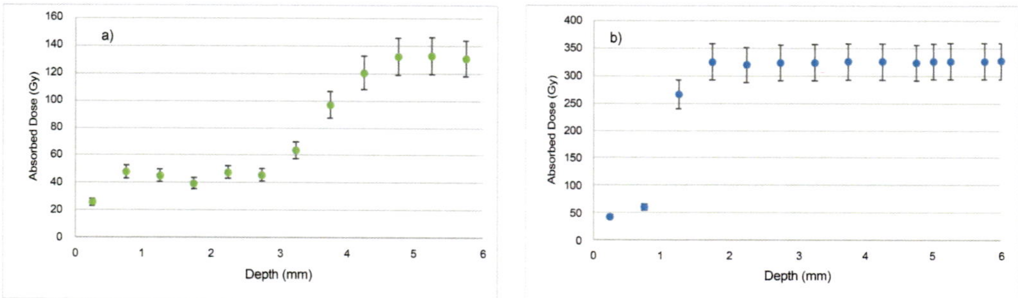

Figure 9.5. Luminescence dating of the rock surface method.

Source: Galli, A., Panzeri, L., Rondini, P., Poggiani, K. R., & Martini, M., (2020). Luminescence dating of rock surface. The case of monoliths from the megalithic sanctuary of Ossimo-Pat (Valle Camonica, Italy). *Applied Sciences*, 10(21), 7403. https://doi.org/10.3390/app10217403.

9.4.6. Amino-Acid Racemization

Amino acids are a component of all living things and are arranged in the geometric shapes "L" and "D." The "L" form is dominant in living things, but when a tissue dies, a process known as racemization takes place, changing the "L" form into the "D" form until they are in balance. The organism's nature determines how quickly this happens, and temperature has an impact. Racemization can occur within a few 1,000 years in hotter climates, but it can take hundreds or even thousands of years in colder climates. This method is only applicable to materials for which the temperature and racemization rate are known.

9.4.7. Annual Cycles in Nature

This category includes two methods: glacial lake varves and tree rings. Tree growth rates vary seasonally, resulting in rings in the wood, the thickness of which is influenced by the length of the growing season. A tree ring chronology reflecting climate fluctuations has been extended several 1,000 years back into the Holocene by comparing the ring patterns of very old living trees. Climatic variations can be identified in the ring pattern. Their calibration has proven beneficial for the carbon-14 dating technique. Varves are millimeter-scale laminae that form in glacial lake deposits as a result of seasonal silt inflow during summer melt. The age of a lake deposit can be determined by counting these laminae backward from the present (Faggetter et al., 2018).

CASE STUDY

IN SITU ^{14}C EXPOSURE AGE DATING IN ANTARCTICA:

Given that the ice sheets in Antarctica and Greenland contain enough water to raise global sea level by about 65 meters, it is important to understand how ice sheets have changed in the past. Knowing how they have changed, and under what kind of environmental conditions, means we can better understand how they will respond to future changes in climate.

Millions of people around the world live close to sea level, from places like Bangladesh, the Netherlands, and low-lying islands in the Pacific and Indian oceans, to cities like New Orleans, Jakarta, and London. How quickly have ice sheets melted, and added to sea level, in the past, and how can we work this out? One way is through measuring something called cosmogenic nuclides (Figure 1).

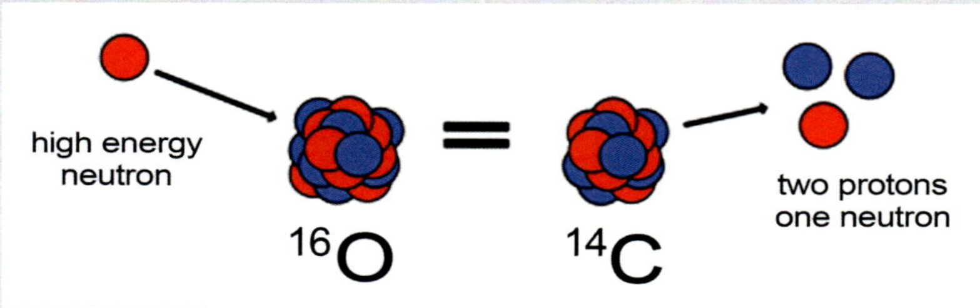

Figure 1. *Cosmic ray (in this case a high energy neutron) interacts with an oxygen atom in quartz to produce in situ ^{14}C, just one of many cosmogenic nuclides.*

Cosmic rays constantly hit the surface of the Earth, producing rare isotopes known as cosmogenic nuclides within minerals in rocks. Cosmic rays contain enough energy to break apart the atoms that make up rocks, leaving behind new cosmogenic nuclides. You can think of a cosmic ray as a cue ball and the atoms in rocks as a triangle of red balls in a game of snooker. When a rock is exposed to the sky, cosmogenic nuclides are constantly being made within it.

- **Glaciers Block Cosmic Rays:** When a glacier grows larger and covers a rock, cosmic rays can no longer reach the rock and production of cosmogenic nuclides pauses. If a glacier later thins and reveals the rock, cosmogenic nuclides start to be made within the rock crystals again.

We can measure the amount of a cosmogenic nuclide contained in a rock. If we know the rate at which the cosmogenic nuclide are produced (the production rate), we can convert this number into an exposure age. The age, usually in thousands of years (ka), tells us how long a surface has been exposed to cosmic rays or, in other words, how long ago in the past a glacier thinned to uncover the rock (Figure 2).

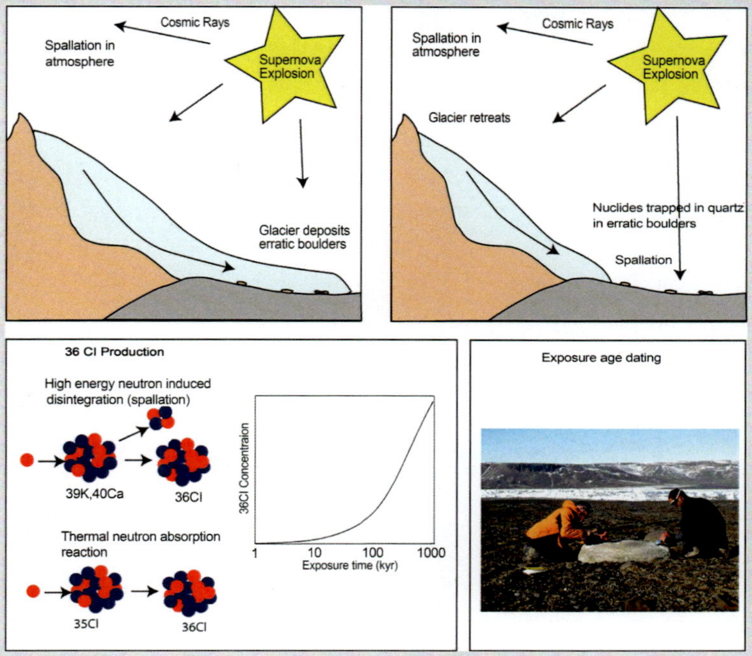

Figure 2. *Cartoon illustrating cosmogenic nuclide exposure ages. A glacier transports an erratic boulder, and then recedes, exposing it to cosmic rays. Spallation reactions occur in minerals in the rocks upon bombardment by cosmic rays. By sampling the rocks and separating certain minerals (such as quartz or pyroxene) and calculating the amount of these minerals (as a ratio to other, stable, minerals), we can work out how long the rock has been exposed on the earth's surface.*

The most commonly-measured cosmogenic nuclide is ^{10}Be, which is produced in the mineral quartz. It is radioactive with a half-life of 1.4 million years, which means for every 1.4 million years that passes, the number of ^{10}Be atoms will decrease by one half. When we convert concentrations to ages, we assume everything we measure was made from the most recent time that a rock was uncovered by a glacier.

- **Glacial Erosion is Needed to 'Reset' the Clock:** We rely on glacial erosion to 'reset' the cosmogenic nuclide concentration of a rock surface and ensure our cosmogenic exposure ages represent one period of exposure.

When a glacier expands, it can erode the landscape beneath, removing cosmogenic nuclides that built up in the past. Evidence of glacial erosion is found in many areas containing glacial landforms; picture the deep U-shaped valleys of Yosemite National Park or the Scottish Highlands.

- **The Case of Minimal Glacial Erosion:** In some places, including Antarctica, glaciers can be cold-based, which means they are frozen to their bed. Rather than eroding the rock beneath them, cold-based glaciers stick to their bed, preserving the landscape beneath, causing problems with exposure dating.

We know the size of ice sheets waxes and wanes with time; they are larger in glacial periods and smaller during interglacials. If a surface is covered by cold-based ice many times over a long timescale, cosmogenic nuclides from previous interglacials can be preserved.

The assumption that everything we measure today was only produced during the most recent time a sample was uncovered by ice is not met, resulting in exposure ages that are much older than the true time that a surface was most recently exposed (Figure 3).

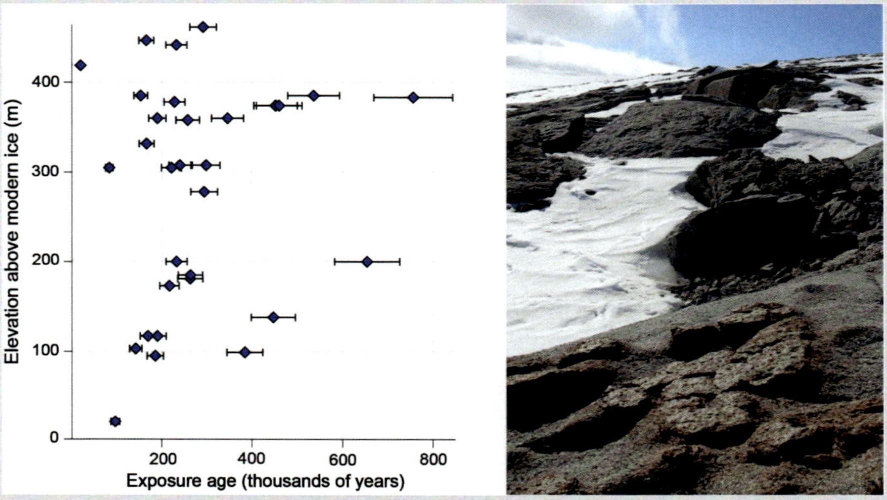

Figure 3. *Left: Exposure ages. Right: Bedrock at the Lassiter Coast, Southern Antarctic Peninsula.*

- **Exposure Age Dating in Antarctica:** The graph above shows ^{10}Be exposure ages from pieces of bedrock collected by Johnson et al. (2019) from the Lassiter Coast, an area on the southern Antarctic Peninsula. These samples were collected to study the history of the West Antarctic Ice Sheet. Most of the ages are very old, with the majority over 100,000 years.

There are two ways we might interpret the exposure ages. Firstly, the ages are accurate and ice has not covered these rocks for hundreds of thousands of years. However, we know that the West Antarctic Ice Sheet was much larger only 20,000 years ago, during a time called the Last Glacial Maximum (LGM) (e.g., RAISED Consortium, 2014). So, instead, the very old ages might tell us something else; the rocks have been buried and uncovered by cold-based ice many times, including during the LGM. With no glacial erosion during ice cover, the long half-life of ^{10}Be means it can be preserved in rock for millions of years.

- **Using ^{14}C to Unravel Complex Exposures:** Another cosmogenic nuclide, *in situ* ^{14}C, can help us answer this question. This slightly heavier, radioactive isotope of carbon is used in radiocarbon dating. With radiocarbon dating, we measure ^{14}C in the remains of organisms to work out how long ago they died.

With exposure age dating we are instead measuring ^{14}C made in the mineral quartz. Quartz contains oxygen (SiO_2), and cosmic rays interact with these oxygen atoms to make ^{14}C (see diagram above). To prevent confusing exposure dating using ^{14}C in rocks with traditional radiocarbon dating of organics, we refer to the former as *in situ* ^{14}C.

So why do we measure in situ ^{14}C when we can already measure ^{10}Be? In situ ^{14}C has a very short half-life of 5,730 years. The short half-life of ^{14}C, compared to the half-life of ^{10}Be (1.4 million years), means that even with no glacial erosion, the ^{14}C concentration in a rock will rapidly decrease through radioactive decay. The animation below shows how the ^{10}Be and in situ^{14}C concentrations change in a sample collected from the surface of a mountain poking out of an ice sheet (Figure 4).

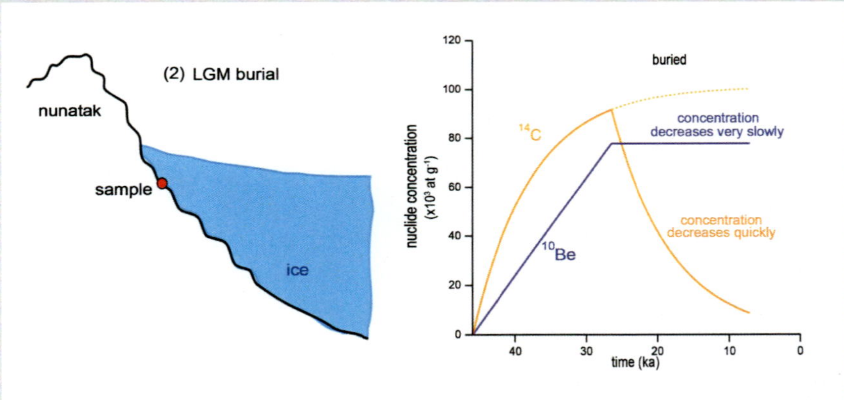

Figure 4. *Animation showing the buildup of two cosmogenic nuclides, ^{10}Be (purple) and in situ ^{14}C (yellow), in a sample (red circle). The dashed line shows the in situ ^{14}C concentration if the sample was not covered.*

The sample is exposed to cosmic rays 46,000 years ago and the concentration of both nuclides increases. When the ice sheet is thicker at the LGM (26,000 years ago), it covers the sample. The amount of ^{10}Be in the sample decreases very slowly, whilst the concentration of *in situ* ^{14}C decreases rapidly. The samples are exposed again 7,000 years ago.

The ^{10}Be concentration at the end of the animation contains a lot of ^{10}Be that was made both before and after the sample was covered by ice.

On the other hand, the vast majority of ^{14}C at the end of the animation was made after the rock was covered by ice. The resulting ^{10}Be age will be much older than the true exposure age of 7,000 years, whilst the *in situ* ^{14}C age will be much closer to it (Figure 5).

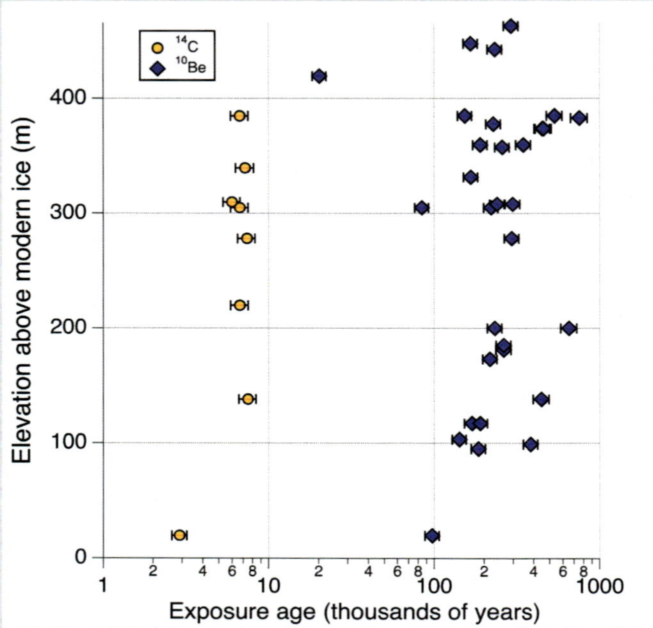

Figure 5. *^{10}Be (purple) and in situ ^{14}C (yellow) exposure ages from the Lassiter Coast.*

- **Using ^{14}C to Determine the Glacial History of the Lassiter Coast:** To shed light on the glacial history of the Lassiter Coast, Johnson et al. measured the in situ ^{14}C concentration in some of the same rocks that were already measured for ^{10}Be. All in situ ^{14}C ages (above, yellow) are younger than the LGM, contrasting with the older ^{10}Be ages (purple).

With young *in situ* ^{14}C and very old ^{10}Be exposure ages from samples collected from ~400 m above the modern Antarctic Ice Sheet, we now know that (i) the ice sheet was at least ~400 m thicker at the LGM than it is now, and (ii) the Lassiter Coast has been covered by cold-based ice multiple times over many hundreds of thousands of years.

- **Using ^{14}C in Data-Calibrated Computer Simulations:** So why does this matter? This information on how much thicker the Antarctic Ice Sheet was in the past, how it has changed since, and under what conditions, will be used to test numerical ice sheet models.

Models that can reproduce past changes in ice sheets can be used to model the response of ice sheets to future climate scenarios with greater confidence. Perhaps most importantly, these models can provide vital information on how quickly the Antarctic Ice Sheets will contribute to sea level rise.

- **Drawbacks of *In Situ* ^{14}C for Exposure Age Dating:** Because ^{14}C is absorbed by living organisms, we can find ^{14}C in organic chemicals used in laboratories. When using *in situ* ^{14}C exposure age dating, as well as traditional radiocarbon dating, is it especially important to avoid contamination of sample material with modern ^{14}C.

Different cosmogenic nuclides are useful for studying different periods of time. Whilst the short half-life of ^{14}C provides advantages described above, it also limits the time that we can use ^{14}C in exposure age dating studies. Note the dashed yellow line in the animation above. The concentration of ^{14}C is flattening-out because a balance between radioactive decay and production from cosmic rays is being achieved.

After about 35,000 years (6 half-lives) of continuous exposure, the in situ ^{14}C concentration in a rock will reach what is known as a saturation concentration. If we measure the saturation concentration in a sample, we know that the rock has been exposed for at least 35,000 years. Whilst this can be a useful result in itself, it also means we cannot measure exposure ages for events older than this time using this cosmogenic nuclide. On the other hand, ^{10}Be can provide exposure ages approaching 6 million years.

Another drawback of in situ ^{14}C, which will hopefully change in the future, is that very few laboratories worldwide have the capability to extract carbon from quartz. Laboratory equipment is expensive, and extraction requires close monitoring and use of a lot of liquid nitrogen.

CLASS ACTIVITY

RELATIVE DATING AND STRATIGRAPHY:

The deposition of sedimentary rocks can be interrupted when no deposition happens, erosion is intense, or tectonic activity happens. The result of this are unconformities.

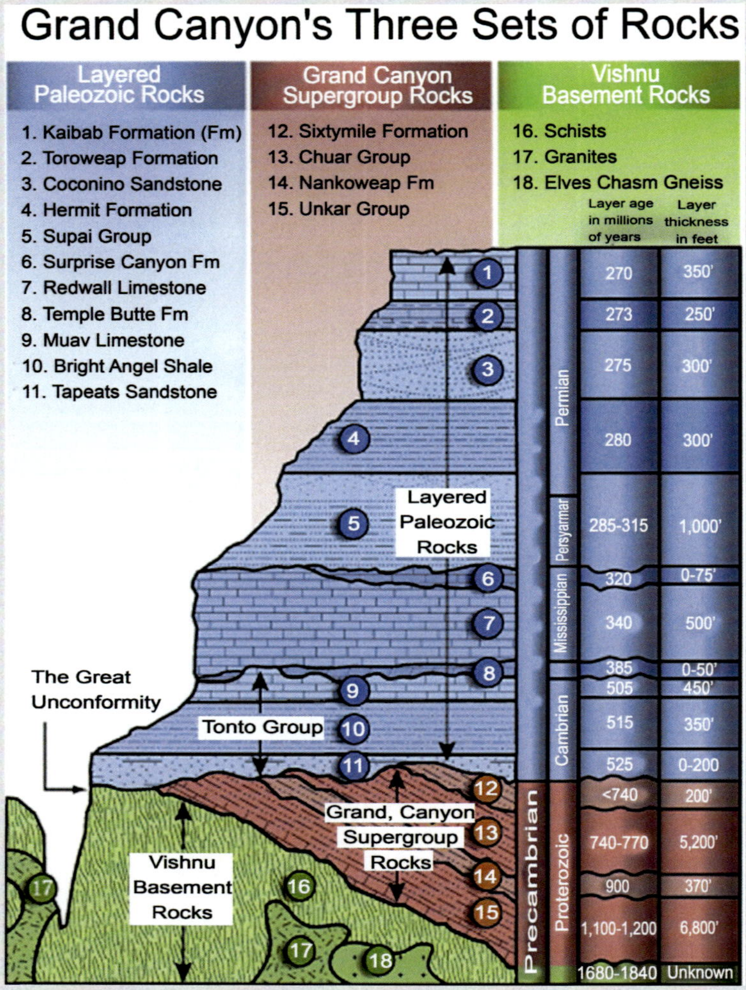

Source: By Mathis, A., (2006). Grand Canyon Association, Grand Canyon, Arizona, USA. https://www.nps.gov/articles/images/pop-up-yardstick.jpg, Public Domain, https://commons.wikimedia.org/w/index.php?curid=3696938.

Describe in words what is seen in the three unconformity diagrams above.

SUMMARY

- Dating techniques are procedures used by scientists to determine the age of rocks, fossils, or artifacts.
- Radiometric dating techniques have been available to geologists since the discovery of radioactivity in the early part of the 20th century and they have provided an absolute scale of millions and billions of years for events in Earth history.
- Radiometric dating is a technique that is used to date materials such as rocks or carbon, in which trace radioactive impurities were selectively incorporated when they were formed.
- For most radioactive nuclides, the half-life depends solely on nuclear properties and is essentially constant.
- The closure temperature or blocking temperature represents the temperature below which the mineral is a closed system for the studied isotopes.
- Uranium–lead radiometric dating involves using uranium-235 or uranium-238 to date a substance's absolute age. This scheme has been refined to the point that the error margin in dates of rocks can be as low as less than 2 million years in two-and-a-half billion years.
- Samarium–neodymium dating is a radiometric dating method useful for determining the ages of rocks and meteorites, based on the alpha decay of the long-lived samarium isotope (^{147}Sm) to the stable radiogenic neodymium isotope (^{143}Nd).
- Potassium–argon dating, abbreviated K–Ar dating, is a radiometric dating method used in geochronology and archaeology.
- The rubidium-strontium dating method (Rb-Sr) is a radiometric dating technique, used by scientists to determine the age of rocks and minerals from their content of specific isotopes of rubidium (^{87}Rb) and strontium (^{87}Sr, ^{86}Sr).
- Carbon-14 is a radioactive isotope of carbon, with a half-life of 5,730 years (which is very short compared with the above isotopes), and decays into nitrogen.
- Strontium is chemically similar to calcium and is found in small quantities in many limestones. There are two common isotopes of strontium, ^{86}Sr and ^{87}Sr, and analysis of carbonates through the Phanerozoic has indicated that the ratio between these two isotopes in seawater has changed over time.
- Thermochronology is a process of determining the age at which a rock was at a particular temperature and is most widely used as a means of calculating the uplift and denudation history of a body of rock.
- A polarity chron, or in context chron, is the time interval between polarity reversals of Earth's magnetic field. It is the time interval represented by a magnetostratigraphic polarity unit.
- Carbon-14 or radiocarbon dating is probably the best-known of all radiometric dating methods because it is widely used in archaeology for the dating of materials such as bone and charcoal.

REVIEW QUESTIONS

1. How is radiometric dating used to determine the age of rocks?
2. Why is it more accurate to use radioactive dating?
3. What is the half-life of samarium-neodymium dating?
4. How accurate is potassium-argon dating? Discuss.
5. What type of rock is commonly dated with rubidium-strontium dating?
6. How to date with decay products of short-lived extinct radionuclides.
7. What is chemostratigraphy, and how is it used in geological studies?
8. How is the magnetic field recorded in rocks?

REFERENCES

1. Adams, E. W., Schröder, S., Grotzinger, J. P., & McCormick, D. S. (2004). Digital reconstruction and stratigraphic evolution of a microbial-dominated, isolated carbonate platform (terminal Proterozoic, Nama Group, Namibia). Journal of Sedimentary Research, 74, 479–497. https://doi.org/10.1306/122903740479.
2. Ahlberg, P., et al. (2009). Cambrian high-resolution biostratigraphy and carbon isotope chemo stratigraphy in Scania, Sweden: First record of the SPICE and DICE excursions in Scandinavia. Lethaia, 42, 2–16. https://doi.org/10.1111/j.1502-3931.2008.00105.x.
3. Becker-Kerber, B., Pacheco, M. L. a. F., Rudnitzki, I. D., Galante, D., Rodrigues, F., & De Moraes Leme, J. (2017). Ecological interactions in Cloudina from the Ediacaran of Brazil: Implications for the rise of animal biomineralization. Scientific Reports, 7, 5482. https://doi.org/10.1038/s41598-017-05753-8.
4. Bernabei, M., & Bontadi, J. (2012). Dendrochronological analysis of the timber structure of the Church of the Nativity in Bethlehem. Journal of Cultural Heritage, 13, 54–60. https://doi.org/10.1016/j.culher.2011.05.003.
5. Bernard-Griffiths, J., & Groan, G. (1989). The samarium–neodymium method. In E. Roth & B. Poty (Eds.), Nuclear Methods of Dating (pp. 53–72). Springer Netherlands. https://doi.org/10.1007/978-94-015-7837-1_4.
6. Birgel, D., Meister, P., Lundberg, R., Horath, T. D., Bontognali, T. R. R., Bahniuk, A. M., De Rezende, C. E., Vasconcelos, C., & McKenzie, J. A. (2015). Methanogenesis produces strong 13C enrichment in stromatolites of Lagoa Salgada, Brazil: A modern analogue for Paleo-/Neoproterozoic stromatolite? Geobiology, 13, 245–266. https://doi.org/10.1111/gbi.12128.
7. Boaretto, E. (2015). Radiocarbon and the archaeological record: An integrative approach for building an absolute chronology for the Late Bronze and Iron Ages of Israel. Radiocarbon, 57, 207–216. https://doi.org/10.1017/S0033822200058665.
8. Cai, Y., Hua, H., & Zhang, X. (2013). Tube construction and life mode of the late

Ediacaran tubular fossil Gaojiashania cyclus from the Gaojiashan Lagerstätte. Precambrian Research, 224, 255–267. https://doi.org/10.1016/j.precamres.2012.09.022.

9. Cao, R., & Xue, Y. (1983). Vadose pisolites of the Tongying Formation (Upper Sinian System) in southwest China. In T. M. Peryt (Ed.), Coated Grains (pp. 538–547). Springer Berlin Heidelberg. https://doi.org/10.1007/978-3-642-68869-0_44.
10. Chen, Y., Chu, X., Zhang, X., & Zhai, M. (2015). Carbon isotopes, sulfur isotopes, and trace elements of the dolomites from the Dengying Formation in Zhenba area, southern Shaanxi: Implications for shallow water redox conditions during the terminal Ediacaran. Science China: Earth Sciences, 58, 1107–1122. https://doi.org/10.1007/s11430-015-5105-5.
11. Cox, G. M., Halverson, G. P., Stevenson, R. K., Vokaty, M., Poirier, A., Kunzmann, M., Li, Z. X., Denyszyn, S. W., Strauss, J. V., & Macdonald, F. A. (2016). Continental flood basalt weathering as a trigger for Neoproterozoic Snowball Earth. Earth and Planetary Science Letters, 446, 89–99. https://doi.org/10.1016/j.epsl.2016.04.029.
12. Cui, H., Grazhdankin, D. V., Xiao, S., Peek, S., Rogov, V. I., Bykova, N. V., Sievers, N. E., Liu, X. M., & Kaufman, A. J. (2016a). Redox-dependent distribution of early macro-organisms: Evidence from the terminal Ediacaran Khatyspyt Formation in Arctic Siberia. Paleogeography, Paleoclimatology, Paleoecology, 461, 122–139. https://doi.org/10.1016/j.paleo.2016.08.011.
13. Cui, H., Kaufman, A. J., Xiao, S., Zhou, C., & Liu, X. M. (2017a). Was the Ediacaran Shuram Excursion a globally synchronized early diagenetic event? Insights from methane-derived authigenic carbonates in the uppermost Doushantuo Formation, South China. Chemical Geology, 450, 59–80. https://doi.org/10.1016/j.chemgeo.2017.01.014.
14. Faggetter, L. E., Wignall, P. B., Pruss, S. B., & Sun, Y. (2018). Sequence stratigraphy, chemo stratigraphy and facies analysis of Cambrian Series 2–Series 3 boundary strata in northwestern Scotland. Geological Magazine, 155(1), 1–13. https://doi.org/10.1017/S0016756817000703.
15. Mattinson, J. M. (2005). Zircon U–Pb chemical abrasion ("CA-TIMS") method: Combined annealing and multi-step partial dissolution analysis for improved precision and accuracy of zircon ages. Chemical Geology, 220(1–2), 47–66. https://doi.org/10.1016/j.chemgeo.2005.03.011.
16. McCulloch, M. T., & Wasserburg, G. J. (1978). Sm-Nd and Rb-Sr chronology of continental crust formation. Science, 200(4345), 1003–1011. https://doi.org/10.1126/science.200.4345.1003.
17. Pommé, S., Stroh, H., Altzitzoglou, T., Paepen, J., Van Ammel, R., Kossert, K., Nähle, O., Keightley, J. D., Ferreira, K. M., Verheyen, L., & Bruggeman, M. (2018). Is decay constant? Applied Radiation and Isotopes, 134, 6–12. https://doi.org/10.1016/j.apradiso.2017.11.002.
18. Wilken, R. D., & Diehl, R. (1987). Strontium-90 in environmental samples from Northern Germany before and after the Chernobyl accident. Radiochimica Acta, 41(4), 157–162. https://doi.org/10.1524/ract.1987.41.4.157.

CHAPTER 10

Subsurface Stratigraphy and Sedimentology

LEARNING OBJECTIVES

After studying this chapter, you will be able to:

- Understand basic principles;
- Explain seismic reflection data;
- Describe borehole stratigraphy and sedimentology;
- Discuss geophysical logging;
- Know subsurface facies and basin analysis

KEY TERMS FROM THIS CHAPTER

Dynamite	Exploration
Geochemical analysis	Geological mapping
Geologists	Geology
Gravity	Hydrocarbons
Independent Simultaneous Sweeping	Ocean bottom nodes
Petroleum reservoirs	Reflections
Refracted rays	S-wave displacement

INTRODUCTION

The primary goal of the development of subsurface geology research techniques has been to meet the demands of the hydrocarbon industries. A branch of geology devoted to the study of subsurface structure, stratigraphy, and sedimentology has emerged as a result of exploration for coal, oil, and gas. The techniques mostly use geophysical methods, such as making seismic reflection profiles and measuring subsurface layer properties with instruments lowered through boreholes. Drill cuttings and core are also used to sample the drilled rocks. In certain regions, oil companies' subsurface exploration has yielded a plethora of data and improved knowledge of sedimentary basin stratigraphy. These activities have led to a significant increase in knowledge, particularly regarding the geology of offshore areas on continental shelves. Sequence stratigraphy's principles and applications originated in subsurface research and were subsequently applied to outcrop geology.

10.1. BASIC PRINCIPLES

While outcrop relationships provide the foundation for most geologists' understanding of sedimentology and stratigraphy, many will work with subsurface data if they are employed as professional geoscientists. When miners discovered mineral-rich beds or layers of coal at the surface, they began to follow them underground by digging tunnels. This was the beginning of the exploitation of mineral resources. A variety of methods are used in modern exploration, especially in the search for hydrocarbons, to determine what is beneath the surface. The majority of exploration uses less direct methods to investigate the strata hundreds or thousands of meters below the surface. In certain cases, this will involve drilling a hole and bringing pieces of rock back to the surface. These methods, which are known as geophysical techniques, entail measuring the physical characteristics of the rocks.

Sometimes used as very general indicators of the nature of the subsurface are surveys of regional variations in the Earth's magnetic field and measurements of gravity, which vary with the density of the rock below ground. However, the creation of seismic reflection profiles throughout a region is typically the first thorough method in subsurface exploration. These offer insights into the strata's structural and stratigraphic relationships as well as some indication of the lithologies that are present. By analyzing these data, boreholes can be drilled to collect cores or to perform additional geophysical measurements of the strata's characteristics. The goal is to develop a picture of the subsurface geology, which includes a sense of the large-scale stratal relationships and the distribution of various facies.

10.2. SEISMIC REFLECTION DATA

Active-source geophysical techniques called seismic-reflection methods were created with oil and gas exploration in mind. Recent technological advancements have increased the ability of seismic-reflection methods to produce high-resolution, two-dimensional images of the subsurface, leading to a greater use of these methods as near-surface investigatory tools. The best applications for seismic-reflection techniques are those that visualize subsurface environments with roughly horizontal layering. Seismic-reflection methods can be useful in many near-surface environmental and hydrogeological studies because the shallow subsurface is often stratified.

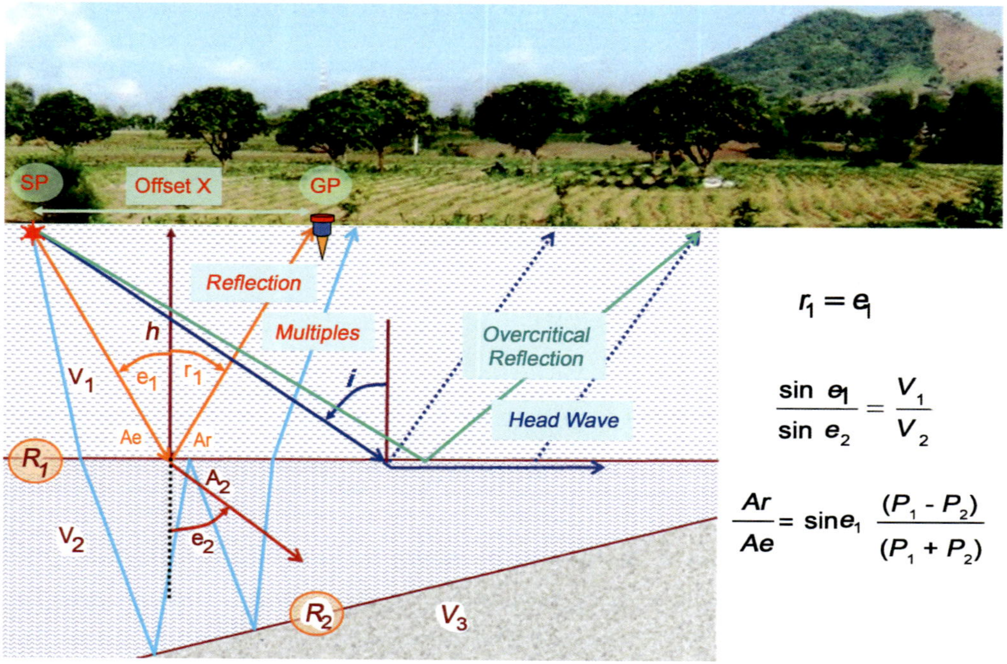

Figure 10.1. Sketch of seismic reflection energy propagation in the subsurface.

Source: By BacLuong at Vietnamese Wikipedia – Transferred from vi.wikipedia to Commons., Public Domain, https://commons.wikimedia.org/w/index.php?curid=75514184.

A seismic-reflection survey employs an above-ground sound source and receiver array to, respectively, transmit sound waves and measure their reflections off acoustic discontinuities (i.e., reflecting interfaces). Depths to acoustic discontinuities are calculated using a relationship between seismic velocity and seismic-wave travel time. Acoustic discontinuities are caused by differences in acoustic impedance. The most popular application for seismic-reflection data is the creation of two-dimensional subsurface cross sections that show the structures and depths of the reflecting interfaces found during the survey (Figure 10.1).

10.2.1. History

Seismic wave recordings from earthquakes were the first to show reflections and refractions of seismic waves at Earth's geologic interfaces. The fundamental model of the Earth's deep interior is predicated on observations of seismic waves produced by earthquakes and propagated through the interior of the planet (Vanda Grubisić & Mirko Orlić, 2007). Not too long afterward, the petroleum industry in particular contributed to the development of the application of human-generated seismic waves for the detailed mapping of the upper few kilometers of the Earth's crust (Figure 10.2).

Figure 10.2. Seismic testing in 1940.

Source: By Robert Yarnall Richie – Flickr: [Workers performing seismic tests, Seismic Explorations, Inc.], No restrictions, https://commons.wikimedia.org/w/index.php?curid=22879698.

The technique of seismic refraction exploration, which was employed to locate oil linked to salt domes, gave rise to seismic reflection exploration. German mine surveyor Ludger Mintrop created a mechanical seismograph in 1914, which he utilized to locate salt domes in his country. In 1919, he filed for a German patent, which was granted in 1926. He established Seismos in 1921, and the company was hired to carry out seismic exploration in Texas and Mexico. In 1924, the first oil discovery using the refraction seismic method was made for commercial purposes. Seismic refraction research along the Gulf Coast surged after the Orchard salt dome in Texas was discovered in 1924. However, by 1930, the majority of the shallow Louann salt domes had been discovered using this method, and the refraction seismic method began to decline.

Following World War I, individuals such as Mintrop, Reginald Fessenden, John Clarence Karcher, William P. Eckhardt. Haseman and McCollum Burton. Haseman, Eckhardt, McCollum, and Karcher established the Geological Engineering Company in 1920. Haseman, I. Karcher, June 1921. Perrine and W. C. In the vicinity of Oklahoma City, Kite recorded the first exploration reflection seismograph.

Many in the oil industry were skeptical of early reflection seismology. An early supporter of the approach said:

"The senior writer personally attempted to implement the method into standard consulting practice, and he or she can vividly remember numerous instances in which reflections were not given the same consideration as the divining rod, since the latter had a more established historical background."

The Geological Engineering Company closed its doors when the price of oil plummeted. However, in 1925, oil prices rose again and Karcher played a role in founding the Geophysical Research Corporation (GRC) with Amerada. Karcher later left GRC in 1930 to help establish Geophysical Service Incorporated (GSI), which became a leading seismic contracting company for over 50 years. GSI was also the parent company of Texas Instruments. Another former GSI employee, Henry Salvatori, went on to found Western Geophysical in 1933. Since the invention of reflection seismology, numerous companies have been formed for hydrocarbon exploration, hydrology, engineering studies, and other applications. More recent major service providers in seismic services include CGG, ION Geophysical, Petroleum Geo-Services, Polarcus, TGS, and WesternGeco. However, the oil price crash in 2015 has led to financial challenges for companies like Polarcus. Some former industry leaders like CGG and WesternGeco have shifted focus away from seismic acquisition and towards managing existing data libraries and providing non-seismic oilfield services.

10.2.2. Theory

Acoustic waves, or seismic waves, are waves that cause oscillations in subsurface earth materials and interact with them. The two seismic body waves that move through the interior of geologic bodies are compressional (P) and shear (S), and these are the ones that reflection methods typically deal with. The way these waves interact with the materials they pass through allows them to be distinguished from one another. P-waves move material in a direction that is parallel to their path of travel, while S-waves move material in a normal (i.e., perpendicular) to the direction of their travel.

A seismic wave travels semispherically in homogeneous media, meaning that the instantaneous extent of seismically influenced material (i.e., the radius of the wavefront) grows with propagation. Each ray of a wave travels perpendicular to the wavefront and arrives at an angle of incidence (Θ_i) at an acoustic interface, describing the wave. This angle, which is between 0° and 90°, is formed by an arriving (i.e., incident) ray and normal to the interface.

There are three possible routes via which a seismic ray can go from a surface-source point to a receiver. Direct beams move in the direction of the receiver just below the surface. Rays that have been refracted reach reflecting interfaces, i.e., at a critical angle of incidence (Θic), are diverted from the normal path and proceed along the lower surface of the reflector. All incident angles of reflected rays reach reflectors, whereupon wave energy is instantly diverted to the surface (Mussett & Khan, 2000).

Interfaces made of materials with a sufficient contrast in acoustic impedance reflect seismic radiation. The density and seismic velocity of the material determine the acoustic impedance, which is a measurement of resistance to sound energy (i.e., seismic wave travel speed that is dependent on both waves and materials). Rays are partially reflected and then continue on toward deeper reflectors, depending on the degree of acoustic impedance contrast. Seismic-reflection techniques can thus image systems with multiple layers.

Seismic velocity data is used to convert two-way travel times (TWT) to depths, which creates a true vertical cross-section of subsurface reflectors. The time it takes for a reflected ray to travel from source to recipient is known as the TWT. The ray travels the least after reflecting off of a single horizontal reflector beneath the source, and it takes the rays a longer amount of time to reach each subsequent geophone. This additional time lag (i.e., normal moveout (NMO)) according to Mussett & Khan (2000) is used to calculate seismic velocity and layer thickness.

10.2.3. Applications

Many domains employ reflection seismology extensively, and its applications can be divided into three groups based on the depth of study they each require:

- Near-Surface Applications: An application used for coal and mineral exploration, engineering, and environmental surveys that seek to understand geology at depths of up to about 1 km. Geothermal energy surveys are a more recent application of seismic reflection, though the depth of the investigation in this case can reach up to 2 km.
- Hydrocarbon Exploration: Utilized to produce a high-resolution map of acoustic impedance contrasts at subsurface depths of up to 10 km, for the benefit of the hydrocarbon industry. This can be utilized in conjunction with other exploration geophysics tools and seismic attribute analysis to assist geologists in creating a geological model of the region of interest.
- Mineral Exploration: While geological mapping, geochemical analysis, and aerial and ground-based potential field methods have been the traditional approaches to near-surface (<300 m) mineral exploration, especially for greenfield exploration, in recent decades, reflection seismic has emerged as a viable method for exploration in hard-rock environments.
- Crustal Studies: Exploration of the composition and formation of the Earth's crust at depths of up to 100 km, all the way to the Moho discontinuity and beyond.

Ground-penetrating radar, or GPR, is a technique that is comparable to reflection seismology but uses electromagnetic waves rather than elastic waves and has a lower penetration depth.

10.2.3.1. Hydrocarbon Exploration

Petroleum geologists and geophysicists use reflection seismology—widely known as "seismic reflection" or simply "seismic" in the hydrocarbon industry—to map and interpret prospective petroleum reservoirs. Since the late 20th century, there have been notable advancements in computer power that have coincided with an increase in the size and scope of seismic surveys. As a result, the seismic industry began routinely obtaining large-scale, high resolution 3D surveys in the 1980s, instead of laboriously and infrequently acquiring small 3D surveys. Over time, the techniques have changed somewhat, but the fundamental ideas and objectives have not.

Three main environments—land, transition zone, and marine—are used for seismic hydrocarbon exploration.

1. Land: Nearly every kind of terrain on Earth is included in the land environment, and each presents unique logistical challenges. The arctic tundra, jungle, desert, forest, urban areas, mountain ranges, and savannah are a few examples of this type of environment.
2. Transition Zone (TZ): The transition zone, which is defined as the area where land meets the sea, presents particular difficulties because the water is too deep for conventional onshore acquisition techniques but too shallow for large seismic vessels. River deltas, marshes and swamps, coral reefs, beach tidal zones, and the surf zone are a few examples of this type of habitat. To acquire a comprehensive map of the subsurface, transition zone seismic crews frequently work on a single project on land, in the transition zone, and in the shallow water marine environment.
3. Marine: The marine zone is either in deep water regions typically associated with seas and oceans (such as the Gulf of Mexico) or in shallow water areas (water depths of less than 30 to 40 meters would normally be considered shallow water areas for 3D marine seismic operations).

10.2.3.1.1. Seismic Data Acquisition

The three distinct stages of seismic exploration are seismic interpretation and seismic data processing, with seismic data acquisition coming first.

National and international oil companies usually design seismic surveys, and they are acquired by service providers like CGG, Petroleum Geo-Services, and WesternGeco. The data is then processed by a different business, though frequently this is the same business that purchased the survey. The completed seismic volume is then given to the oil company for geological interpretation.

10.2.3.1.2. Land Survey Acquisition

Large-scale land seismic surveys typically involve hundreds of tons of equipment and several 100 to several 1,000 workers who are spread out over a long period of time. In a land survey, there are several alternatives for a controlled seismic source. Two popular options are dynamite and vibroseis. Vibroseis is a cheap, non-impulsive source that works best on level ground, which makes it more challenging to use in undeveloped areas. The process involves lowering a steel plate to the ground using one or more heavy, all-terrain vehicles. The plate is then vibrated with a particular frequency distribution and amplitude. Although the massive weight of a Vibroseis truck can have negative environmental effects of its own, it produces a low energy density that makes it suitable for use in cities and other populated areas where dynamite would cause significant damage. Although it produces an almost perfect impulse function, dynamite is an impulsive source that is thought to be the perfect geophysical source despite having clear environmental disadvantages. Prior to weight dropping being introduced in 1954, it was the only seismic source available for a considerable amount of time. This allowed geophysicists to weigh the trade-off between environmental damage and image quality. Since dynamite must be inserted into each source point after it has been drilled, it is also less operationally efficient than Vibroseis.

Land geometries are not constrained by narrow paths of acquisition, in contrast to marine seismic surveys. As a result, a wide range of offsets and azimuths are typically acquired, with the biggest challenge being accelerating the rate of acquisition. It goes without saying that the speed at which the source—in this case, Vigroseis—can be fired and then move on to the next source location determines the rate of production. To improve survey efficiency, attempts have been made to use multiple seismic sources simultaneously. One effective example of this technique is Independent Simultaneous Sweeping (ISS).

Significant logistical support is needed for a land seismic survey. In addition to the daily seismic operation, there is need for support with resupply, medical care, camp and equipment maintenance, security, personnel crew changes, and waste management for the main camp. Additionally, some operations might run smaller "fly" camps that are set up in remote locations where it is too far to travel back to the main camp every day. These camps will frequently require logistical support.

10.2.3.1.3. Marine Survey Acquisition (Towed Streamer)

Utilizing specialized seismic vessels, tow streamers—cables with groups of hydrophones, also known as receiver groups—just below the surface for a maximum of 5 to 15 meters, depending on the project specifications—is how marine seismic surveys are carried out. Many streamers that can be fastened to underwater wings—also referred to as doors or vanes—that enable several streamers to be towed out widely to the port and starboard sides of a vessel are typically towed astern by modern streamer vessels. With the use of modern technology, such as that found on the PGS-operated Ramform series of ships constructed between 2013 and 2017, the total number of streamers on these vessels has increased to 24. A streamer stretched from "door to door" across the stern of a vessel with this kind of capacity is frequently more than one nautical mile. The client's intended geological area of interest beneath the sea floor will determine the exact streamer configuration for each project, including streamer length, streamer separation, hydrophone group length, offset, and distance between the source center and the receivers.

The high-pressure air gun arrays on streamer vessels, which fire at 2000 psi to produce a tuned energy pulse into the seabed from which the reflected energy waves are recorded on the streamer receiver groups, are the main source of high energy that the vessels tow. Gun arrays are tuned, meaning that the number, arrangement, and individual volumes of the guns in a given array can alter the frequency response of the air bubble that is released when the gun is fired. Guns can be found scattered across an array or grouped together to form clusters. Source array volumes typically range from 2000 to 7000 cubic inches, however, exact volumes will vary based on the geology of the survey region. Because of the size and towing capacity of contemporary towed streamer vessels, marine seismic surveys produce a high volume of data.

A Narrow-Azimuth Towed Streamer (also called NATS or NAZ) is a seismic vessel that has two sources and is towing a single streamer. It was acknowledged by the early 2000s that this kind of acquisition was good for preliminary research but not suitable for development and production, where precise well placement was required. As a result, the Multi-Azimuth Towed Streamer (MAZ) was created in an attempt to overcome the constraints imposed by a NATS survey's linear acquisition pattern by acquiring a combination of NATS surveys at various azimuths. This effectively produced a higher signal to noise ratio and more subsurface illumination.

Because salt attenuates seismic waves and has difficult-to-image overhangs, its seismic properties present an additional challenge for marine seismic surveys. The wide-azimuth towed streamer, also known as WAZ or WATS, is another variation on the NATS survey type that resulted from this. It was initially tested in the Mad Dog field in 2004. For this kind of survey, two different vessels were used to tow seismic sources at the beginning and end of the final receiver line, and one vessel was used to tow a set of eight streamers. The receiver vessel moved farther away from the source vessels during each of the four "tiles" in this configuration, which ultimately produced the impression of a survey with four times as many streamers. As a result, seismic imaging was revolutionized with a seismic dataset that covered a wider range of azimuths. These three forms of marine towed streamer seismic surveys are currently in use.

10.2.3.1.4. Marine Survey Acquisition (Ocean Bottom Seismic (OBS))

In addition to using seismic vessels, cables of geophones and hydrophones can be laid on the sea floor in a manner akin to that of a land seismic survey, using a different source vessel for each survey. This technique was first created out of an operational need to allow seismic surveys to be carried out in places where there are obstacles, like production platforms, without compromising the quality of the images that are obtained. In areas where a seismic vessel cannot be used, such as shallow marine environments (water depth <300 m) and transition zone environments, ocean bottom cables (OBC) are also widely used. In deep water, where repeatability is important, OBCs can be deployed by remotely operated underwater vehicles (ROVs). Conventional OBC surveys use dual-component receivers, combining a pressure sensor (hydrophone) and a vertical particle velocity sensor (vertical geophone), but more recent developments have expanded the method to use four-component sensors, i.e., a hydrophone and three orthogonal geophones. The benefit of using four-component sensors is that they can record shear waves, which don't pass through water but can nevertheless provide insightful data.

Apart from its operational benefits, OBC offers geophysical advantages over a traditional NATS survey due to its wider azimuth range and increased fold associated with the survey geometry. Large-scale OBC survey capabilities are severely constrained, and the wider azimuths and increased fold come at a cost similar to a land survey.

Ocean bottom nodes (OBN) were a collaboration between BP and Fairfield Geotechnologies that began in 2005. OBN is an extension of the OBC method that uses battery-powered cableless receivers positioned in deep water. Because these nodes are lighter and smaller than OBC cables, they can be placed more freely and are simpler to store and deploy.

10.2.3.1.5. Marine Survey Acquisition (Ocean Bottom Nodes (OBN))

The advancement of ocean bottom cable technology directly led to the development of node technology, i.e., the capacity to use towed streamer technology to eliminate the seafloor to hydrophone sea water space by placing a hydrophone in direct contact with the seafloor. The idea of an ocean bottom hydrophone is not new; it has been used for many years in scientific research. However, its application as a quick method of gathering data for oil and gas exploration is relatively new.

Three horizontal and vertical axis orientation sensors and a hydrophone make up the self-contained four-component nodes. Nodes' physical dimensions vary based on the manufacturer and design requirement, but generally speaking, they weigh more than 10 kg each to compensate for buoyancy and reduce the likelihood of seafloor movement caused by currents or tides.

To ensure the safe navigation of node vessels and before the deployment of nodes, a bathymetry seabed survey is typically carried out of the survey area using side-scan technology to map the seabed topography in detail. This is because nodes can be used in areas where streamer vessels may not be able to enter. In addition to identifying any problems with node deployment, such as submerged obstructions, wrecks, oilfield infrastructure, or abrupt changes in water depth from underwater cliffs, canyons, or other locations where nodes might not be stable or make a good connection to the seabed, this will also identify any potential hazards that could affect the safe navigation of node and source vessels.

Ocean bottom cables must be physically attached to a recorder vessel to record data in real-time, in contrast to OBC operations, where a node vessel does not connect to a node line. There is no real-time quality control component to a node's operating status because they are self-contained and not connected to any system once they are deployed, so until the nodes are recovered and the data is reaped from them (reaping is the industry term used to remove data from a recovered node when it is placed within a computerized system that copies the hard drive data from the node), it is assumed that the data will be recorded. Since the technology is now well-established and dependable, there is a high level of confidence that a node will function as intended once it and its battery system have met all setup requirements. Technical pauses throughout node initiatives, i.e., when compared to the total number of deployed nodes, individual node failures during deployment typically occur in single digits.

Rechargeable internal Lithium-ion battery packs or replaceable non-rechargeable batteries power nodes; the type of battery utilized depends on the node's specifications and design. The solid-state hard drive's inability to store collected data and the loss of all data recorded since the node was deployed on the sea floor make the battery life of a node unit an essential factor to take into account when designing a node project. As a result, a node with a 30-day battery life needs to be set up, record data, recover, and harvest all within that time frame. This is also related to the number of nodes that will be deployed, since the two have a close relationship with battery life. Batteries can run out and data can be lost if too many nodes are deployed and the OBN crew does not have enough resources to recover them in time, or if unfavorable weather restricts recovery operations. Because batteries must be transported to and from an operation and their drained batteries must be disposed of by a licensed contractor ashore, disposable or non-rechargeable batteries can also pose a serious waste management challenge.

Timing synchronization of individual node clock units with an internal clock drift correction is another crucial factor to take into account. Unusable data may result from any mistake in the proper synchronization of nodes prior to their deployment. Since node acquisition frequently occurs in multiple directions and from multiple sources at the same time throughout a 24-hour period, it is essential that all nodes operate on the same clock time to process data accurately.

The deployment and recovery modes, as well as the node handling system design, will be determined by the node type and specification. There are currently two popular methods: ROV operations and node on a rope.

10.2.3.1.6. Node on a Rope

The node must be fastened to a high-grade rope or steel wire to use this method. Depending on the prospect design, each node will be evenly spaced along a rope that has unique fittings every 50 meters to securely attach the node to the rope. Next, a specialized node vessel uses an node handling system to lay this rope, usually with dynamic positioning along a pre-established node line. The nodes are "landed" onto pre-plotted positions that have an agreed-upon and allowable error radius; for instance, a node has to be positioned within a radius of 12.5 meters from the pre-plot positions for navigation. They are frequently accompanied by pingers, which are tiny transponders that each node on deployment can precisely pinpoint its sea floor position by using an underwater acoustic positioning transducer to detect them. Pingers may be found on every node or every third node, for example, depending on the terms of the contract. The term "pinging and pinging equipment" refers to the application of USBL, or Ultra-short baseline, acoustic positioning systems that interface with navigational equipment for vessel-based differential GPS, or Differential Global Positioning System.

Node lines are often recovered by dragging them back onto the vessel using an anchor or grapple hook. The specific design of handling systems on node vessels is determined by the node design and is utilized for the purposes of storing, deploying, and recovering nodes. While larger nodes are automatically moved, stored, recharged, and reaped by robotic systems, smaller nodes may have a manual handling component. Rope bins and spoolers are two more tools that node vessels use to store and manage the many kilometers of rope that are frequently carried on board.

When a prospect has shallow water, such as less than 100 meters or a transition zone area near a beach, a node on a rope is typically utilized. A dynamic positioning vessel is used for operations in deeper water to guarantee precise node deployment; however, the safe operating depth of these larger vessels is limited; the typical cutoff is between 15 and 20 meters of water depth, depending on the vessel and its in-water equipment. Then, in water as shallow as one to three meters, nodes can be deployed and recovered using specialized shallow water boats. To create a consistent seismic line transition from water to land, these shallow water nodes can then be used to tie in with geophones on the shore.

There are several vulnerabilities in this approach that could lead to damage or loss on a project, all of which must be carefully assessed and mitigated. Nodes connected on a rope resting on the sea floor are at risk of being moved by strong currents, getting caught on obstructions, being dragged by vessel anchors, or getting caught by fishing vessels. These potential hazards should be identified and evaluated during the project planning phase, especially in oilfield locations where contact with subsea structures must be avoided. To prevent accidents, exclusion zones are typically established. Since node lines can be displaced after deployment, it is crucial to accurately position them during both deployment and recovery as a standard navigation quality control check. In some cases, node lines may need to be repositioned if they have moved outside of the specified contract parameters.

10.2.3.1.7. ROV Deployment

To handle and position nodes at their pre-planned locations, this technique makes use of remotely operated underwater vehicle (ROV) technology. With this kind of deployment and recovery technique, a basket filled with nodes is submerged in water. After establishing a connection with the appropriate node basket, a ROV will remove each node from a tray in the prescribed order. Every node will be positioned on the pre-plot where it has been assigned. When a node needs to be recovered, the procedure is carried out in reverse: the ROV picks it up and places it into the node basket tray, filling it up until it is lifted back to the surface. The nodes are extracted from the basket and harvested after the basket is recovered onto the node vessel.

Typically, ROV operations are employed for deep water node projects, frequently in open ocean waters as deep as 3,000 meters. Nonetheless, there are a few things to think about when it comes to ROV operations. Because deep water ROV operations are particularly complex, production may be impacted by the need for periodic maintenance. Repairs to ROVs that need onshore or outside expert assistance will put an end to a node project. Umbilicals and other high-tech spare parts for ROVs can be very costly. Extreme water depths will likely limit ROV operations in open ocean areas due to weather and sea conditions, and node deployment and recovery production rate will be significantly lower due to the time required for node basket transit from surface to seafloor. Resupply, bunkering, and crew change activities can pose logistical challenges when supporting operations offshore.

10.2.3.1.8. Time Lapse Acquisition (4D)

Time lapse or 4D surveys, so named because they are 3D seismic surveys that are repeated after some amount of time, refer to the fourth dimension—in this case, time. Time-lapse surveys are obtained to track changes in reservoir characteristics during production and pinpoint locations with flow obstructions that may not be visible with traditional seismic. A baseline survey and a monitor or repeat survey, obtained after the field has been in production, make up time-lapse surveys. Since NATS surveys are less expensive to obtain and most fields have previously had a baseline NATS survey, the majority of these surveys are repeats of NATS surveys. Ocean-bottom cables are used for some of these surveys because, once removed, they can be precisely repositioned in their original location. Increased repeatability and higher signal-to-noise ratios are the results of precisely repeating the source and receiver locations. Additionally, a number of 4D surveys have been set up over fields where ocean bottom cables have been permanently installed through purchase. This technique is sometimes referred to as permanent reservoir monitoring (PRM) or life of field seismic (LoFS).

Since the goal of a 4D survey is to replicate the baseline or original survey as precisely as possible, 4D seismic surveys employing towed streamer technology can be extremely difficult. The accuracy with which such a survey can meet that repeatability objective can be greatly impacted by factors such as weather, tides, currents, and even the season. Another excellent method for precisely reproducing a seismic acquisition is OBN. Year 2009 saw the acquisition of the world's first 4D survey using nodes over the Atlantis Oil Field. A ROV deployed the nodes in a water depth of 1,300–2,200 meters, up to a few meters from their original location in 2005.

10.2.3.1.9. Seismic Data Processing

Seismic data processing involves three primary procedures: migration, common-midpoint (CMP) stacking, and deconvolution. Deconvolution is an attempt to extract the Earth's reflectivity series based on the idea that a seismic trace is nothing more than the Earth's reflectivity series convolved with distorting filters. By collapsing the seismic wavelet, this method increases temporal resolution; however, it is not unique unless additional data, such as well logs, are obtained or new assumptions are made. It is possible to perform deconvolution operations in a cascade, where each deconvolution is tailored to eliminate a specific kind of distortion.

A reliable method called CMP stacking makes use of the fact that a specific subsurface location will have been sampled multiple times at various offsets. This enables the creation of a Common Midpoint Gather, a collection of traces with various offsets that all sample the same subsurface location, by a geophysicist. After that, the average amplitude is computed over a time sample. This considerably reduces random noise but also eliminates all useful information regarding the correlation between seismic amplitude and offset. Statics and normal moveout correction are less important procedures that are applied just prior to the CMP stack. Land seismic data must be adjusted for the elevation differences between the shot and receiver locations, in contrast to marine seismic data. This correction, known as a statics correction, takes the form of a vertical time shift to a flat datum. However, because the near-surface velocity is not precisely known, additional corrections will be required later in the processing sequence. A residual statics correction is the name given to this additional adjustment.

The process of geometrically relocating seismic events in space or time to the subsurface location of the event instead of the surface location where it was recorded results in a more accurate subsurface image. This is known as seismic migration.

10.2.3.1.10. Seismic Interpretation

Seismic interpretation aims to produce a logical geological narrative from the processed seismic reflection map. Seismic interpretation, in its most basic form, consists of following and correlating continuous reflectors across the 2D or 3D dataset, then using these to inform the geological interpretation. The goal of this is to create structural maps that show how different geological layers vary in depth over space. These maps can be used to locate hydrocarbon traps and to build subsurface models that enable volume calculations. Seismic datasets, however, seldom provide a sufficiently clear picture to accomplish this. The primary cause of this is the vertical and horizontal seismic resolution, but noise and processing issues frequently contribute to a lower-quality image as well. This is the reason why there is never 100% accuracy in seismic interpretation and why multiple solutions may fit the same dataset. More information will be required in this situation to limit the solution, such as additional seismic acquisition, borehole logging, or gravity and magnetic survey data. Like a seismic processor, a seismic interpreter is typically urged to maintain optimism to promote continued work rather than leaving the survey area. Both geologists and geophysicists perform seismic interpretation; the majority of seismic interpreters are knowledgeable in both disciplines.

The elements that comprise a petroleum reservoir—the source rock, the reservoir rock, the seal, and the trap—are the characteristics that the interpreter is especially attempting to distinguish in hydrocarbon exploration.

10.2.3.1.11. Seismic Attribute Analysis

To improve information that might be more subtle in a traditional seismic image and provide a better geological or geophysical interpretation of the data, seismic attribute analysis entails extracting or deriving a quantity from seismic data that can be analyzed. A few characteristics that can be examined are coherency, amplitude versus offset, and mean amplitude, which can be used to identify bright and dim areas. Direct hydrocarbon indicators are characteristics that have the ability to reveal the presence of hydrocarbons.

10.2.3.2. Crustal Studies

Initiating deep seismic exploration in other nations like BIRPS in Great Britain and ECORS in France, organizations like the Consortium for Continental Reflection Profiling (COCORP) pioneered the use of reflection seismology in studies of tectonics and the Earth's crust in the 1970s. The North Sea's oil and hydrocarbon exploration led to the establishment of the British Institutions Reflection Profiling Syndicate (BIRPS). It became evident that the tectonic processes that had created the sedimentary basins and geological structures under exploration were poorly understood. The endeavor yielded noteworthy outcomes, demonstrating that marine seismic surveys can be utilized to profile characteristics like thrust faults that pierce through the crust and reach the upper mantle.

10.2.4. Environmental Impact

Seismic reflection surveys have some effect on the Earth's natural environment, as do all human activities. To learn more about these effects, environmental organizations and the hydrocarbon industry conduct research.

10.2.4.1. Land

Building roads may be necessary for the transportation of workers and equipment during a land-based seismic survey, and vegetation may need to be removed to make room for the equipment. Significant habitat disturbance may happen if the survey is conducted in an area that is still developing, and many governments have strict regulations governing environmental destruction, such as banning the use of dynamite as a seismic source. Seismic lines can use pre-existing non-straight tracks and trails, or they can veer around natural obstacles thanks to seismic processing techniques. This can significantly lessen the impact of a land seismic survey on the environment with careful planning. By enabling survey lines to be wound between trees, the more recent use of inertial navigation instruments rather than theodolites for land survey reduced the impact of seismic activity.

Before beginning any seismic survey on land, it is crucial to assess and effectively manage the potential impact it may have. In well-regulated environments, Environmental and Social Impact Assessment (ESIA) or Environmental Impact Assessment (EIA) reports are typically required before work can commence. Additionally, project planning should consider the long-term impact that may be left behind once the project is completed. It is the responsibility of both contractors and clients to adhere to the remediation plan outlined in the contract and in accordance with the laws of the project location.

Land seismic operations can have a substantial physical footprint and local impact depending on the size of the project. This is particularly the case when living accommodations, workshops, maintenance facilities, general and seismic vehicle parking areas, camp utilities, waste management facilities (including black and grey water management), and storage facilities are needed. Interaction with the local population may result in potential disruptions to their daily routines, such as increased traffic, noise, and 24-hour operations, all of which need to be evaluated and minimized. In addition to the importance of archeological considerations, project planning needs to take into account the legal, cultural, and social requirements that will need to be taken into account. To reduce their impact and stop damage, specialized techniques can be used to determine safe working distances from buildings and archaeological structures.

10.2.4.2. Marine

Since whales, porpoises, and dolphins are among the cetaceans that primarily communicate with sound, the primary environmental concern for marine seismic surveys is the possibility that noise from the high-energy seismic source will disturb or harm wildlife. While lower-level noise can temporarily shift one's hearing threshold, obstruct sounds essential to marine life, or disrupt behavior, high-level and prolonged sound can cause physical harm, including hearing loss.

According to a study, humpback whales that are migrating will leave a minimum of 3 km between themselves and an operational seismic vessel. Pods of resting whales that include cows will leave an even larger gap, ranging from 7 to 12 km. On the other hand, the study discovered that male humpback whales were drawn to a lone working airgun because it was thought that the low-frequency sound confused them with whale breaching behavior. When a nearby seismic source was detected, not only did whales exhibit alarming and evasive behavior, but also fish and squid. Comparing reports on how seismic survey noise affects marine life is challenging since methods and units are frequently not well documented.

Where there is seismic testing, the gray whale will stay more than 30 km away from its typical migratory and feeding grounds. Likewise, it has been observed that the breathing of gray whales is faster, which suggests that the whale is in discomfort or even panic. Though investigation into these issues is still ongoing, circumstantial evidence like this has led researchers to surmise that avoidance and panic may be to blame for an increase in whale beachings. Even so, airguns are shut down only when cetaceans are seen at very close range, usually under 1 km.

An alternative viewpoint is presented in a joint paper published by the International Association of Oil and Gas Producers (IOGP) and the International Association of Geophysical Contractors (IAGC), which contend that seismic noise produced by marine surveys is similar to seismic noise produced by natural sources.

The Joint Nature Conservation Committee, or JNCC as it is more widely known, is a government organization in the United Kingdom. The public organization that provides advice on national and worldwide nature conservation to the UK government and devolved administrations. "has long had a stake in understanding how seismic or geophysical surveys affect the marine environment. At the government level, it was recognized as early as the 1990s that monitoring and research were necessary to determine the effects of the sound energy generated by seismic surveys. The "JNCC guidelines for minimizing the risk of injury to marine mammals from geophysical surveys (seismic survey guidelines)" from 2017 are just one example of the international references that are still in use today as a potential baseline standard for surveys in seismic contracts worldwide.

The size and scope of seismic surveys being conducted into the 21st century complicates the issue of seismic sound energy as a disruptive factor to marine mammals. Seismic surveys used to typically last a few weeks or months and be limited to a small area of the ocean. However, with the advent of OBN technology, surveys can now span thousands of square kilometers and last for years, continuously adding sound energy to the ocean from various energy sources. One such instance is the 85,000 square kilometer mega seismic survey contract that the national oil company of Abu Dhabi, ADNOC, signed in 2018. The survey is expected to cover a variety of deep-water, coastal, island, and shallow-water locations and is expected to last until 2024. Determining the long-term effects of these massive operations on marine life may prove to be extremely challenging.

In 2017, IOGP recommended that, to avoid disturbance whilst surveying:

- Every operation uses protective measures to address site-specific environmental conditions to prevent marine mammals from being harmed by vessel traffic and sound exposure.
- Surveys are designed to stay away from known sensitive locations and times, like areas used for breeding and feeding. To further shield marine life from any potentially harmful effects of sound, exclusion zones are usually created around the seismic source.
- Usually, the exclusion zone encircles the sound source in a circle with a minimum radius of 500 meters.
- Before any sound-producing operations start, the area is visually and auditorily checked for marine mammals and other

protected species using trained observers and listening devices. These observers assist in ensuring that safety procedures are followed during operations, and the local governments receive detailed reports from them that include information on the biodiversity of the survey area.

- At the start of the seismic lines, sound production usually starts with a "soft-start" or "ramp-up," which is a gradual increase in sound level from the air gun source from a very low level to full operational levels – usually over 20 to 40 minutes. The purpose of the soft-start technique is to give any animal that might be nearby enough time to get away as the noise gets louder.

The regulatory framework in which the seismic survey is being conducted is the second factor. The legal requirements in highly regulated areas, like the North Sea or the Gulf of Mexico, will be spelled out in detail in the contract, and both the contractor and the client will abide by it because breaking the rules can have serious repercussions, like significant fines or the removal of exploration block permits. On the other hand, some nations with rich and diverse marine biomes also have lax environmental regulations and nonexistent or ineffective regulators. In situations where state-owned oil and gas companies hold a dominant position in a country, and the regulator is also a state-owned and operated entity, undermining its perceived independence, attempts to protect marine environments can be seriously compromised by a weak regulatory framework.

10.3. BOREHOLE STRATIGRAPHY AND SEDIMENTOLOGY

An understanding of potential stratigraphic and structural relationships in the subsurface can be gained through the interpretation of seismic reflection profiles. These sources of data can offer some indications of the subsurface lithologies, but only the inclusion of lithology and facies data can yield a complete geological picture. This can be obtained by boring holes through the succession and collecting samples of the rocks or by measuring the properties of the rocks in great detail using geophysical instruments. There are several methods for gathering data from the subsurface when a borehole is drilled; these are briefly discussed below.

10.3.1. Borehole Cuttings

A fluid is pumped down to the drill bit during the process of drilling a deep borehole to lubricate it, remove cut rock (cuttings), and balance subsurface formation fluid pressures. Fluids (water, oil, and gas) trapped in porous and permeable strata will be under pressure due to the weight of the rocks above, and in the absence of an object to counteract that pressure, they will rush to the surface up the borehole. As a result, the drilling fluid is typically a "mud," consisting of a blend of water or oil and powdered material to increase the fluid's density. Powdered barite ($BaSO_4$) is a common ingredient because of its 4.48 density. To balance the pressure in the subsurface formations, the density of the drilling mud is changed. Pumping the drilling mud down the inside of the drill string (pipe) and back up the outside is how it is recirculated; as it is a thick, viscous fluid, it will carry the cuttings with it when it reaches the surface. To create a record of the drilled strata, the cuttings are sieved out of the mud and then cleaned. These cuttings, which usually have a diameter of 3–5 mm, are separated from the drilling mud at the surface by sieving. Although details like sedimentary structures are not preserved, information about the rock types of the strata that the borehole has penetrated can be obtained by recording the lithology of these drill chips, a process known as "mud-logging." Cuttings can be used to recover microfossils, including palynomorphs, nanofossils, and foraminifera, for use in biostratigraphic analysis. Since the fluid usually mixes some material from various layers as it ascends the borehole, the most important factor is usually the depth at which a lithology or fossil first appears.

10.3.2. Core

It is possible to design a drill bit so that it removes an annulus of rock, exposing a core—a cylinder in the center—that can be raised to the surface. Drilling stops when coring is being done, and the section of core is raised to the surface inside the hollow drill string in a sleeve. Every section of core that is raised to the surface is put into a box that has a label indicating the depth interval from which it was recovered. Recovery is frequently insufficient, with the core possibly breaking apart during drilling and only a portion of the succession drilled remaining intact. Afterwards, the core is often cut vertically to produce a smooth-surfaced rock slab that, depending on the borehole's width, is normally 90 to 150 mm across.

This type of core cutting yields a great deal of information about the lithologies that are present, as well as small-scale sedimentary structures, body and trace fossils. Cores are cut through "target horizons" in the process of searching for oil and gas and developing fields for the production of hydrocarbons. These are sections of the succession that have been determined to be either reservoir bodies or, more crucially, likely source rocks based on the interpretation of seismic data. Only these sections of the stratigraphy are typically cored and recovered; the remainder of the succession must be interpreted using geophysical wireline logs. On the other hand, continuous cores can be cut through successions that are too complex to interpret with geophysical data alone.

This can happen when the characteristics of the rock units make it difficult to distinguish between different lithologies with wireline logging equipment. Coal and mineral exploration, in contrast to oil and gas exploration, typically entails inserting the entire core through the drilled section. The core is not split vertically and is typically only 40 mm wide when it is cut. While the continuous core offers good vertical coverage of the drilled succession, its small size and curved surface may make it more challenging to identify sedimentary structures than in the traditional, larger, split core used in oil and gas exploration.

10.3.3. Core Logging

Creating a graphic sedimentary log of a succession exposed in the field is very similar to the process used to record the specifics of the sedimentary rocks in a core. Field logging sheets and core logging sheets have a similar format and record the same kinds of data (lithology, bed thickness, bed boundaries, sedimentary structures, biogenic structures, and so forth). Typically, the scale is 1:20 or 1:50. In certain respects, field description is more difficult to record than information about strata from the core.

A nearly full record of the succession, including the finer-grained lithologies, will exist if the core recovery is strong. Although weaker, finer-grained beds will frequently fracture more during drilling, weathering of mudrocks in the field typically means that they are less well preserved than the coarser beds. However, in core this tends to be less of an issue.

The primary constraints stemming from the core's width. Only fragments of larger scale sedimentary structures are preserved, and it is difficult to see the lateral geometry of the beds and identify features like channels. However, the smooth, cut surface of a core may make it easier to see the details of ripple-scale features. Only when the orientation of the core was noted during the drilling process can paleocurrent data be extracted from sedimentary structures, and this is not always the case. Although it is unlikely to be as interesting to work in as a field location in an exotic place, the geologist can record data in the relatively comfortable setting of a core store. This is another, not insignificant, difference between core and outcrop. Not every core crosses the strata perpendicular to the bedding. A vertical drill core will cut through the beds at an angle if the strata are tilted, resulting in all bed boundaries and sedimentary structures seen in the core being inclined. Drilling is frequently directed along pathways (directional drilling), which can be at any angle, including horizontal, during the development phase of oil and gas extraction. For this reason, knowledge of the well's angle is necessary for the interpretation of inclined and nearly horizontal cores.

10.4. GEOPHYSICAL LOGGING

Accurate information about the physical characteristics of groundwater and geologic units inside the borehole environment can be obtained through geophysical borehole logging. Borehole logging is a cost-effective method of obtaining detailed information that would otherwise require performing and analyzing multiple cores. Data from borehole logging is commonly utilized to characterize geology, fluid flow, fracture patterns, and geologic structural properties.

Video, resistivity, natural gamma, electromagnetic induction, 3-arm caliper, spontaneous potential, borehole deviation, and temperature are examples of common borehole logging techniques that can be quickly and affordably implemented and interpreted. These programs are frequently used to make the most of the data gathered from geotechnical borings. Other tools like ground penetrating radar (GPR), downhole seismics, heat pulse flow meters, optical and acoustic televiewing, and so forth are available, but they are typically saved for more in-depth subsurface investigations.

10.4.1. Wireline Logging

Various industries, including mining, oil, and gas, utilize wireline logging to gather a continuous record of a formation's rock properties, as well as groundwater consultants. Wireline logging is the process of acquiring and analyzing geophysical data based on well bore depth, along with providing related services. It's important to note that wireline logging and mud logging are not the same, but they are closely connected through the integration of the data sets. Measurements are taken about True Along Hole (TAH) depth, which can then be used to determine additional properties such as hydrocarbon saturation and formation pressure, aiding in making informed decisions regarding drilling and production activities.

Wireline logging involves lowering a 'logging tool' – or a string of multiple instruments – on a wireline into an oil well or borehole to record petrophysical properties using various sensors. These tools can measure natural gamma ray, electrical, acoustic, stimulated radioactive responses, electromagnetic, nuclear magnetic resonance, pressure, and other properties of the rocks and fluids in the well over time.

A printed record or electronic presentation known as a "well log" is given to the client along with an electronic copy of the raw data once the data is recorded to an electronic data format, either at the surface (in real-time mode) or in the hole (memory mode). To obtain real-time information about the formations the well is penetrating, well logging operations can be carried out either during the drilling process or after the well has reached Total Depth, at which point the borehole's entire depth can be recorded.

Data is recorded in real-time directly correlating with measured cable depth. Memory data is recorded over time and depth data is measured at the same time. The two datasets are combined using a common time base to produce an instrument response versus a depth log. Recorded memory depth can also be corrected the same way as real-time corrections, ensuring consistent accuracy in TAH measurements.

The measured cable depth can be obtained in a variety of ways, but it is typically recorded using one of two methods: either a wheel counter that has been calibrated, or (more precisely) magnetic markings that offer calibrated cable length increments. Then, the measurements need to be adjusted for temperature and elastic stretch.

Wireline logs can be classified based on their function or the technology used. "Open hole logs" are performed before casing the well, while "cased hole logs" are done after casing. Logs can be grouped by the physical properties they measure.

10.4.1.1. Electrical Logs

10.4.1.1.1. Resistivity Log

The ability to obstruct the flow of electric current is known as subsurface electrical resistivity, and resistivity logging quantifies it. This makes it easier to distinguish between formations that are filled with hydrocarbons, which are poor conductors of electricity, and those that are filled with salty waters. Water saturation is computed through measurements of resistivity and porosity. Resistivity is expressed in ohms. meter (Ω·m), and is frequently charted on a logarithm scale versus depth because of the large range of resistivity. Depending on the tool, the current's penetration depth from the borehole can range from a few centimeters to a meter.

10.4.1.1.2. Borehole Imaging

The term "borehole imaging" refers to the methods used to create detailed images of the borehole wall and surrounding rocks at a centimeter scale. These techniques are commonly used in open hole drilling, but some tools can also be used in cased-hole situations. Borehole imaging technology has rapidly advanced in wireline well logging and has a wide range of applications, including reservoir description, reservoir performance analysis, and hydrocarbon recovery optimization. Specific uses include identifying fractures, analyzing sedimentological features, evaluating net pay in thinly bedded formations, and detecting breakouts in the borehole wall. The subject area of borehole imaging can be divided into four categories:

- Optical imaging;
- Acoustic imaging;
- Electrical imaging; and
- Methods that draw on both acoustic and electrical imaging techniques using the same logging tool.

10.4.1.2. Porosity Logs

Porosity logs assess the amount of pore volume within a rock formation, typically reported as a fraction or percentage. The two main technologies used for porosity logs are acoustic and nuclear. Acoustic logs analyze sound wave properties within the well-bore, while nuclear logs rely on nuclear reactions either in the downhole instrument or within the formation itself. Nuclear logs can include density, neutron, and gamma-ray logs for correlation purposes. The basic idea behind using nuclear technology is that a neutron source placed near the formation being measured will interact with hydrogen atoms in the formation fluid, providing an accurate porosity measurement. This is different from electrical resistivity measurements, which reflect the conductive properties of the formation fluid instead. By comparing neutron porosity with electrical porosity measurements, it is possible to detect the presence of hydrocarbons in the formation fluid.

10.4.1.2.1. Density

By subjecting a formation to a radioactive source and counting the gamma rays that remain after Compton Scattering and photoelectric absorption, the density log calculates the bulk density of the formation. The porosity can then be ascertained using this bulk density.

10.4.1.2.2. Neutron Porosity

For the neutron porosity log to function, a formation must be bombarded with high-energy epithermal neutrons. These neutrons lose energy through elastic scattering to levels that are almost thermal before being absorbed by the formation atoms' nuclei. Either the gamma-ray of capture, scattered thermal neutrons, or scattered, higher energy epithermal neutrons are detected, depending on the specific type of neutron logging equipment. The amount of hydrogen atoms in a given formation, which typically equates to rock porosity, is the main factor that affects the neutron porosity log.

Boron's high capture cross section for thermal neutron absorption is known to result in anomalously low neutron tool count rates. A rise in the hydrogen content of clay minerals influences the count rate in a comparable way.

10.4.1.2.3. Sonic

The formation interval transit time, which depends on lithology, rock texture, and especially porosity, can be obtained from a sonic log. Two or more receivers and a minimum of one piezoelectric transmitter make up the logging tool. An interval transit time is the amount of time it takes for a sound wave to travel a set distance between two receivers.

10.4.1.3. Lithology Logs

10.4.1.3.1. Gamma Ray

An API-unit-measured log of the formation's inherent radioactivity along the borehole is especially helpful for telling sands from shales in a siliclastic environment. This is because shales naturally contain radioactive elements due to adsorbed uranium and thorium and potassium isotopes found in clays, while sandstones are typically composed of nonradioactive quartz.

Certain rocks, especially carbonate rocks, have a significant and irregular uranium contribution that can lead people to believe the carbonate is shale. The carbonate gamma ray is a more accurate measure of the shale content in this instance. The uranium contribution to the gamma-ray log has been subtracted to create the carbonate gamma ray log.

10.4.1.3.2. Self/Spontaneous Potential

Without any applied current, the natural or spontaneous potential difference between the borehole and the surface is measured by the Spontaneous Potential (SP) log. When a single potential electrode was lowered into a well and a potential was measured in relation to a fixed reference electrode at the surface, one of the earliest wireline logs was discovered.

The electrochemical potential is the most valuable aspect of this potential difference, as it can create a noticeable shift in the SP response when encountering permeable beds. The extent of this shift is influenced by the difference in salinity between drilling mud and formation water, as well as the clay content of the permeable bed. As a result, the SP log is frequently utilized to identify permeable beds, estimate clay content, and determine formation water salinity. Additionally, the SP log can differentiate between impermeable shale, permeable shale, and porous sands.

10.4.1.4. Miscellaneous Logs

10.4.1.4.1. Caliper

An apparatus that uses two or four arms to mechanically measure the borehole's diameter or high-frequency acoustic signals. The caliper log can show where logs are potentially compromised because the borehole is either under-gauged (like mudcake buildup) or over-gauged (due to washout) because most logs rely on borehole regularity to record accurately.

10.4.1.4.2. Nuclear Magnetic Resonance

Nuclear magnetic resonance (NMR) logging provides a continuous record along the borehole's length by directly determining a formation's porosity and permeability using its NMR response. The moveable fluid volume (BVM) of a rock is the main use of the NMR instrument. This is the pore space that does not include irreducible water (IRI) or clay-bound water (CBW) (BVI). These volumes are difficult to see on older logs because neither of them is movable in the sense of NMR. After converting the relaxation curve to the porosity domain, both CBW and BVI are frequently visible in the signal response on contemporary instruments. It should be noted that not all of the fluids that are considered moveable (BVM) in the NMR context are truly moveable in the context of an oilfield. Though they might seem mobile to the NMR precession measurement, bitumen, heavy oil, and residual oil and gas do not always flow into a well bore.

10.4.1.4.3. Spectral Acoustic Logging

An acoustic measurement method called spectral acoustic logging is used in oil and gas wells for hydrodynamic reservoir characterization, production and injection interval identification, and well integrity analysis. The acoustic energy produced by fluid or gas flowing through a reservoir or by leaks in downhole well components is recorded by spectral acoustic logging.

The petroleum industry has been using acoustic logging tools for many years. An acoustic detector was first proposed in 1955 for use in casing hole identification during well integrity analysis. Downhole acoustic logging tools have shown to be useful for many years in a variety of applications, including leak detection, cross-flow location behind casing, inflow and injectivity profiling of operational wells, and even composition analysis of reservoir fluids.

10.4.1.4.4. Corrosion Well Logging

Calipers and thickness gauges are used to check the integrity of the steel and cemented column (casing and tubing) over the course of the wells' lifetime. Non-destructive technologies like magnetic, electromagnetic, and ultrasonic transducers are used in these cutting-edge technical techniques.

10.4.2. Logging While Drilling

Logging while drilling (LWD) was a novel approach to wireline logging that was first introduced in the 1970s. While the measurements are made in real-time while the well is being drilled, this method offers well information comparable to traditional wireline logging. Rather than having sensors lowered into the well at the end of wireline cable, the sensors are integrated into the drill string. This makes it possible for geologists and drilling engineers to swiftly acquire data on porosity, resistivity, hole direction, and weight-on-bit. With this data, they can promptly decide on the well's future and the course of the drilling.

In LWD, pressure pulses in the mud fluid column of the well are used to transmit measured data in real-time to the surface. Although drilling through rock is a fairly slow process, data compression techniques mean that this mud telemetry method's bandwidth of less than 10 bits per second is sufficient for real-time information delivery. At bit changes, a higher sample rate of data is retrieved from memory when the drillstring is removed. High-definition subsurface and downhole data can be obtained via networked or wired drillpipes that provide real-time, memory-quality data (Ali et al., 2008).

10.4.2.1. Memory Log

Instead of sending "Real Time" data to the surface, this data acquisition method records the sensor data into a downhole memory. This memory solution has a few benefits and drawbacks.

- The instruments can be transported into wells where the path is bent or extended past the capacity of traditional Electric Wireline cables. This may involve the electric cable's weight-to-strength ratio over its extended reach. In these situations, coil or pipe tubing can be used to convey the memory tools.
- The available sensor types are restricted when compared to those on the Electric Line, and they are typically used in the well's cased hole, production stage. Even so, a few memory "Open Hole" compact formation assessment tool combinations have recently been developed. These instruments are capable of being deployed, carried downhole, and "Pumped" out the bottom to start logging. They are kept safe from harm while running in the hole by being hidden inside a drill pipe. To cut expenses and operating time, additional fundamental open hole formation evaluation memory tools are available for use in "Commodity" markets on Slickline.
- A "Slick Line" intervention unit is typically present in cased hole operations. This operates the well-bore completion system by manipulating or performing other tasks using a solid mechanical wire (0.072–0.125 inches in OD). It is more common to perform memory operations on this Slickline

conveyance rather than dispatching a fully functional Electric Wireline unit.

- Because the results are not revealed until the equipment is brought back to the surface, it is not possible to monitor any real-time changes in well dynamics. This hinders the ability to accurately modify or adjust the downhole production conditions during memory logging by altering the surface production rates, as is commonplace in Electric Line operations.
- It is unknown what went wrong during recording until the memory tools are recovered. This data loss can be a serious problem in large, costly offshore locations. On the ground (e.g., South Texas, US), where logging is frequently done without the rig infrastructure, is home to the so-called "Commodity" oil service sector. This is less troublesome, and running the logs again is frequently trouble-free.

10.4.3. Coring

The process of extracting a real sample of a rock formation from a borehole is called coring. There are two primary methods of coring: "sidewall coring," which involves taking multiple samples from the side of the borehole after it has passed through a formation, and "full coring," which involves taking a sample of rock using a specialized drill bit as the borehole is first penetrating the formation. The primary benefits of sidewall coring over full coring are its lower cost (since drilling doesn't have to stop) and its ease of obtaining multiple samples. The main drawbacks are the possibility of sample failure and uncertainty regarding the depth at which the sample was obtained.

10.4.4. Mudlogging

Mud logs are well logs created by recording soil or rock shavings that are raised to the surface by the mud flowing through the borehole. In the oil industry, the operating company typically hires a mud logging company to prepare them. The formation gas (gas units or ppm) is one parameter that is typically displayed in a mud log. The various gas-detector manufacturers define arbitrary gas units differently, so the gas recorder is typically scaled in terms of these units. In actuality, only the relative changes in the detected gas concentrations are given significance. In addition to real-time drilling parameters like rate of penetration (ROP), lithology, gas hydrocarbons, flow line temperature (the temperature of the drilling fluid), and chlorides, the standard mud log used by the oil industry also sometimes contains information on mud weight, estimated pore pressure, and corrected d-exponent (corrected drilling exponent) for pressure pack logs. Directional data (deviation surveys), weight on bit, rotary speed, pump pressure, pump rate, viscosity, drill bit information, casing shoe depths, formation tops, and mud pump information are a few other items that are typically noted on a mud log.

10.4.5. Information Use

In the oil industry, the operating company receives the well and mud logs in "real-time." It uses these logs to make decisions about the well's operations, correlate formation depths with nearby wells, and interpret the amount and quality of hydrocarbons present. Log analysts are experts in the interpretation of well logs.

10.5. SUBSURFACE FACIES AND BASIN ANALYSIS

It should be clear from the above that a thorough stratigraphic and sedimentary analysis of a subsurface succession of rocks can be conducted using data interpreted from seismic reflection data, core, cuttings, and wireline logging tools. Subsurface resources contain the great majority of the world's oil and gas reserves. Whether data are gathered from outcrops or below ground, the principles of stratigraphic analysis and sedimentary facies interpretation are the same. With the development of 3-D seismic cubes of data, seismic reflection data can now be used to obtain information about large-scale structural and stratigraphic relationships, which provide a more comprehensive picture of the geology than scattered outcrops at the surface. Seismic reflection data can be used to help correlate, and although data from boreholes may be fragmented and one-dimensional, they offer a more comprehensive record of the sedimentary history than is available from surface exposures (Mussett & Khan, 2000).

CASE STUDY

HYDROGEOLOGICAL CHARACTERIZATION AND CONTROLS ON CONTAMINANT TRANSPORT AT THE HANFORD FACILITY, RICHLAND, WA:

The Hanford 300 Area is situated in Richland, Washington, at the southernmost point of the Hanford Site, on the west bank of the Columbia River. Groundwater in the area became contaminated with uranium due to the underground storage of spent nuclear waste between 1943 and 1975. Through a zone of surface water-groundwater interaction beneath and next to the river, this contaminated groundwater discharges into the Columbia River. At this location, one of the main scientific challenges is figuring out how lithologic variability and river stage variations interact to control the exchange of contaminated groundwater with river water. The upper Hanford Formation, which houses the unconfined aquifer from which groundwater flows, and the underlying, semi-confining Ringold Formation make up the two main formations of the site geology.

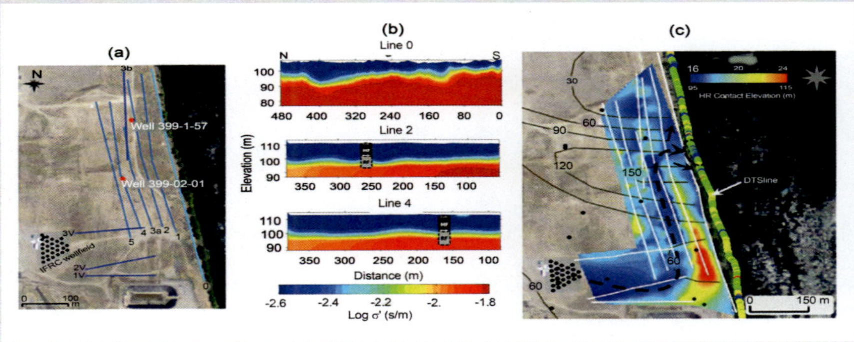

Figure 1. *Resistivity imaging at the 300 Area of the Hanford Facility, Richland, WA. (a) location of 2D resistivity survey lines. (b) selected 2D resistivity profiles (locations in part a) showing imaging of variations in depth to the Hanford-Ringold contact. (c) Map view of depth to the Hanford-Ringold contact estimated from the electrical resistivity data, showing evidence of a paleochannel controlling groundwater discharge into the river.*

Source: Modified from Mwakanyamale et al. (2012); https://www.enviro.wiki/index.php?title=Geophysical_Methods_-_Case_Studies.

The interface that governs the vertical flow and transportation of contaminated groundwater into the Columbia River is the hydrogeological contact between the relatively impermeable Ringold Formation and the permeable Hanford Formation. To enhance groundwater flow modeling and the transportation of uranium from the aquifer to the river, it is essential to identify spatial variability in the elevation of this interface throughout the site. The upper and lower portions of the Ringold Formation are primarily composed of mud and gravel, respectively. The Hanford Formation, which sits on top of it, is mainly made up of flood deposits with permeable paleochannels that could act as preferred flow paths for contaminated groundwater and locally help mix it with river water. It is possible that these paleochannels are locally incised beneath the Hanford-Ringold regional contact (Figure 1).

The site's 2D electrical resistivity imaging shows how geophysical techniques help to comprehend the movement of contaminants (Figure 1). Over 400–500 m along each line, variations in the depth to the Hanford–Ringold contact are easily captured by the surface-acquired resistivity images (Figure 1(b)), with findings verified by in-person borehole observations at two sites (Figure 1(b), grey symbols). Strong evidence of a paleochannel (paleo = old) carved into the Ringold Formation can be seen in the image (Figure 1(c)). The Columbia River receives most of its uranium discharge from this paleochannel. A zone of enhanced groundwater-surface water exchange identified by temperature monitoring on the riverbed is spatially correlated with the location where the paleochannel directs groundwater discharge into the river.

CLASS ACTIVITY

SANDWICH STRATIGRAPHY

Using a sandwich model, students will determine the relative ages of different "sedimentary rock layers" and experiment with "folding" and "faulting."

Objectives:

Students will create a model to:

- Determine the relative age of a series of "sedimentary rock layers" by applying the Law of Superposition; and
- experiment with "folding" and "faulting."

Materials:

Each student or student group will need a paper plate, a slice of white bread, a slice of light brown bread, a slice of dark brown bread, a spreader (knife or tongue depressor), 1 tablespoon of grape jelly, 1 tablespoon of soft chunky peanut butter, paper towels.

Getting Ready:

Gather the materials listed above. Have students wash their hands and wipe off their work area with a paper towel.

Setting the Stage:

1. Introduce each vocabulary term as it comes up in the activity.
2. Write the following information on the board:
   - white sandstone = white bread;
   - light brown sandstone = light brown bread;
   - brown shale = dark brown bread;
   - purple limestone = grape jelly;
   - conglomerate = peanut butter.
3. Tell students that each food item represents a specific sedimentary rock as noted.

Doing the Activity:

1. Determining Relative Age Using the Law of Superposition:
   - Distribute the materials to each student.
   - Tell students to:
     - o Deposit the "white sandstone" on the paper plate.
     - o Spread a layer of "purple limestone" on top of it.
     - o Place the "brown shale" on top of the "limestone."
     - o Spread the "conglomerate" on top of the "brown shale."
     - o Deposit a layer of "light brown sandstone" on top.
   - Discuss the following questions as a class:
   - Which is the oldest layer? Why? The white sandstone, because it was deposited first. It is on the bottom; everything else came after it.
   - Which is the youngest layer? Why? The light brown sandstone is the youngest because it was put on top and was the last layer to be deposited.
   - What is the relative age of the brown shale layer compared to the other layers? It is older than the light brown sandstone and younger than the white sandstone.
   - Tell students that this "sandwich" illustrates the Law of Superposition: A given sedimentary rock layer is younger than any of the sedimentary rock layers below it and older than any of the sedimentary rock layers above it.

Faulting and Folding:

- Tell students to "fold" the "sedimentary rock layers" to form a "valley" and "mountain slope."
- Have students eat the highest parts of the folded sandwich ("mountain") until the top is "eroded" away. Are the newly exposed surface layers older or younger than the surface of the valley? *Older*
- What happens when the sandwich is torn apart? *Faulting*
- Now students can enjoy what's left of their sandwiches!

SUMMARY

- Geologists usually learn the principles of sedimentology and stratigraphy from outcrop relationships in the field, but many will work with subsurface data if they are employed as professional geoscientists.
- Seismic-reflection methods are active-source geophysical methods that were developed for oil and gas exploration applications.
- A seismic-reflection survey employs an above-ground sound source and receiver array to, respectively, transmit sound waves and measure their reflections off acoustic discontinuities (i.e., reflecting interfaces).
- Reflections and refractions of seismic waves at geologic interfaces within the Earth were first observed on recordings of earthquake-generated seismic waves. The basic model of the Earth's deep interior is based on observations of earthquake-generated seismic waves transmitted through the Earth's interior.
- Seismic waves are waves of acoustic energy that induce oscillations within, interact with and are influenced by subsurface earth materials. Reflection methods are most commonly concerned with the compressional (P) and shear (S) seismic body waves, which travel within the interior of geologic bodies. These waves can be differentiated by how they affect the materials through which they travel. P-waves displace material in the direction parallel to their travel direction, whereas S-wave displacement is normal (i.e., perpendicular) to their travel direction.
- The interpretation of seismic reflection profiles provides a model for the stratigraphic and structural relationships that may exist in the subsurface. Data from these sources can provide some indicators of the lithologies in the subsurface, but a full geological picture can be obtained only by the addition of information on lithology and facies.
- In the course of drilling a deep borehole, a fluid is pumped down to the drill bit to lubricate it, remove the rock that has been cut (cuttings) and counteract formation fluid pressures in the subsurface. Due to the weight of the rocks

above, fluids (water, oil, and gas) trapped in porous and permeable strata will be under pressure, and without something to counteract that pressure they would rush to the surface up the borehole.

- A drill bit can be designed such that it cuts an annulus of rock away leaving a cylinder in the center, a core, that can be brought up to the surface. Where coring is being carried out the drilling is halted and the section of core is brought up to the surface in a sleeve inside the hollow drill string. As each section of core is brought to the surface it is placed in a box, which is labeled to show the depth interval it was recovered from.
- The procedure for recording the details of the sedimentary rocks in a core is very similar to making a graphic sedimentary log of a succession exposed in the field. Core logging sheets are similar in format to field logging sheets, and the same types of information are recorded (lithology, bed thickness, bed boundaries, sedimentary structures, biogenic structures, and so on).
- Geophysical borehole logging can provide accurate data on the physical properties of geologic units and groundwater within the borehole environment. Borehole logging is a time and money-saving approach to gaining detailed information that is otherwise only obtainable from performing and analyzing numerous cores.
- Wireline logging can be defined as being "The acquisition and analysis of geophysical data performed as a function of well-bore depth, together with the provision of related services."
- In the 1970s, a new approach to wireline logging was introduced in the form of logging while drilling (LWD). This technique provides similar good information to conventional wireline logging but instead of sensors being lowered into the well at the end of wireline cable, the sensors are integrated into the drill string and the measurements are made in real-time, whilst the well is being drilled.
- Coring is the process of obtaining an actual sample of a rock formation from the borehole. There are two main types of coring: 'full coring,' in which a sample of rock is obtained using a specialized drill-bit as the borehole is first penetrating the formation and 'sidewall coring,' in which multiple samples are obtained from the side of the borehole after it has penetrated through a formation.
- Mud logs are well logs prepared by describing rock or soil cuttings brought to the surface by mud circulating in the borehole. In the oil industry, they are usually prepared by a mud logging company contracted by the operating company.
- From the foregoing, it should be apparent that a combination of information from the interpretation of seismic reflection data, core, cuttings and wireline logging tools can be used to carry out a full stratigraphic and sedimentary analysis of a subsurface succession of rocks.

REVIEW QUESTIONS

1. What is seismic refraction data?
2. How to interpret seismic reflection data?

3. What are seismic reflection profiles used for?
4. What are the three types of boreholes?
5. What is a borehole structure?
6. What is geophysical logging?
7. What are geophysical well logging techniques?

REFERENCES

1. Birkelo, B. A., Steeples, D. W., Miller, R. D., & Sophocleous, M. (1987). Seismic reflection study of a shallow aquifer during a pumping test. Groundwater, 25(6), 703–709. https://doi.org/10.1111/j.1745-6584.1987.tb02211.x.
2. Bruno, P. P., Improta, L., Castiello, A., Villani, F., & Montone, P. (2010). The Vallo di Diano Fault System: New evidence for an active range-bounding fault in southern Italy using shallow, high-resolution seismic profiling. Bulletin of the Seismological Society of America, 100(2), 882–890. https://doi.org/10.1785/0120090210.
3. Büker, F., Green, A. G., & Horstmeyer, H. (1998). Shallow seismic reflection study of a glaciated valley. Geophysics, 63(4), 1395–1407. https://doi.org/10.1190/1.1444441.
4. Dehghannejad, M., Juhlin, C., Malehmir, A., Skyttä, P., & Weihed, P. (2010). Reflection seismic imaging of the upper crust in the Kristineberg mining area, northern Sweden. Journal of Applied Geophysics, 71(4), 125–136. https://doi.org/10.1016/j.jappgeo.2010.06.002.
5. Grubišić, V., & Orlić, M. (2007). Early observations of rotor clouds by Andrija Mohorovičić. Bulletin of the American Meteorological Society, 88(5), 693–700. https://doi.org/10.1175/BAMS-88-5-693.
6. Haines, S. S., & Ellefsen, K. J. (2010). Shear-wave seismic reflection studies of unconsolidated sediments in the near surface. Geophysics, 75(2), 1MA-Z39. https://doi.org/10.1190/1.3340969.
7. Hunter, J. A., Pullan, S. E., Burns, R. A., Gagne, R. M., & Good, R. L. (1984). Shallow seismic reflection mapping of the overburden–Bedrock interface with the engineering seismograph—Some simple techniques. Geophysics, 49(8), 1140–1395. https://doi.org/10.1190/1.1441766.
8. Miller, R. D., Steeples, D. W., & Brannan, M. (1989). Mapping a bedrock surface under dry alluvium with shallow seismic reflections. Geophysics, 54(12), 1528–1534. https://doi.org/10.1190/1.1442620.
9. Mussett, A. E., & Khan, M. A. (2000). Looking into the Earth: An Introduction to Geological Geophysics. Cambridge University Press.
10. Mwakanyamale, K., Slater, L., Binley, A., & Ntarlagiannis, D. (2012). Lithologic imaging using complex conductivity: Lessons learned from the Hanford 300 Area. Geophysics, 77(6), E397–E409. https://doi.org/10.1190/geo2011-0407.1.
11. Pugin, A. J. M., Pullan, S. E., & Duchesne, M. J. (2013). Regional hydro stratigraphy and insights into fluid flow through a clay aquitard from shallow seismic reflection data. The Leading Edge, 32(7), 725–856. https://doi.org/10.1190/tle32070742.1.

Index

A

Acidic environment 9
Aeolian processes 176
Aeolian sandstones 1, 3
Alluvial 52
Alluvial fans 139
Alluviated intermontane basins 180
Amino acid racimization 309
Ammonoids 267, 282
Analytical Procedures 329
Animal species 274
Anticline cores 228
Aquatic animals 267
Aquatic habitat 113
Aragonite structures 76
Archaeological stratification 2
Argonite skeletal grains 76
Assemblage biozones 267
Asymmetrical ripples 108, 134
Atmospheric agents 8
Atmospheric elements 8
Authigenic minerals 309

B

Ball-and-pillow structures 223
Bank gravel 48
Beach depositional environments 178
BEAST (Benthic Environmental Assessment Sediment Tool) 117
Bedload transport 107, 112, 113, 137, 138
Bedrock 41, 54
Bedrock weathering 48
Bed thicknesses 141
Bench gravel 48
Biochemical sedimentary rocks 43, 44
Biochemical sedimentation 43
Biogenic sediments 73
Biogenic structures 61
Biological activity 73
Biological weathering 8, 9
Biostratigraphic correlation 267, 268, 276, 309
Biostratigraphic data 268, 286
Biostratigraphic unit 269, 276, 286, 305
Biostratigraphic zonation techniques 268
Biostratigraphy 1, 24, 25, 37, 267, 268, 275, 276, 277, 278, 279, 281, 282, 283, 284, 285, 287, 288, 292, 293, 302, 306, 307
Bioturbated mudstone 162, 163, 165
Bioturbation 108, 129, 139
Bituminous coal 46
Boulder-sized gravels 42
Breccia 41, 44
Breccio-conglomerate 50
British Institutions Reflection Profiling Syndicate (BIRPS) 359

Broken-down rock 8
Bryozoan grains 76

C

Calcareous green algae 76
Calcite 41, 45
Calcite compensation 81
Carbonate facies 139
Carbonate minerals 309
Carbonate rock 75, 80
Cenozoic 73, 81
Cephalopoda 267, 273
Chemical alteration 108
Chemical modification 108
Chemical reactions 8
Chemical sedimentary rocks 43
Chemical sedimentation 43
Chemical weathering 8, 9, 38, 39
Chemostratigraphy 309, 325
Chert 46
Chlorine-36 Dating Method 321
Chronological framework 268
Chronostratigraphy 1, 27
CHUR Model 317
Civil engineering 111
Clast Composition 51
Clastic deposits 139
Clastic detritus 175
Clastic dikes 223
Clastic rocks 3
Clastic sediment 175
Clastic sedimentary rocks 43
Clastic sedimentation 41, 43
Clastic terrigenous sedimentary rocks 42
Clay-bound water (CBW) 369
Claystone 41, 45
Climate change 10
Closure temperature 309
Coal 41
Coarser sediment particles 113
Coastal depositional environments 115
Coastal erosion 10
Cobble conglomerate 52
Common-midpoint (CMP) 358
Compaction 108
Conglomerate 44, 47, 49, 52, 53, 69
Consortium for Continental Reflection Profiling (COCORP) 359
Continental drift 180, 183, 190
Continental environments 140
Continental landform 175
Continental landscapes 176
Continental land surface 175
Continuous succession 275
Convolute bedding 223
Convolute lamination 233, 263
Coquina 73, 78
Coralline algae 76
Creamy semi-liquid condition 238
Cross-bedding 108, 129, 134, 135, 136
Cross-cutting relationships principle 5
Crystalline limestone 46

D

Deep marine basins 111
Deepwater Marine 52
Deforestation 10
Deltaic deposits 115
Deltaic environments 241
Deltaic systems 139
Dense limestone 76
Density instabilities 243
Denudational processes 180, 183
Denudational system 175
Depleted-Mantle Model 317
Deposition 2, 3, 7, 8, 9, 11, 12, 14, 16, 17, 19, 20, 22, 23, 37
Depositional environments 139, 268, 277, 278, 279, 280, 282, 283, 306
Diagenesis 108, 129
Diluted hydrochloric acid 77, 141
Dish structures 223, 246, 248, 249, 264
Dolomitic limestone 75
Dolomitization 80

Dolostone 46
Dorag dolomitization 80
Dunham Classification 78
Dynamic continental environments 176
Dynamite 345

E

Echinoderms 179, 267, 282
Ecological imbalance 10
Electron spin resonance (ESR) 332
Entropy 181, 187, 191, 194
Environmental and Social Impact Assessment (ESIA) 360
Environmental engineering 111
Environmental Impact Assessment (EIA) 360
Eogenesis 41
Eolian depositional environment 178
Equilibria pattern 275
Erosion 9, 175, 186, 206, 207, 209, 210, 212, 213, 214
Evaporative basins 74
Exfoliation 10
Exogenic processes 1, 9
Exploration 345, 351, 352

F

Facies analysis 140, 162
Facies interpretation 162
Facies models 139
Fecal pellets 76, 92
Fertile soils 73
Field sedimentologists 140, 171
Field sedimentology 139, 140, 172
Fine-grained silt 115
Fine gravel 49
Fission Track Dating Method 320
Flame structures 223
Floods 10
Fluidization 225, 226, 227, 228, 229, 249
Fluidized deformation 228
Fluidized sediment 226
Fluid velocity 123
Fluvial 52
Fluvial depositional environments 177
Folk Classification 78
Foraminifera 73
Foraminiferida 267, 293, 306
Fossil organisms 268, 273
Fossils 1, 20, 25, 26

G

Gastropods 267, 282
Gemstones 4
Genetic basis 272
Geochemical analysis 345
Geochemistry 5, 29
Geochronology 311, 315, 318, 341
Geographical cycle 186, 187
Geographical isolation 274, 279
Geological alterations 109
Geological map 142
Geological mapping 345
Geological record 268, 275, 299
Geological strata 2
Geological time 268, 270, 279, 287
Geologic time 311, 314
Geomorphology 111
Geophysical Research Corporation (GRC) 350
Geophysical Service Incorporated (GSI) 350
Glacial meltwater 175
Glaciers 10, 11, 41
Glaciers pulverize 177
Glacier terminus 177
GPS (Global Positioning Satellite) 141
Graded bedding 5
Gradual process 225
Grain size distribution 139
Granular gravel 47, 48
Graphic log 145, 146, 154, 171
Graphic sedimentary 139, 141, 142, 143, 144, 145, 163, 164, 171, 172
Gravel 41, 47, 48, 50, 69
Gravitational instability 238, 239, 242, 244,

254
Gravity 345
Ground penetrating radar (GPR) 366

H

Heterogeneous fluids 107, 120
Hierarchical classification system 271
High-concentration dispersion 228
Hydraulic engineering 107, 111
Hydrocarbon exploration 1
Hydrocarbons 345
Hydroplastic condition 226
Hydroplastic deformation 228
Hydroplastic sediments 238

I

Ice crystallization 8
Independent Simultaneous Sweeping 345, 353
Industrial minerals 4
Interbreeding test 274
International Association of Geophysical Contractors (IAGC) 361
International Association of Oil and Gas Producers (IOGP) 361
Interval biozones 267
Intraformational conglomerate 51
Ion concentration 41
Ionizing radiation 331
Isochron plot 315

L

Lacustrine environments 117
Lag gravel 49
Laminar flows 107
Land degradation 10
Landform evolution 181, 182, 183, 186, 187, 190, 195, 196, 197
Land planation 188
Landscape dynamics 176
Landscape evolution theory 181
Landslides 107, 119
Life of field seismic (LoFS) 358
Limestone 73, 75, 76, 78, 82, 83, 103, 104
Limestone formations 75, 81, 82, 95
Linnaean scheme 271
Linnaean System 271
Liquefaction 225, 226, 227, 228, 236, 242, 250, 251, 254, 255
Liquidization 226, 227, 228, 236, 239, 245, 258
Lithification 1, 41
Lithofacies analysis 139
Lithologies 175
Lithostratigraphy 1, 22, 37
Load casts 223, 231
Logging while drilling (LWD) 370
Low-energy reflective beaches 115
Lower-density mud 239
Low-lying coastal areas 115
Luminescence Dating Methods 321
Lycopods 175
Lyell's gradualism concept 186

M

Magnetic material 309
Magnetostratigraphic Correlation 327
Magnetostratigraphy 310, 326, 327
Marine depositional environments 115
Marine environments 140
Marine macrofossils 267
Marine organisms 73
Mass spectrometer 315, 321
Meltwater streams 176
Mesozoic 73, 81, 91
Metadeposition 240
Metamorphic rock 48, 59
Metamorphism 73, 309
Metamorphosed limestone 4
Micrite 77
Microcrystalline 46
Microscopic fossil fragments 45
Migliorini attributes convoluted lamination 235
Mineral dolomite 75
Mineral-laden water 74

Mollusks 76, 179
Morphological trends 274
Morphological variability 275, 295
Mountainous terrain 180
Mudrock 61
Multi-Azimuth Towed Streamer (MAZ) 354

N

Natural environment 140
Negligible yield strength 225, 226
Normal moveout (NMO) 351
Nuclear magnetic resonance (NMR) 369
Numerous contemporary plant 274

O

Ocean bottom nodes 345, 355
Ocean Bottom Seismic (OBS) 354
Ocean circulation 118
Oceanic crust 81
Olivine crystals 65
Oolitic limestone 76, 83
Optical stimulating luminescence (OSL) 331
Organic matter 73, 95, 114, 115
Organic sedimentary rocks 43
Orogenic belts 175
Orogenic system 175
Orogenic tectonism 189
Oscillating water 108

P

Paleoclimatology 267
Paleocurrent 145, 148, 149, 152, 155, 156, 172
Paleocurrent data 142, 145, 150, 172
Paleoecology 267
Paleoenvironmental habitat 271
Paleontology 272, 307
Paleozoic rocks 267, 273, 277, 282, 284, 305
Paraconglomerates 51
Pegmatites 48
Peloids 76
Permanent reservoir monitoring (PRM) 358
Petrography 5
Petroleum reservoirs 345
Petrology 5
Petromictic conglomerate 51
Phyletic gradualism 275, 306
Piedmont gravel 49
Pillar structures 223
Plagioclase 54, 55, 59, 65
Planktonic (free-floating) foraminifera 76
Plateau gravel 49
Plate-tectonic subduction system 175
Polarity Chron 328
Polarizing microscope 60
Polydisperse systems 228
Pore fluid 223
Porosity 73, 76
Post-Depositional Convolute Lamination 234, 263
Precambrian 73, 79, 81, 84, 90, 91, 93
Precursor rocks 3
Predominant clast size 52
Predominant grain 3
Pseudoextinction 275
Pseudo-nodules 240, 242

Q

Quartzite 59
Quaternary continental successions 332

R

Radioactive decay constant 313
Radioactivity 310, 311, 313, 331, 341
Radiogenic decay 309
Radiolarians 75, 90, 91, 92
Radiometric dating 309, 311, 341
Radiometric dating techniques 310, 311, 321, 330
Rainfall 10
Rainwater 9
Rate of penetration (ROP) 371
Rayleigh-Taylor instability 229, 236, 239, 243
Recrystallization 73
Reef material 4
Reflections 345, 376

Refracted rays 345
Remotely operated underwater vehicles (ROVs) 354
Re-sedimenting dispersion 245
Ripple marks 139
River floodplains 225
River Formation 82
Riverine landscapes 176
Rock disintegration 9
Rock formation 5, 22
Rock gypsum 46
Rock salt 46
Rock structure 10
Rubidium–Strontium Dating Method 319

S

Saltation 112, 113, 121
Samarium–neodymium dating 341
Samarium–Neodymium Dating 316
Sampling Procedures 329
Sand-mud interface 239
Sand-mud systems 239
Sandstone 54, 57, 69
Sand volcanos 223
Scanning electron microscope (SEM) 62
Secondary dolomite 75
Sedimentary deposits 140, 171
Sedimentary environments 139, 140
Sedimentary geology 111
Sedimentary layers 245
Sedimentary log 141, 142, 143, 158, 163, 171, 172
Sedimentary petrology 139
Sedimentary rocks 2, 3, 4, 5, 7, 8, 12, 14, 16, 19, 20, 21, 27, 36, 37, 38, 39
Sedimentary sequence 1
Sedimentary structures 1
Sedimentary successions 4
Sediment consolidation 225
Sediment deposition 108, 129
Sediment formation 176, 221
Sedimentological analysis 141
Sedimentological circumstances 5
Sedimentological data 146
Sedimentological research 2, 225
Sedimentologists study the composition, structure, textures, and fossil content of deposits that were made in different geographical settings 2
Sedimentology 1, 2, 3, 4, 21
Sediment source 176, 177
Sediment transport 2, 10, 11, 107, 110, 111, 112, 113, 115, 117, 119, 120, 128, 132, 133, 134, 135, 136, 137, 138
Sediment transportation 1, 37
Seismic-reflection methods 348, 376
Seismic velocity 351
Seismic waves 376
Seismites 223, 264
Shallow Marine 52
Silica-rich clastic sedimentary 3
Silicate minerals 43, 56, 62, 69, 70
Siliciclastic sedimentary rocks 42
Siltstone 45, 62
Simulator for Particle Erosion (PES) 117
Single sedimentation units 230
Slumps 230, 257
Slump structures 264
Small rock cavities 77
Sm and Nd geochemistry 316
Sm–Nd Radiometric Dating 316
Soft-sediment deformation 224, 263
Soft-sediment deformation structures 223, 225, 229
Soil microorganisms 9
Sparite 77
Spatial distribution 140, 171
Spontaneous Potential (SP) 369
Stratigraphic record 268, 269, 270, 274, 275, 277, 278, 280, 281, 283, 285
Stratigraphy 1, 3, 17, 21, 22, 24, 37
Stress-release fractures 9
Stromatolites 1, 12, 13, 82
Strontium 319, 323, 341, 343
S-wave displacement 345, 376
Symmetrical ripples 108
Syndepositional convolute lamination 233,

234, 235, 240

T

Taxonomy 272
Tectonic effects 10
Tectonism manifests 180
Terrigenous clastic sedimentary rocks 42
Thermochronological Techniques 324
Thermochronology 309, 314, 324, 341
Thermochronometry 309
Thermoluminescence (TL) 331
Thickness variations 81
Tidal mudflat depositional environments 178
Tillites 52
Topography 10
Transitional environments 140
Transport mechanisms 108, 118, 129, 132, 133
Travertine 74, 78, 104
True Along Hole (TAH) 366
True orthoquartzite 60
Turbidite basin 156
Turbidite deposits 240
Turbidity 108
Turbulent eddies 113
Two-way travel times (TWT) 351

U

Uniformitarianism 17, 20
Uniformitarianism principle 4, 20
Uranium–Lead Dating Method 315
Uranium-series dating 320, 331
Uranium–Thorium Dating Method 320

V

Vibroseis 352
Volcanic cones 180, 221
Volcanic eruption 110
Volcanic fragments 65
Volcaniclastic sediments 74, 101
Volcanic sequences 238
Vortex rings 244

W

Water-escape structures 224, 225, 263
Water-saturated mineral-density sediments 239
Weathering 8, 9, 33, 34, 37
Wind-borne sand transportation 113
Wind transport 108

X

X-ray diffractometer (XRD) 63

Y

Younging indicators 5

Z

Zircon, tourmaline, rutile 56